MINING OF MINERAL DEPOSITS

T0076737

Mining of Mineral Deposits

Editors

Genadiy Pivnyak
Rector of National Mining University, Ukraine

Volodymyr Bondarenko
Department of Underground Mining, National Mining University, Ukraine

Iryna Kovalevs'ka
Department of Underground Mining, National Mining University, Ukraine

Mykhaylo Illiashov
PJSC "Donetsksteel", Ukraine

CRC Press
Taylor & Francis Group
Boca Raton London New York

CRC Press is an imprint of the
Taylor & Francis Group, an **informa** business

A BALKEMA BOOK

CRC Press
Taylor & Francis Group
6000 Broken Sound Parkway NW, Suite 300
Boca Raton, FL 33487-2742

First issued in paperback 2019

© 2013 by Taylor & Francis Group, LLC
CRC Press is an imprint of Taylor & Francis Group, an Informa business

No claim to original U.S. Government works

ISBN-13: 978-1-138-00108-4 (hbk)
ISBN-13: 978-0-367-37959-9 (pbk)

Typeset by Olga Malova & Kostiantyn Ganushevych, Department of Underground Mining, National Mining University, Dnipropetrovs'k, Ukraine

This book contains information obtained from authentic and highly regarded sources. Reasonable efforts have been made to publish reliable data and information, but the author and publisher cannot assume responsibility for the validity of all materials or the consequences of their use. The authors and publishers have attempted to trace the copyright holders of all material reproduced in this publication and apologize to copyright holders if permission to publish in this form has not been obtained. If any copyright material has not been acknowledged please write and let us know so we may rectify in any future reprint.

Except as permitted under U.S. Copyright Law, no part of this book may be reprinted, reproduced, transmitted, or utilized in any form by any electronic, mechanical, or other means, now known or hereafter invented, including photocopying, microfilming, and recording, or in any information storage or retrieval system, without written permission from the publishers.

For permission to photocopy or use material electronically from this work, please access www. copyright.com (http://www.copyright.com/) or contact the Copyright Clearance Center, Inc. (CCC), 222 Rosewood Drive, Danvers, MA 01923, 978-750-8400. CCC is a not-for-profit organization that provides licenses and registration for a variety of users. For organizations that have been granted a photocopy license by the CCC, a separate system of payment has been arranged.

Trademark Notice: Product or corporate names may be trademarks or registered trademarks, and are used only for identification and explanation without intent to infringe.

**Visit the Taylor & Francis Web site at
http://www.taylorandfrancis.com**

**and the CRC Press Web site at
http://www.crcpress.com**

Table of contents

Mining of Mineral Deposits – Pivnyak, Bondarenko, Kovalevs'ka & Illiashov (eds)
© 2013 Taylor & Francis Group, London, ISBN: 978-1-138-00108-4

Preface

The present collection of scientific papers is addressed to mining engineers, scientific and research personnel, students, postgraduates and all professionals connected with the coal and ore industry.

The papers published describe topics related to mine workings drivage, optimization of longwall working parameters, modeling of mine support interaction with rock massif, stress strain state of rock massif during mining operations, geomechanical tasks solving, economic aspects and environment protection.

Additional information is provided regarding recovery and utilization of mine methane, borehole underground coal gasification and alternative energy sources development such as gas hydrates.

Genadiy Pivnyak
Volodymyr Bondarenko
Iryna Kovalevs'ka
Mykhaylo Illiashov

Mining of Mineral Deposits – Pivnyak, Bondarenko, Kovalevs'ka & Illiashov (eds)
© 2013 Taylor & Francis Group, London, ISBN: 978-1-138-00108-4

New-generation technique and technology for leakage tests

A. Bulat, O. Voloshyn & S. Ponomarenko
M.S. Polyakov Institute of Geotechnical Mechanics, Dnipropetrovs'k, Ukraine

D. Gubenko
Yuznoye State Design Office named after M.K. Yangel, Dnipropetrovs'k, Ukraine

ABSTRACT: This article describes principle and mathematic model for testing various devices for leakage with the help of fixed-volume method; shows advantages of this method when compared with manometric methods with no pressure chamber; and presents functional arrangement for applying the method under consideration in industries.

1 INTRODUCTION

Energy saving is the key issue of economic development in any country. One of the main types of energy for mineral mining is compressed air which is widely used in the industry thanks to highly safe pneumatic equipment. It is especially important for gassy and dusty mines and, besides, in some cases usage of compressed air is the only possible way when electric power is dangerous to be used for the mineral mining in underground mines under the coal burst and gas release hazard. However, air ducts and pneumatic devices in the active mines are in such condition that requires special measures in order to reduce direct energy inputs and material resources when compressed air is used. Today nonproduction cost of compressed air supplied to the mining equipment is very high, and the problem to cut the cost is a strong business case as tariffs for the energy carrier are growing.

One of the ways to save energy at compressed air production, transportation to and consumption by the mining pneumatic equipment is minimization of the compressed air losses through the improved leak proofness in the pneumatic systems and equipment. With this end in view, the mining companies should improve their standards for testing the pneumatic equipment and air-supply system for leakage in the course of repair and preventive maintenance.

2 RESULTS OF THE STUDY

The most widely spread method of testing pneumatic devices with no pressure chamber is manometric method which determines value of the air pressure drop per time unit. This method is known as pressure drop method. Its main drawback is essential inaccuracy caused by impact of environment parameter gradient on accuracy of the leakage tests. This drawback can be eliminated with the help of fixed-volume method at which:

– two similar vessels (reference vessel and compensating vessel) are used;

– the vessels and device which measures differential pressure between the vessels are located inside the closed thermostat;

– value of factual total leakage in the system or device is determined by the varied gas mass value in the compensating vessel;

– mathematic model which calculates total leakage size takes into account factual variations of gas pressure and temperature in the device under consideration.

Functional arrangement of components for testing device for leakage with the help of the fixed-volume method is shown in the Figure 1.

In this scheme, device 1, which is a compressed-air consumer (volume V_t) in the mine, is connected to the compensating vessel 7 and reference vessel 8 inside the thermostat 6 with the help of pneumatic lines 2, 3 and shutoff valves 4, 5. Differential pressure meter 9 is installed between the vessels. Initial compressed air parameters are fixed in the reference vessel, and current parameters are fixed in the compensating vessel per certain period of time, and thermostat maintains positive temperature balance with the environment.

Below is a sequence of preparatory operations for the leakage testing:

– the device under the test, compensating vessel and reference vessels are filled with compressed air up to the operating pressure;

– operating pressure is aligned throughout the

whole system, and the system is shut off from the compressed air source by the valve 4 (not shown in the Figure 1);

– compressed air temperature in the vessels is stabilized up to specified value, and the reference vessel is shut off from the device under the test; the compensating vessel is still connected to the device, and both vessels are connected to each other through the differential pressure meter.

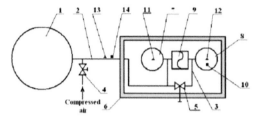

Figure 1. Functional arrangement for testing device for leakage with the help of the fixed-volume method: 1 – device under the test; 2, 3 – pneumatic lines; 4, 5 – shutoff valves; 6 – thermostat; 7, 8 – compensating and reference vessels; 9 – differential pressure meter; 10 – sensor of the gas absolute pressure in the reference vessel; 11, 12, 13 – temperature sensor; 14 – barometric-pressure sensor.

The following measurements are made in real time mode:

– absolute air pressure in the reference vessel – by the sensor 10;

– air temperature in the geometric centers of the reference and compensating vessels – by the sensors 11 and 12;

– environment temperature in the entry to the thermostat – by the sensor 13;

– atmospheric pressure – by the sensor 14 and differential pressure between the compensating and reference vessels – by the sensor 9 .

The layout of the reference vessel 8, compensating vessel 7 and differential pressure meter 9 in the thermostat 6 (Figure 1) allows aligning temperature fields in them and removing temperature disturbance received from the device 1. Thank to this, the compensating vessel receives from the device 1 only disturbance caused by differential pressure of compressed air. Changes of the air parameters can be explained by two factors: leakages in the device 1 and changed environment parameters. The changes occur in accordance with the Clapeyron-Mendeleev equation for the gas state and the law of mass and energy conservation.

The mathematic model of testing the devices for leakage by the fixed-volume method is based on the key laws of molecular-kinetic theory of gas (Ginsburg 1966 & Loistyankiy 1973).

Generally, processes occurring in the "device-compensating vessel" system are described by the following set of equations:

$$
\left.\begin{aligned}
PV_c &= m_{\kappa,n} R T_{\kappa,c}; \\
PV_t &= m_{u,n} R T_{u,i}; \\
\left(P + \Delta P_c\right) V_c &= \left(m_\kappa + \Delta m_\kappa\right)\left(T_{\kappa,i} + \Delta T_{\kappa,c}\right) R; \\
\left(P + \Delta P_c\right) V_t &= \left(m_{u,n} + \Delta m_t\right)\left(T_{u,i} + \Delta T_{u,c}\right) R,
\end{aligned}\right\} \quad (1)
$$

where P – pressure of compressed air at the beginning of testing (boost pressure); V_t and V_c – volumes of the device under the test and compensating vessel; m_κ and Δm_κ – air mass in the compensating vessel at the beginning of testing and its changes in the course of testing; $T_{\kappa,i}$ and $\Delta T_{\kappa,c}$ – initial and current temperature in the compensating vessel; R – universal gas constant; m_t and Δm_t – air mass in the device at the beginning of testing and its changes in the course of testing; $T_{u,i}$ and $\Delta T_{u,c}$ – initial and current temperature in the device; ΔP_c – current differential pressure between the compensating and reference vessels measured by differential pressure meter.

This method of leakage test with the help of compensating and reference vessels is based on the assumption that when there is no compressed air leakage from the system mass of the compensating vessel does not change. This conclusion is proved by the equations of the gas state for the compensating vessel at the beginning and at the end of testing when the given gas volume is in the equilibrium state. Equation 1 and 3 of the set (1) shows that:

$$
\frac{\Delta m_{\kappa,t}}{m_\kappa}\left(1 + \frac{\Delta T_{\kappa,c}}{T_{\kappa,i}}\right) = \frac{\Delta P_c}{P} - \frac{\Delta T_{\kappa,c}}{T_{\kappa,i}}. \quad (2)
$$

From the latter equation, it is obvious that $\Delta m_{\kappa,t} = 0$, provided:

$$
\frac{\Delta P_c}{P} = \frac{\Delta T_{\kappa,c}}{T_{\kappa,i}} \quad \text{or} \quad \frac{\Delta P_c}{\Delta T_{\kappa,c}} - \frac{\Delta P}{T_{\kappa,i}}.
$$

Thus, if there is no leakage in the system ($\Delta m_{\kappa,t} = 0$) though temperature changes due to the heat exchange between the device under the test and environment then the ΔP change caused by this temperature factor is linearly connected with the $\Delta T_{\kappa,c}$ change in the compensating vessel. As pressure in the vessel cannot be less than pressure in the device (it would contradict the Pascal law on the isometry of pressure) the mass can flow only from the compensating vessel into the device and not vice versa. Therefore, in the equation (2) the $\Delta m_{\kappa,t}$ is always ≤ 0.

Consequently, as during the time period needed for estimating leakage proofness of the device we have the following correlation:

$$\frac{\Delta P_c}{\Delta T_{\kappa,c}} = \frac{P}{T_{\kappa,i}} = const,$$

then we can say about complete leakage proofness of the device with accuracy commensurable with the measurement inaccuracy.

Distinctive features of this method (they were confirmed by mathematic model during studying physical process of the compressed air leaking from the device under the test) are the following:

1. Re-distribution of the compressed air parameters in the compensating vessel is determined by the adiabat law with adiabat ratio k, and re-distribution of gas parameters in the device – by polytropy law with polytrope index n:

$$\frac{P + \Delta P_c}{P} = \left(\frac{m_\kappa + \Delta m_{\kappa,t}}{m_\kappa} \right)^k;$$

$$\frac{P + \Delta P_c}{P} = \left(\frac{m_t + \Delta m_{u,i}}{m_t} \right)^n;$$

$$n = \frac{\lg \left. P_{a,c} \middle/ (P + \Delta P_c) \right.}{\lg \left. P_{a,c} \middle/ (P + \Delta P_c) \right. + \lg \left. T_{e,t} \middle/ T_{a,c} \right.},$$

where $P_{a,c}$ – measured current atmosphere pressure; $T_{e,t}$ – mean value of the current compressed-air temperature measured in the compensating and reference vessels; $T_{a,c}$ – current atmosphere pressure measured in the entry into the thermostat.

2. Changed compressed air mass in the device includes air mass flowing out from the device due to the leakage and some quantity of air flowing into the device from the compensating vessel.

3. Mass value Δm_y of factual leakage in the device in the atmospheric condition is determined by the following equation:

$$\Delta m_y = \frac{V_t (P + \Delta P_c) \rho_n T_n}{T_{a,i} P_n} \left[\left(1 + \frac{T_{t,i}(\Delta P_c - \Delta P_{i,c}) - P(T_{e,t} - T_{t,i})}{P T_{e,t}} \right)^{k/n} - 1 \right] -$$

$$- \frac{T_{t,i}(\Delta P_c - \Delta P_{i,c}) - P(T_{e,t} - T_{t,i})}{T_{t,i} T_{e,t}} \frac{V_e}{R},$$

where $V_e = V_c$ – volume of the reference vessel; ρ_n, P_n, T_n – air density, pressure and temperature at normal conditions which are chosen from the reference data; $T_{a,i}$ – initial temperature of atmospheric air measured in the entry into the thermostat; $T_{t,i}$ – initial temperature measured in the compensating vessel; $\Delta P_{i,c}$ – computed correction for pressure variation under the impact of environment temperature.

Mean value of the current compressed-air temperature in the vessels and correction for pressure variation are calculated by the following formulas:

$$T_{e,t} = T_{\kappa,i} + \Delta T_{\kappa,c} - \Delta T_{e,t};$$

$$\Delta P_{i,c} = P\left(\frac{T_{a,c}}{T_{a,i}} - 1 \right),$$

where $\Delta T_{e,t}$ – current change of temperature measured in the reference vessel; $T_{a,c}$ – current temperature of environment.

In spite of impact of external factors (heat exchange between the device under the test and environment and real variations of pressure) mass value of factual leakage in the device Δm_y depends only on re-distribution of the compressed air mass between the device and compensating vessel. In order to determine Δm_y in the formula (3) impact of changed external factors and steadiness of thermostat operation are taken into account in the polytrope index n and values of $\Delta P_{i,c}$, $T_{t,i}$ and $T_{e,t}$.

The mathematic model is based on the following assumptions:
– only quasi-statistic processes are considered;
– relaxation period is essentially less than time periods for which gas-state equations are written;

– time period during which pressure of compressed air in the device changes due to the leakage in the device and heat exchange with the environment is essentially longer than the relaxation period;

– compressed air pressure and its changes are the same in all points of the system according to the Pascal law;

– compressed air temperature and density can differ in different points in the device volume and are inversely dependent on each other;

– compressed air temperature in the pneumatic line in the entry into the thermostat is equal to temperature of environment;

– leakage in the pneumatic line, shutoff valve, compensating and reference vessels and instrumentation equipment is essentially less than measured leakage of compressed air.

Depending on how compressed air mass in the reference vessel has changed factual value of leakage in the device is determined by specially designed fixed-volume method and with taking into account character of changes of environment parameters.

For the purpose of practical testing devices for leakage by the fixed-volume method, the Institute of Geotechnical Mechanics under the National Academy of Science of Ukraine together with the Yuznoye State Design Office named after M.K. Yangel created and tested on the space-rocket hardware a precision Leak Testing Device for Hollow Wares (LTDHW), which consists of the following key structural elements:

– compensating and reference vessels designed as the Dewar spherical container;

– electronic unit for measuring local temperature in gaseous media by quartz frequency thermometers (QFT);

– electronic unit for measuring differential pressure between the vessels which consists of low-limit differential pressure sensor, series LPX/LPM, produced by the "DRUCK Company (UK), and AD converter;

– electronic unit for measuring excess pressure and barometric pressure which consists of two resonance pressure sensors of extra accuracy, RPT series, produced by the "DRUCK Company (UK), and

AD converter. To measure barometric pressure, barometric pressure sensor RPT 410F is used. To measure excess (working) pressure, pressure sensor RPT 200 is used.

The instrumentation equipment has passed incoming metrological inspection and metrological attestation in the National Scientific Center "Metrology Institute".

At working pressure up to 0.3 MPa (3 kgf / cm^2), the LTDHW detects factual leakage in the device under the testing with error not more than 20% from the measured values. The low limit of the leakage size measured by the working example of the LTDHW is 10 l micron HG / s. Total time period needed for testing device for leakage is not more than 8 hours.

3 CONCLUSIONS

The method to test pneumatic equipment in mines for compressed air leakage with the help of the fixed-volume method is a result of complex theoretical and experimental studies of various methods of detecting micro leakages without using indicating gases and barometric equipment. The fixed-volume method improves accuracy and reliance of detecting compressed-air leakage from the wares of any configuration and helps to determine factual size of total leakage in the devices in real testing conditions. Computerized method of the leakage tests on the basis of the proposed fixed-volume method will provide proper test control, automatic measurement of all parameters, computer processing of measuring results with their conversion to any needed dimensions. The method also makes shorter testing time and improves accuracy and reliance of measurement of factual leakage size in the devices.

REFERENCES

Ginsburg, I.P. 1966. *Aerogas dynamics.* Moscow: Vysha shkola.
Loistyankiy, L.G. 1973. *Fluid and gas mechanics.* Moscow: Nauka.

Optimal parameters of wall bolts computation in the united bearing system of extraction workings frame-bolt support

V. Bondarenko & I. Kovalevs'ka
National Mining University, Dnipropetrovs'k, Ukraine

R. Svystun
Mokryanskiy quarry, Zaporozhye, Ukraine

Yu. Cherednichenko
"Fuel-energy company of Donetsk" Donets'k, Ukraine

ABSTRACT: Engineering approach of required rational parameters of wall bolts installation, which provide optimal load on the frame support all over its contour in view of their mechanical interrelation in the united bearing construction using spatial-flexible units is developed.

Conducted researches of frame-bolt support interaction with rock massif around extraction workings, which are exposed to the intensive impact of stoping, stresses into the support, caused by loads of rock massif computation, obtained solutions numerical analysis allowed to work out engineering approach of flexible frame-bolt support rational parameters. It is based on nomograms range creation, which allows to compute parameters efficiently and accurate enough depending on mining environment and mine engineering conditions of extraction working maintenance. Main design parameters of frame-bolt support include: reactions N_j of bolts interaction (equal within their bearing ability), coordinates θ_{N_j} of bolts installation.

Computation of impact required reaction on flexible bolts frame props. Nomogram used to measure required reaction N_1 of bottom bolt, located from the side of the coal seam is shown on Figure 1. Numerical analysis has shown, that required rate of bolts reaction N_1 essentially (more than 10% of N_1 intensity) depends on variables: φ, φ_c, $\dfrac{k}{q_v}$, $\dfrac{m}{r}$, $\dfrac{h}{r}$. Here, φ and φ_c are the angle of internal friction of rocks and coal, respectively; q_v and k – are the vertical load on the frame and the coefficient of its skewness (Bondarenko, Kovalevs'ka & Symanovych 2012), respectively.

Computation of parameter $\dfrac{N_1}{q_v r}$ is conducted in terms of quadrants I-V in accordance with wrench move for following basic data: $\varphi = 30°$, $\varphi_y = 20°$, $\dfrac{h}{r} = 0.3$ and $\dfrac{m}{r} = 0.4$ (which with working arch radius $r = 2.5$ m corresponds to the height of rock bankette $h = 0.75$ m and of the seam $m = 1$ m), $\dfrac{k}{q_v} = 0.4$. From I quadrant's horizontal scale and point $\varphi = 30°$ we drop a perpendicular to meet with the line $\dfrac{h}{r} = 0.3$, from which we contour in quadrant II up to meet with the line $\dfrac{m}{r} = 0.4$; from this point we drop a perpendicular to the lower border of the quadrant II, where we receive point A. From the mark $\varphi_c = 20°$ on the vertical scale of quadrant IV we contour to the point of meeting with line $\dfrac{m}{r} = 0.4$, from which we erect a perpendicular to the top border of the quadrant IV and receive a point B. We connect points A and B with line we receive intermediate parameter value 0.235 on the summarizing scale of the quadrant 3. We put this value on quadrant's V vertical scale. From this point we contour to the point of meeting with the line

$\dfrac{k}{q_v} = 0.4$, from which we erect a perpendicular to the top horizontal scale of quadrant V, where we read the answer $\dfrac{N_1}{q_v r} = 0.285$ m. Bottom bolt reaction magnitude N_1 (for example with $r = 2.5$ m and $q_v = 165$ kPa) builds up $N_1 = 0.285$ m $\times$ 2.5 m $\times$ 165 kPa = 117 kN in terms of one frame installation on one long meter of the working. If the number of frames on one long meter equals one, that means $n = 1$ that the reaction N_1 of the bottom bolts is 117 kN. With other value n of frames installation on working long meter number reaction N_1' of the bolt equals

$$N_1' = \dfrac{N_1}{n}.$$

Nomogram used to measure required reaction N_2 of the top bolt, located from the side of the coal seam is shown on Figure 2. Computation is conducted in accordance with wrench move over the quadrants I-V by analogy with nomogram of reaction N_1 computation. Parameter value $\dfrac{N_2}{q_v r}$ for the top bolt (located from the side of the coal seam) builds up 0.20 m. for previously shown basic data, and reaction magnitude N_1 (with $r = 2.5$ m and $q_v = 165$ kPa) equals $N_2 = 0.20$ m $\times$ 2.5 m $\times$ 165 kPa = 83 kN for one long meter of the working.

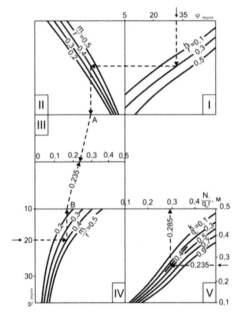

Figure 1. Nomogram used to measure required reaction N_1 of the bottom bolt, located from the side of the coal seam.

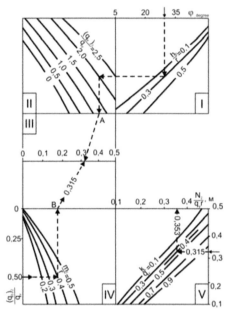

Figure 2. Nomogram used to measure required reaction N_2 of the top bolt, located from the side of the coal seam.

Nomogram used to measure required reaction N_4 of the bottom bolt, located from the side of the goaf is shown on Figure 3. Parameter $\dfrac{N_4}{q_v r}$ dimensioning is carried out in accordance with wrench move over the quadrants I-V for basic data: $\varphi = 30°$, $\dfrac{h}{r} = 0.3$, $\dfrac{(q_b)_3}{q_v} = 2.0$, $\dfrac{(q_b)_2}{q_v} = 0.5$, $\dfrac{m}{r} = 0.4$, $\dfrac{k}{q_v} = 0.4$ and

is run in the following order. From the mark $\varphi = 30°$ on the horizontal scale of quadrant I we drop a perpendicular to meet the line $\dfrac{h}{r} = 0.3$, from which we contour in quadrant II to the line $\dfrac{(q_b)_3}{q_v} = 2.0$ and from the point of meeting we drop a perpendicular to the bottom border of the quadrant II, where we receive point A. From the value $\dfrac{(q_b)_2}{q_v} = 0.5$ on the vertical scale of quadrant IV we contour to the point of meeting with the line $\dfrac{m}{r} = 0.4$ and erect a perpendicular to the top border of the quadrant IV, where we receive point B. Then we connect points A and B, and compute value of the intermediate parameter on the horizontal scale of the quadrant III that equals 0.315. We put this value on the vertical scale of quadrant V, and contour to the point of meeting with the line $\dfrac{k}{q_v} = 0.4$, from which we erect a perpendicular to the horizontal scale of the quadrant V, where we read the answer $\dfrac{N_4}{q_v r} = 0.353$ m. Bottom bolt reaction magnitude N_4 (for example with $r = 2.5$ m and $q_v = 165$ kPa) builds up $N_4 = 0.353$ m $\times$ 2.5 m $\times$ 165 kPa = 145 kN on one long meter of the working.

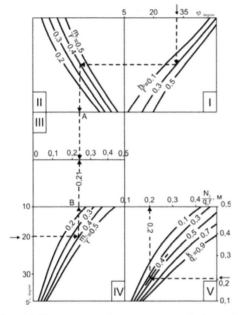

Figure 3. Nomogram used to measure required reaction N_4 of the bottom bolt, located from the side of the goaf.

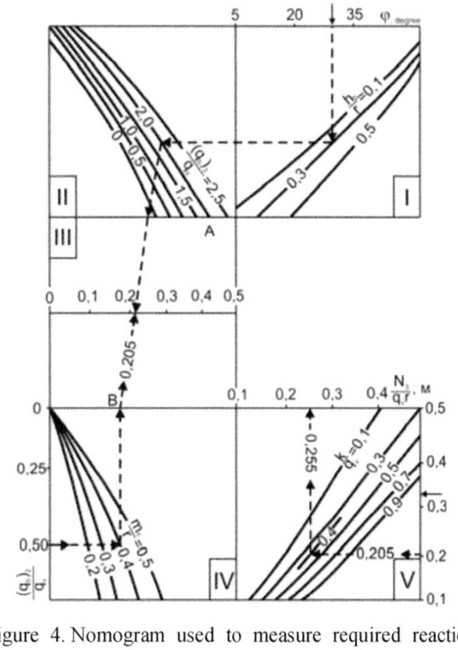

Figure 4. Nomogram used to measure required reaction N_3 of the top bolt, located from the side of the goaf.

In order to compute top bolt required reaction N_3, located from the side of the goaf, nomogram was designed (Figure 4). Usage rules of it are similar to the previous example. A value $\dfrac{N_3}{q_v r} = 0.255$ m is received for previously shown basic data, and reaction N_3 magnitude built up $N_3 = 105$ kN per one long meter of the working.

Units of flexible bolts connection on frame support props installation place computation. Except of the required bolts reaction stresses N_j, rational parameters of their installation also include coordinates of their location over the working contour.

Angular coordinate θ_{N_j} ($j = 1,...,4$) is used as such parameter. According to the analysis, it essentially depends on existing variables: angle of internal friction of rocks φ; ratio of $\dfrac{h}{r}$ natural bankette height to the working arch radius; ration $\dfrac{m}{r}$ of seam height to the radius of working arch; angle of internal friction of φ_c coal seam; ratio of $\dfrac{k}{q_v}$ load increment k in zone of stoping influence to the vertical load q_v on the support out of this zone.

Following variables need to be considered during the computation θ_{N_j} for the other pair of bolts ($j = 3,4$), that are installed from the side of the goaf: ratio of $\dfrac{(q_b)_3}{q_v}$ wall load $(q_b)_3$ along the length of natural bankette (under security element) to the vertical one q_v; ratio $\dfrac{(q_b)_2}{q_v}$ of wall load $(q_b)_2$, acting along the height of the security element to the vertical one; ratio $\dfrac{k}{q_v}$, angle of internal friction of rock φ; ratio of $\dfrac{h}{r}$ and $\dfrac{m}{r}$.

Nomogram used to compute rational coordinates θ_{N_1} of bottom bolt installation over the quadrants I-V is presented on Figure 5 in accordance with wrench move for following basic data: $\varphi = 30°$, $\varphi_y = 20°$, $\dfrac{k}{q_v} = 0.4$, $\dfrac{h}{r} = 0.3$, $\dfrac{m}{r} = 0.4$.

From the mark $\varphi = 30°$ on the vertical scale of quadrant I we contour to the point of meeting with the line $\dfrac{h}{r} = 0.3$; from this point we drop a perpendicular to quadrant II till it meets the line $\dfrac{k}{q_v} = 0.4$, from which we contour to the quadrant II, where we receive value 1.77 of accessory parameter. According to this value we build a curve in quadrant III. From the mark $\varphi_c = 20°$ in quadrant IV we contour to the line $\dfrac{m}{r} = 0.4$, from which we drop a perpendicular to quadrant III to the built by us line 1.77. From the point of meeting we contour to the right vertical scale of quadrant III, where we read the answer $\theta_{N_1} = 90°$. As the result, it is the most rational to install the bottom bolt near the frame bolt arch footing for the given example.

Rational coordinate θ_{N_2} of top bolt installation computation from the side of the coal seam is conducted according to the nomogram (Figure 6) similarly to above described example. As a result of computation for the same basic data we discover, that the top bolt must be installed at the angle of $\theta_{N_2} = 38.6°$ to the working vertical axis.

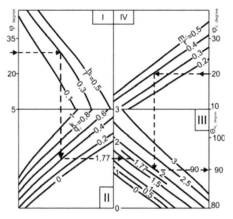

Figure 5. Nomogram used to measure rational coordinate θ_{N_1} of the bottom bolt, located from the side of the coal seam.

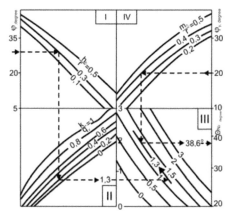

Figure 6. Nomogram used to measure rational coordinate θ_{N_2} of the top bolt, located from the side of the coal seam.

Nomogram used to Figure rational value θ_{N_3} of the top bolt installation, located from the side of the goaf is shown on Figure 7. Computation is conducted in accordance with wrench move for following basic data: $\dfrac{(q_b)_3}{q_v} = 0.8$, $\dfrac{(q_b)_2}{q_v} = 0.4$; $\dfrac{k}{q_v} = 0.4$; $\varphi = 30°$;

$\dfrac{h}{r} = 0.3$; $\dfrac{m}{r} = 0.4$. The other parameters do not have such significant impact on the angular coordinate θ_{N_3}. Method of computation is following: from the mark $\dfrac{(q_b)_3}{q_v} = 0.8$ of the horizontal scale of quadrant and we drop a perpendicular to the line $\frac{h}{r} = 0.3$; then we contour in quadrant II to the line $\frac{(q_b)_2}{q_v} = 0.4$; after that we drop a perpendicular to quadrant III to the line $\frac{m}{r} = 0.4$; then we contour from this mark and read on the vertical scale of quadrant III the answer for intermediate parameter. It equals 1.08. After that we contour from the mark $\varphi = 30°$ on the

vertical scale of quadrant IV to the line $\frac{k}{q_v} = 0.4$, there we erect a perpendicular in quadrant V; we read the answer – $\theta_{N_3} = 27°$ from the point of meeting of vertical wrench with the line 1.08 in quadrant V on its vertical scale. We compute bottom bolt $\theta_{N_4} = 95°$ coordinate using nomogram shown on Figure 8 by analogy with the previous one.

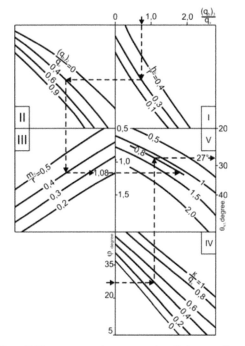

Figure 7. Nomogram used to compute installation coordinate θ_{N_3} of the top bolt, located from the side of the goaf.

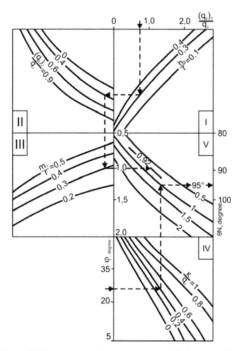

Figure 8. Nomogram used to compute installation coordinate θ_{N_4} of the bolt, located from the side of the goaf.

Thus, as the result of wall bolts and frames unification by means of mechanical bonds into the united bearing system it is possible to create frame-bolt supports in terms of high wall loads. They are notable for reduced material capacity and sufficient bearing ability.

REFERENCES

Bondarenko, V., Kovalevs'ka, I., Symanovych, G. and other. 2012. *Process operator guide of extraction working maintenance and security at flat seams.* Research and practice guide. Dnipropetrovs'k: NMU: 98.

Pillars sizing at magnetite quartzites room-work

M. Stupnik, V. Kalinichenko & S. Pismennyi
Kryvyi Rig National University, Kryvyi Rig, Ukraine

ABSTRACT: The methods of the inclined pillars parameters calculation at level room-work of thick complex-structural magnetite quartzite deposits in Kryvyi Rig iron ore basin is given.

1 SCIENTIFIC AND PRACTICAL TASKS

In Kryvyi Rig iron ore basin rich and poor iron deposits are generally mined by room-work or sublevel caving methods (Table 1).

The main efficiency indices of mining methods are production costs, which is largely determined by the level of ore losses, degree of waste rock dilution and the specific volume of the access workings (Table 2).

Table 1. Mining methods applied in mines.

Enterprise	Mine	Mining depth, m	Mining methods
PJSC "Krivbaszhelezorudkom"	Rodina	1315	sublevel ore caving
	Octyabrskaya	1190	sublevel ore caving; room and pillar caving
	Lenin mine	1275	
	Gvardeiskaya	1270	
PJSC "ArcelorMittal Kryvyi Rig"	#1 Artem mine	1135	sublevel ore caving
PJSC "Evraz Sukha Balka"	Yubileinaya	1260	sublevel ore caving; room and pillar caving
	Frunze mine	1135	

Table 2. Technical and economic indices of mining methods in Kryvyi Rig iron ore basin.

Name	Mining methods		
	Level room-work	Sublevel room-work	Sublevel caving
Specific weight in annual production, %	35.0	20.0	45.0
Specific volume of development and access workings m / th. t	1.9-3.0	2.5-4.5	3.0-5.0
Ore losses, %	5.0-10.0 17.4-25.0	7.0-12.0 16.9-20.0	14.7-18.0
Ore dilution, %	4.0-7.0*⁾ 13.0-16.0*⁾	4.0-6.0*⁾ 11.4-14.0*⁾	16.5-18.0
Iron content reduction in ore output, %	0.5-2.0	0.3-1.5	1.5-3.0

Note: * – without pillar and ceiling mining.

Table 2 represents that the sublevel caving methods reduce iron ore content in ore output almost by two times in comparison with the room-work (Hivrenko 2001 & Development of technological... 2012). Taking into account that magnetite quartzite deposits are composed of very thick hard rocks technological advancement of their room-and-pillar methods is rather essential.

2 PAPER ANALYSIS

Iron ore deposits in Kryvyi Rig basin, reach the horizontal area of more than 1500 m² and strike length of more than 700 m, of ore bodies ranging in size from 50 to 500 m² and strike length from 10 to 75 m. The share of large deposits is 80% from the ore area in the basin. Their thickness varies from 20 to 150 m and more. The ore bodies are extended in the north-east direction and lie at angle from 20 to 80 degrees with the grade of ore from 36 to 64%. Physical and mechanical properties of Kryvyi Rig iron ore basin vary widely. Some mine fields have one or two parallel iron deposits containing about 70% of the reserves of the mine field, others have more than 20 separate ore bodies having a strike length from 150 to 500 m with the grade of ore from 58 to 64% (Development of technological... 2012).

According to the occurrence iron ore deposits are divided into homogeneous and heterogeneous (Development of technological... 2012 & Bizov 2001). There are inclusions of barren area or ores with low grade quality in heterogeneous deposits. The thickness of barren areas varies from 3.2 to 6 m in some areas to 10.6 m. The specific area of barren inclusions within the level (sublevel) is 10...15-18%. The deposits with the presence of barren area, are usually mined by complete mining, see Table 1.

3 THE PROBLEM STATEMENT

Application of traditional mining methods with ore complete mining at ore deposits mining, including barren area inevitably leads to a decline of the ore grade quality from 3 to 10%, which significantly af-

fects the sale price of commercial products and increases the cost of extraction, transportation, hoisting of extracted rock mass and its dressing.

Thereby, the development of improved version of mining methods for deposits with barren area inclusions, allowing to increase the quality of mined ore mass, is an important scientific and technical task for mines.

4 MATERIAL PRESENTATION AND RESULTS

Ore deposits of Kryvyi Rig iron ore basin according to their structure can be divided into five types: 1 – without barren area inclusions; 2, 3 and 4 – mining ore area has single; double and triple barren area inclusions; 5 – ore area has combined barren area inclusions, Figure 1.

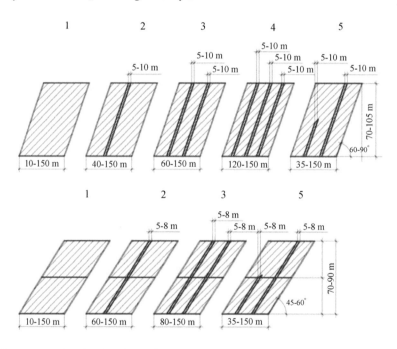

Figure 1. The structure of ore deposits of Kryvyi Rig iron ore basin.

The first type includes all single and parallel and contiguous deposits that don't contain barren area inclusions or the thickness of barren area inclusions between the ore deposits is more than 15 m. In this case, it should be noted that the parallel and contiguous deposits are mined separately. Deposits which have one barren area inclusion with the thickness of not more than 10 m belong to the

second type. The third and the fourth type are ore deposits having two or more barren area inclusions, the distances between barren area inclusions vary from 15 to 35 m and more. The fifth type is ore deposits with barren area inclusions of irregular shape.

Based on researches, the classification of ore deposits of Kryvyi Rig iron ore basin tend to be mined by room and pillar systems is given (Table 3).

Table 3. Morphological classification of ore deposits of Kryvyi Rig iron ore basin.

Name	Without barren area inclusions	Single barren area inclusions		Doubled by barren area inclusion		Tripled by barren area inclusion	Combined barren area inclusions	
Deposit type	1	2		3		4	5	
Dip angle of ore deposits, degree	45-90	45-60	60-90	45-60	60-90	60-90	45-60	60-90
Thickness of ore deposits, m	10-150	60-150	40-150	80-150	60-150	120-150	35-150	35-150
Dip angle of barren area inclusions, degree	—	45-90	60-90	45-60	60-90	60-90	45-70	60-90
Thickness of barren area inclusions, m	—	5-8	5-10	5-8	5-10	5-10	5-8	5-10
Rigidity of ore body	+/-	+/-	+/-	+/-	+/-	+/-	+/-	+/-
Rigidity of hanging wall rock	+	+	+/-	+	+	+	+	+/-
Rigidity of bottom wall rock	+	+/-	+	+/-	+	+	+/-	+
Rigidity of rock inclusions	—	+	+/-	+	+	+	+	+/-

Note: + hard ores or rock; – soft ores or rock.

For mining of iron ore deposits with barren area inclusions (type 2-5) it is necessary to use selective mining, leaving barren area inclusions in the waste area (Bizov 2001). This can be achieved by using level (sublevel) room and pillar systems with caving or leaving pillars and crowns between rooms. However, their use has a number of boundary conditions, which include: the minimum allowable thickness of barren area and ore deposit, the amount of mining panels, the thickness of inclined dirt area inclusion (Storchak 2003).

The minimum allowable thickness of barren area inclusion is conditioned by inclined pillar integrity support, normal conditions of ore crashing and determined by

$$m_n \geq 1.5 \cdot W ,$$ (1)

where m_n – minimum allowable thickness of barren area inclusion, m; W – line of least resistance at longhole work, m.

The minimum allowable thickness of ore body limited by the barren area inclusion depends on the underground mining technology, height of level (sublevel) and is determined by

$$m_p \geq (0.1...0.3) \cdot h \geq m_n ,$$ (2)

where m_p – minimum allowable thickness of ore body which is situated near barren area, m, h – height of level, m.

The amount of mining panels in the stope limited across by barren area inclusions is determined by

$$N = \frac{M}{n} + 1 ,$$ (3)

where N – amount of mining areas in the stope limited across by barren rock inclusions; M – horizontal thickness of ore deposit, m, n – amount of barren area inclusions the thickness of which are ranged from 5 to 8-10m.

The thickness of the inclined barren area inclusion that will ensure its stability for a period of the panel mining is determined by the conditions of the longitudinal compressive forces P_l in which there is no integrity. Side forces P_s, are directed towards the previously mined room filled with caved rocks (Storchak 2003). The design formula for determining the width of the inclined interstall pillar is

$$b = \frac{P_l \cdot K_d \cdot \xi \cdot \sqrt{\sigma_t \cdot h}}{n_c \cdot \sigma_{com} \cdot \sqrt{K_f \cdot \gamma}} \geq m_s ,$$ (4)

where P_l – longitudinal compressive forces work along the inclined pillar; K_d – ratio depending on the tensile stress and rock deformation; ξ – ratio of rock creeping; σ_t – rock tensile strength, kPa; n_c – amount of longitudinal pillars per one room; σ_{com} – rock compressive strength; K_f – inclined pillar stability factor; γ – specific weight of rock, forming the inclined pillar, kg / m³.

In the case when there is no tensile stress and deformation in the pillar K_d is 1.15-1.41, when inclined pillar subjected to maximum deformation without affecting its integrity K_d is 1.41-1.73, in the laminated fractured ground with possible or partial pillar caving K_d is 1.63-2.0, and at crack initiation with the following caving K_d is 2.0-2.44 (Slesarev 1948).

So, the width of the inclined barren pillar defined by the expression (4) should be 1.5 times greater than the thickness power of barren area inclusion. As a result of researches an improved version of the level room mining methods with pillars and crown caving is developed.

A distinctive feature of the proposed version of the room mining method shown in Figure 2, from the traditional is the following. Mining section is divided into mining panels according to the thickness.

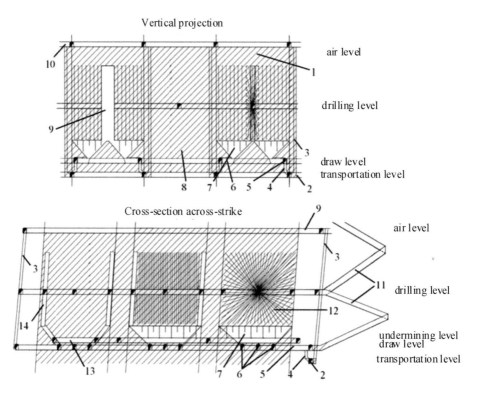

Figure 2. Level room development with dirt inclusions leaving in the section: 1 – primary stope (room); 2 – hauling roadway; 3, 4 – raise, air and passageway, ore-pass; 5, 10 – draw entry; 6 – loading rooms; 7 – ditch undercutting; 8 – stope of the second turn (room fender); 9 – vertical compensation room; 11 – spiral ramp; 12 – rings of block caving deep holes; 13 – cutoff ditch ort; 14 – cut raise.

The first section is limited by hanging wall rock and hanging wall of barren area inclusion, the last one is limited by hanging wall of barren area inclusion and bottom wall. Section mining is carried out by mining panels from hanging to bottom wall.

The panels are mined separately by level, (sublevel-) room and pillar system with the formation of a separate compensation space, drilling and receiving levels. Between the mining panels the inclined pillar consisting of barren inclusion is left. Interstall inclined pillars (barren inclusions) are not mined but remain unaffected between panels. Ore pillars and crowns mining are carried out according to the traditional technology. The results of calculation of improved mining method application compared with traditional technologies are shown in Table 4.

Table 4. Technical and economic indices of mining methods in deposits with barren area inclusions.

Name	Room and pillar mining without pillar caving	Ore and cover caving methods	Proposal room and pillar mining method
Block (panel) parameters			
Block strike length, m	50	50	50
Mining thickness, m	100	100	80
Level height, m	90	90	90
Barren inclusion thickness, m	10	10	-
Dip angle of ore deposits, degree	80	80	80
Barren inclusions amount, pieces	2	2	-
Mined blocks (panels) amount, pieces	1	1	3
Ore volume weight, t / m³	2.8	2.8	2.8
Barren inclusions volume weight, t / m³	2.2	2.2	-
Ore mass reserve in the block, th. t	1206	1206	1008
– ore reserve in the block, th. t	1008	1008	1008
– barren inclusion reserve in the block, th. t	198	198	-
Grade, %:			
– in ore	46.0	46.0	46.0
– in rock	24.0	24.0	24.0
– in dirt inclusions	16.0	16.0	-
Specific rate of preliminary development and access working, m / th.	2.8	3.6	3.8
Ore output per hole meter run, ton / m	21	20	25
Output per man-shift, t / per shift	136.2	154.8	155.72
Grade of mining block (panel), %	40.0	40.0	46.0
Ore loss, %	10.0	16.0	10.0
Ore dilution, %	7.0	15.0	7,0
Ore mass amount, th. t	1084.5	1191.8	975.5
Grade of mined ore, %	38.9	37.6	44.5

5 CONCLUSIONS

It is determined that the use of inclined pillars consisting of barren area inclusions allows to increase the iron content in the mined ore from 37.6-38.9% to 44.5%, and to reduce drilling, output and minerals processing costs. Thus, the ore output is reduced by 10-18%, which significantly reduces the rock processing and haulage costs.

The given method of pillars determination is applicable when the calculated width of inclined pillars is equal to or less than the barren inclusion thickness. In the case when the calculated width of the inclined pillar is more than the barren inclusion thickness, the traditional ore and cover caving method is applied.

REFERENCES

Hivrenko, V. 2001. *Technological classification of complicated structural deposits.* Ore deposits development. Kryvyi Rig: KTU, #76: 26-29.

Development of technological opening schemes, preparation and longwall mining for complicated structural deposits at further development on big depth: Report from research work. 2012. #GR 0109U002336. Kryvorizhs'kyi national university, #30-84-11: 306.

Bizov, V., Storchak, S., Sirichko, V., Cherednichenko, O., Garkusha, A., Vitryak, V., Plothikov, V., Repin, O., Hivrenko, O., Schelkanov, V. & Andreev, B. 2001. *Patent № 37982A E 21 C41/16 UA. Method of steep ore bodies development that consist waste rock insertion".* Publish 15.05.2001. Bulletin #4.

Storchak, S., Chelkanov, V., Karamanic, F., Andreev, B., Korzh, V. & Pismennyi, S. 2003. *Patent. 62168 UA, MKI E21C41/06. Method of steep deposit development of mineral resources.* Publisher 02.01.2003; Publish 15.12.2003; Bulletin #12.

Slesarev, V. 1948. *Rock mechanics and mine support.* Moscow: Coal Publisher: 45.

The calculation scheme of mathematical modeling of displacement process of a terrestrial surface by working out of coal layers

M. Antoshchenko, L. Chepurnaya & M. Filatyev
Donbass State Technical University, Alchevs'k, Ukraine

ABSTRACT: On the basis of the carried-out theoretical research and analysis of experimental data it's been developed method of forecasting final subsidence of terrestrial surface which uses trajectory displacement of points on the earth surface and which takes into account the mechanical properties of rocks, mining depth and geometry sizes of extraction sites.

Determination of regularities of process of displacement of a terrestrial surface at a side job its clearing developments is one of the main objectives while working off of coal layers. The authentic forecast of parameters of displacement of a terrestrial surface promotes the successful decision of others, not less important, mining tasks. To them, except protection of objects on a terrestrial surface, the choice of the location of excavations and rational ways of their protection from influence of mountain pressure, the gas emission forecast from undermining sources, justification of rational schemes of airing of extraction sites, calculation of bearing ability support also many other things belong. By the solution of the specified tasks essential value has establishment of dynamics of displacement process and allocation of its characteristic stages.

Duration of process is considered the period of time during which the terrestrial surface is in a condition of displacement owing to influence of clearing works. The general duration divides into three stages: initial, active and subside stage. Establishment of the specified stages according to the normative document (The rule of undermining of buildings, 2004) is made rather conditionally and in modern conditions of big depths of development to their definition it is impossible to declare existing approach completely correct (Kulibaba 2010). Its main shortcoming is lack of accurate and unambiguous methods of definition of a temporary framework of course, both all process of displacement of a terrestrial surface, and its separate stages. The most perspective direction in the solution of a considered problem it's offered the approach represented by Professor Gavrilenko Yu.N. (Gavrilenko 2007).

Division of displacement process of a terrestrial surface on separate stages is offered to be made by means of characteristic points of the mathematical function describing development of subsidence of a terrestrial surface in time. By way of such points it is offered to use extremes of the first three derivatives on time from the main equation describing the change of subsidence of a point of a terrestrial surface in the process of displacement (2-4).

By the mathematical models (2-4) it's supposed the formation of a flat bottom displacement trough on a terrestrial surface (η_0) by carrying out clearing works within one extraction site. It is recommended (Gavrilenko 2011) to take a time point as the end of the process when the current subsidence (η_0) reaches $0.97 \div 0.99$ its final value. At such approach it is possible to consider that depth of a flat bottom displacement trough (η_0) is equal to final displacement of a terrestrial surface (η_κ), taken for one of main parameters of mathematical models (2-4).

The analysis of known experimental data (Borzych, 1999) showed that there are mining-and-geological conditions in which the flat bottom displacement trough on a terrestrial surface isn't formed even by working out of several extraction sites. It testifies that mathematical models (2-4) adequately describe processes of displacement of a terrestrial surface only for the greatest possible extent of development of clearing works after formation of a flat bottom displacement trough. By their use for forecasting of course of process it's still unknown the value of final subsidence of a terrestrial surface (η_κ). The offer (Gavrilenko 2011) to define (η_κ) according to (The rule of undermining of buildings... 2004) is insufficiently reasonable for the reasons given earlier in work (Kulibaba 2010). Besides this it has been established (Filatyev 2011) that criteria of formation of a flat bottom displacement trough by working out of anthracitic layers significantly differ from recommended (The rule of undermining of buildings... 2004). For the specified reasons in this work the purpose to develop the scheme of subsidence of a terrestrial surface before achievement of its full side job during removal of a clearing face from the cutting furnace is set. Such approach will allow to open more fully features of course of process of displacement of a terrestrial surface and to predict value at any extent of development of clearing works with removal of a clearing face from the cutting furnace on distance of L.

Processes of displacement of a terrestrial surface are considered by mathematical model (Gavrilenko 2011) in time for approximately identical speed of a moving of a clearing face. Such condition of application of model is noted in work (Kulibaba 2010). The author of the model (Gavrilenko 2011) offers also for the description of development of process of displacement instead of time on abscissa axis to use distance concerning a projection of the line of a clearing face to a terrestrial surface to a supervision point. Use of geometrical parameters, in our opinion, is more expedient as they allow coordinating the extent of development of clearing works, the change of corners of full displacements in the undermining rocks and the maximum subsidence of a terrestrial surface. Confirmation of validity of such approach are also almost functional dependences of the maximum subsidence of a terrestrial surface (η_m) in specific mining-and-geological conditions by changing of one of the geometrical amount of clearing development (Filatyev 2011) that is explained by constancy of thickness of developed layer (m), depth of carrying out works (H) and strength properties of undermining rocks. In the conditions of one mining layers, owing to the specified reasons, it is possible to apply mathematical dependences, both with absolute parameters, and with relative ones (Filatyev 2011). Relative parameters (η_m / m, L/H) it is expedient to use for generalization of the results received in different mining-and-geological conditions (Filatyev 2011).

By developing the scheme of formation of trough parameters of displacement of a terrestrial surface (fig. 1) it has been used modern ideas of the geomechanical processes happening in undermining rocks by development of clearing works. They consist of the following:

– the beginning of displacement of a terrestrial surface occurs in a point A during removal of a clearing face from the cutting furnace on some distance L_{H}, which is defined by strength properties of

rocks (f) and depth of carrying out works (H). The scheme of change of a ratio between boundary corners and corners of full displacement in process of development of clearing works is given in the reference (Filatyev 2010);

– the maximum subsidence of a terrestrial surface (η_m^1, η_m^2, η_m^3 ... η_m^i) to its full undermining by layers of a flat bedding take place approximately over the middle of the developed space. Dependence $\eta_m = \varphi_1(L)$ is described be curve 4 (Figure 1). Final subsidence (η_κ^i) for the concrete position of a clearing face (the amount of clearing development) is characterized by the maximum subsidence $\eta_\kappa^i \approx \eta_m^i$;

– the full undermining of a terrestrial surface is observed during removal of a clearing face from the cutting furnace on distance more L_K. In this case the final, most possible value of subsidence of a terrestrial surface (η_K) is approximately equal to depth of of a flat bottom trough (η_0);

– after formation of a flat bottom trough displacement of any point on a terrestrial surface doesn't depend any more on distance of its projection to the cutting furnace, and is connected only with a further moving of a clearing face. The description of process of displacement of a terrestrial surface during this period of development of clearing works completely corresponds to mathematical models (2-4).

The developed scheme of trough formation of displacement of a terrestrial surface during removal of a clearing face from the cutting furnace is confirmed both direct measurement of some parameters, and their calculation with use of experimental data about the processes which are caused by displacement of undermined rocks and indirectly characterizing their condition.

For example, fixing distances between a clearing face and the cutting furnace and observing dynamics of methane emission in vent wells (excavations), it is possible to calculate the change of corners of unloading (full displacement). In specific conditions the sizes of these corners changed in process of development of clearing works from 35 to 65° (Antoshchenko 2013) that practically corresponds to their final values (The rule of undermining of buildings... 2004).

One of important points is definition of withdrawal of a clearing face from the cutting furnace (L_{H}), by which displacement of a terrestrial surface begins. For specific mining-and-geological conditions it can be defined, having statistically processed experimental data of directly proportional dependences $\eta_m = \varphi_1(L)$ or $\eta_m / m = \varphi_2(L/H)$. Such type of dependences is caused by an active stage of course of process of displacement of rocks during this period of time of clearing works. The point of intersection of these dependences with abscissa axis (A) defines required value $L_{_H}$. The example of such definition is given in the work (Antoshchenko 2012) when at $H = 97 \div 114$ m value $L_{_H}$ made 21 m.

The key moment for the adequate description of processes by means of mathematical models is establishment of a trajectory of movement of points with the maximum value of subsidence η_m towards a moving of a clearing face. In the scheme offered by us the trajectory of movement of these points corresponds to a curve of 4 (Figure 1). In the

scheme (Antipenko 2001) dependence $\eta_m = \varphi_1(L)$ is accepted rectilinear. For specification of a type of this dependence by a method of the smallest squares made statistical processing of the experimental data known from references obtained for the last fifty years. The empirical dependences characterizing change of relative maximum subsidence of a terrestrial surface (η_m / m), are received for different mining-and-geological and mining conditions. Initial grouping of basic experimental data were made on strength properties of containing rocks. The rocks containing anthracitic layers are referred to the strongest. The intermediate group on durability is represented by layers with coals of average degree of a metamorphism. Less strong are rocks of the Western Donbass.

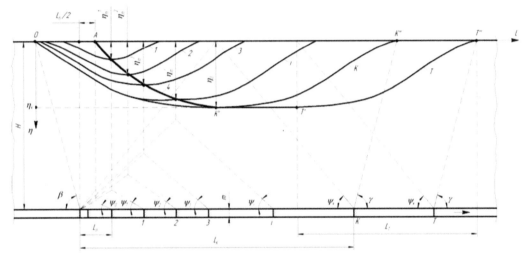

Figure 1. Scheme of formation of trough parameters of displacement of a terrestrial surface during removal of a clearing face from the cutting furnace: H – depth of carrying out clearing works; m – thickness of developed layer; 1, 2, 3, ... i, ... κ, ... T – the position of a clearing face at its withdrawal from the cutting furnace and trough of subsidence of a terrestrial surface corresponding to them; β, γ – boundary corners; ψ_0 – the corner of full displacement corresponding to the beginning of subsidence of a terrestrial surface, ψ_1, ψ_2, ψ_3, ... ψ_i – the corners of full displacement corresponding to 1, 2, 3, ... i – positions of a clearing face; ψ_{κ} – final value of a corner of full displacement, η_m^1, η_m^2, η_m^3, ... η_m^i – maximum subsidence of the terrestrial surface, corresponding to 1, 2, 3, ... i – positions of a clearing face; 4 – trajectory of movement of points with the maximum value of subsidence η_m towards a moving of a clearing face; η_{κ} – final subsidence of a terrestrial surface approximately equal to depth of a flat bottom trough of displacement η_0; $L_{_H}$ - distance from a clearing face to the cutting furnace at which displacement of a terrestrial surface begins; L_{κ} – distance from a clearing face to the cutting furnace at which there is a full undermining of a terrestrial surface; $\longrightarrow$ – direction of a moving of a clearing face.

Except strength properties of containing rocks during processing experimental data the depth of carrying out works (H) and the amount of clearing developments were considered (L_1, L_2). In the

course of experiments one of the geometrical sizes changed, and the second remained constant. If the single lava of variable length L_1 was object of supervision, the change η_m during removal of a clearing face from the cutting furnace on a distance L_2 was considered. By working out of several extraction sites when the full side job of a terrestrial surface wasn't reached, length of an extraction column was invariable. In this case value η_m was defined after discrete increase of the second size of the developed space at the size of length of the next fulfilled lava.

Such approach allowed receiving the empirical equations (1-3) considering two sizes of clearing development:

– for conditions of working out of anthracitic layers

$$\eta_m / m = \frac{0.67}{1 + 9.83 \cdot exp\left(-2.16\frac{L_1}{H} \cdot \frac{L_2}{H}\right)} ; \qquad (1)$$

– by extraction of layers with coals of average degree of a metamorphism

$$\eta_m / m = \frac{0.78}{1 + 11.31 \cdot exp\left(-3.14\frac{L_1}{H} \cdot \frac{L_2}{H}\right)} ; \qquad (2)$$

– for the Western Donbass

$$\eta_m / m = \frac{0.92}{1 + 23.97 \cdot exp\left(-4.79\frac{L_1}{H} \cdot \frac{L_2}{H}\right)} . \qquad (3)$$

The logistic equations of type (1-3) are usually used for modeling of processes of transition from one stable condition in another. In relation to the description of subsidence of a terrestrial surface the numerator characterizes the greatest possible value η_m / m. In the case under consideration it corresponds to depth of a flat bottom trough of displacement of a terrestrial surface at a stage of carrying out clearing works. Empirical coefficients of a denominator define the position of a curve concerning abscissa axis and width of an average site (an active stage). Dependences (1-3) almost functionally describe processes of displacement of a terrestrial surface and correspond to their physical sense at values $\frac{L_1}{H} \cdot \frac{L_2}{H} \geq 0.3 \div 0.5$. Smaller values of argument $(\frac{L_1}{H} \cdot \frac{L_2}{H})$ characterize processes of the beginning of displacement of a terrestrial surface and demand the separate studying.

After removal of a clearing face from the cutting furnace on distance L_κ, a site ($K' - T'$) of curve

dependence $\eta_m = \varphi(L)$ becomes almost parallel to abscissa axis (Figure 1) that testifies to formation of a flat bottom displacement trough. Subsidence of a terrestrial surface on a site between points T' and T'' also is defined by only the current position (T) of a clearing face. The distance from a projection of points of a terrestrial surface to the cutting furnace in this case has no any more practical impact on processes of displacement and consolidation of the undermined rocks.

The developed scheme, coordinating development of clearing works and processes of displacement of a terrestrial surface, allows expanding a scope of mathematical modeling by means of characteristic points. On the basis of the carrying out researches the important practical conclusions have been don for mining science:

– in mathematical models instead of temporary parameter more expedient is to use the geometrical sizes of clearing development (the developed space). It will allow developing the general mathematical models for the description of processes of displacement by removing of a certain group of coal layers. In this case it is possible to go to time factor, setting values of speed of a moving of a clearing face;

– three stages of processes of displacement of a terrestrial surface (initial, active and attenuations) have the features connected with development of clearing works. By an incomplete side job of a terrestrial surface it is necessary to consider mathematical models with use of the parameters characterizing the geometrical sizes of clearing developments (the developed spaces). After achievement of a full side job processes of displacement of points of a terrestrial surface depend only on their position in relation to a clearing face;

– for mathematical modeling it is conditionally possible to consider that formation of a flat bottom trough of displacement of a terrestrial surface is one of criteria of the end of processes at a stage of carrying out clearing works;

– final displacement of a terrestrial surface needs to be defined taking into account the sizes of earlier fulfilled extraction sites;

– in the conditions of one mining layer at approximately constant values of depth of carrying out works, the thickness of developed layer and strength properties of containing rocks can be used mathematical dependences both with absolute parameters, and with the relative ones. Relative parameters are recommended to be applied to generalization of the experimental data obtained in different mining-and-geological conditions;

– movement of a trajectory of points towards a moving of a clearing face during its removal from the cutting furnace happens to the maximum value of subsidence of a terrestrial surface generally on curvilinear dependence that is confirmed by statistical processing of experimental data. The beginning of displacement in this case is determined by a point of intersection of the specified dependences on abscissa axis;

– some parameters of displacement of rocks and a terrestrial surface can be determined on the experimental curve characterizing dynamics of gas emission from underlined sources;

– the choice of mathematical functions for the adequate description of process of subsidence of a terrestrial surface at all stages of development of clearing works by means of characteristic points demands the further analysis and development of recommendations about their application in specific mining-and-geological conditions.

REFERENCES

The rule of undermining of buildings, constructions and natural objects at coal mining by underground way. Publishing house is official. 2004. GSTU 101.00159226.001-2003. Sectoral standard of Ukraine. Kyiv: Ministry of Energy of Ukraine: 128.

Kulibaba, S.B., Rozhko, M.D. & Hokhlov's, B.V. 2010. *Nature of development of process of displacement of a terrestrial surface in time over moving clearing face.* Science works UKRNDM NAN of Ukraine, 7: 40-54.

Gavrilenko, Yu.N. 2007. *The mathematical description of dynamics of process of displacement on coal mines of Donbass.* International Society for Mine Surveying, XIII International Congress. Budapest, Hungary. Report 032: 6.

Gavrilenko, Yu.N. 2011. *Forecasting of displacement of a terrestrial surface in time.* Coal of Ukraine, 6: 45-49.

Borzych, A.F. & Gorovoi, E.P. 1999. *Influence of width of the developed space on activization of displacement of carboniferous massif.* Coal of Ukraine, 9: 26-30.

Filatyev, M.V., Antoshchenko, N.I. & Syatkovsky, S.L. 2011. *About the maximum displacement of a terrestrial surface by working out of coal layers.* Coal of Ukraine, 2: 37-40.

Filatyev, M.V. 2011. *Influence of extent of development of clearing works on the maximum subsidence of a terrestrial surface.* Coal of Ukraine, 4: 12-16.

Filatyev M.V, Antoshchenko, N.I. & Syatkovsky, S.L. 2010. *Necessary conditions of formation of a flat bottom trough of displacement of a terrestrial surface after removing of coal layers.* Sb. scientific works of DonSTU. Vyp. 31. Alchevs'k: 41-49.

Antoshchenko, N.I., Kulakova, S.I. & Filatyev, M.V. 2013. *The gas emission forecast from the underlined coal layers.* Coal of Ukraine, 1: 44-49.

Antoshchenko, N.I., Chepurnaya, L.A. & Filatyev, M.V. 2012. *Quantitative assessment of parameters of displacement of the underlined rocks and a terrestrial surface by removing of coal layers.* Sb. scientific works of DonSTU. Vyp. 38. Alchevs'k: 17-24.

Antipenko, G.A. & Nazarenko, V.A. 2001. *About some terms and definitions of process of displacement of a terrestrial surface.* Coal of Ukraine, 9: 44-45.

Mining of Mineral Deposits – Pivnyak, Bondarenko, Kovalevs'ka & Illiashov (eds)
© 2013 Taylor & Francis Group, London, ISBN: 978-1-138-00108-4

Changes of overburden stresses in time and their manifestations in seismic wave indices

A. Antsyferov, A. Trifonov, V. Tumanov & L. Ivanov
UkrSRMI of the NAS of Ukraine, Donets'k, Ukraine

ABSTRACT: During one and a half – two months, due to the redistribution of the stress state, the energy indicator of low-velocity low-frequency waves of channel nature can increase and decrease two-threefold and the energy indicator of high-frequency component of the refracted waves can change by a factor of four.

Coal-producing areas are characterized by manifestations of widespread geodynamic phenomena. Among these are rock bursts, induced earthquakes, ground surface subsidence and caving and others. Unpredictability or poor predictability leads to emergency and catastrophic situations. Occurrence of these phenomena is closely connected with the changes in the stress state of rocks in time. For that reason research into the changes in time of the stress state conducted in coal-producing areas is of immediate interest.

Changes in the stress state are caused by two main groups of factors: natural and human-induced. The main natural factors are recent tectonic fault movements, and undermining of rock mass is human-induced factor.

Estimation of the stress-deformation state of rock mass based on the data of ground surface movement is traditionally used for long-lasting processes. Monitoring of geophysical fields provides information on the changes in the stress state during comparatively short time determined by months, days and hours. In recent times estimation of rock stresses employing seismic sounding techniques growths rapidly (Antsyferov, Tirkel and other 2009; Antsyferov, Kiseleev and other 2012; Trifonov, Arkhipenko and other 2010; Antsyferov, Trifonov and other 2009; Trifonov, Tirkel and other 2009; & Antsyferov, Tirkel and other 2008 & Antsyferov, Tirkel and other 2007). Prospects of these techniques are specified by parametric abundance of seismic waves and capability of simultaneous use of several types of waves with physically different principles of generation and propagation.

Changes in seismic wave dynamic indices due to a joint impact of fault and rock mass undermining have been detected at the mine field of the Pokrovskoe Mining Unit in the Krasnoarmeisk coal province of the Donets Coal Basin. Surveys were conducted at two areas (area #1 and area #2 located

within the overburden exposure of fault zone of the Kotlinsky overlap fault. The bottom of the overburden lies at the depth of about 40 m. Overlap fault throws in the range of the first tens to the first hundreds meters. Its recent activity is confirmed by geophysical observations conducted in combination with geophysical survey. Both areas are located within subsidence profile, closer to its axial line. Rock mass have been undermined down to the depth of 750 m for one and a half year before surveying began. Taking into account that rock movement has been occurring for several years, the rock mass under consideration is in conditions of the changes in the stress-deformation state.

Sounding was conducted down to the depth of the overburden with the offset recording range of 60 m. Seismic signal generation and recording were carried out in shallow (about 1.0 m) wells with multiple (up to 100 times) stacking based on seismic energy excitation. Three cycles of observations were conducted at two mine field areas repeated in 1.5-2 months (07.08.12; 26.09.12; 20.11.12).

Spectra of the recorded seismic signals are shown in Figure 1. These spectra reflect several types of waves: high-intensity surface waves, refracted and low-velocity low-frequency waves. It is determined that during three cycles of observations changes in the spectrum of seismic signals take place. Sufficiently expressed differences are typical for its high-frequency region. At the area #1 they are confined to the frequency range of 50-70 Hz, and at the area #2 to 45-60 Hz (Figure 2).

With regard to the sounding conditions under examination the described changes in the spectrum is the manifestation of high-frequency components of refracted waves propagating in the deep dense part of the overburden. It is determined that energy index of the spectrum high-frequency components (as the sum of the amplitudes of frequency components of spectrum ranges) changes considerably during monitoring. At first (cycles 1-2) this index decreases, and then (cycles 1-2) increases. Its change takes place concurrently at two areas and achieves four-fold value (see Figure 2).

Seismic signals as absolute values of the amplitudes reduced by energy index of the first signal phase with averaging the energy index by sliding window method (width = 50 ms, interval = 25 ms) are shown in Figure 3.

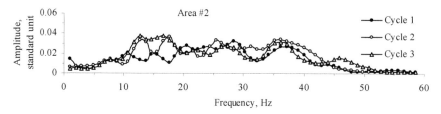

Figure 1. Seismic signal spectra based on three cycles of observations.

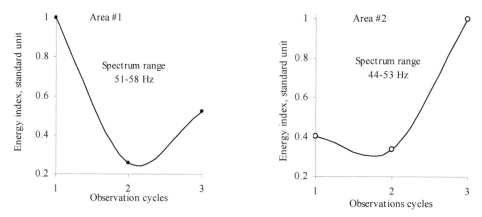

Figure 2. High-frequency part of seismic signal spectra and energy indices based on three cycles of observations.

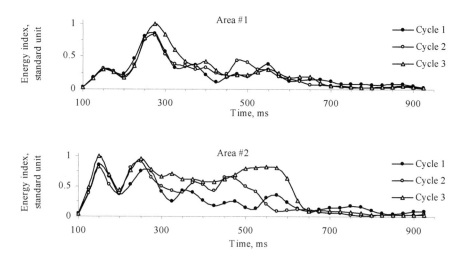

Figure 3. Seismic signals as absolute values of the amplitudes reduced by the energy index of the first signal phase with averaging the signal energy index in 50 ms time window and navigating the window based on the signal duration with 25 ms interval.

The difference of these signals by observation cycles enables us to identify three time bands: 100-160 ms, 300-450 ms and 500-950 ms. The first time band is typical for refracted waves (which propagation velocity is 400-500 m / s), the second one – for surface waves (150-200 m / s), and the third one – for low-velocity low-frequency waves (70-120 m / s). Energy index of the signals for three observation cycles based on the mentioned time bands is shown in Figure 4.

We can see that in the first time band (100-160 ms) the energy index practically does not change, or, as at the area #2, changes a little (up to 20%). Therefore refracted waves within their basic frequency (of the order of 30 Hz) shall be considered as the less sensitive of the analyzed wave types. The reason for this is apparently the long wavelength at this frequency (about 15 m) at which the sensitivity of refracted waves is much lower that that of the examined high-frequency components of these waves.

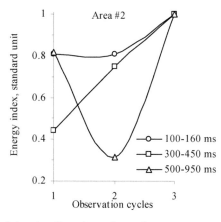

Figure 4. The energy index of seismic signals of different time bands based on three observation cycles.

The second time band is characterized by highly apparent increase in the energy index of the waves from cycles 1-2 and 3. At the area #1 this index increases by half, and at the area #2 it increases twofold. During monitoring there was no precipitation that influences the propagation of surface waves. Therefore, taking into consideration location of the areas with regard to the active overlap fault and subsidence profile being formed, the reason of the changes in the waves under examination is the redistribution of stress state of rock mass.

The third time band differs by the character and magnitude of changes in the energy index of the waves with time. This index decreases at first (from cycle 1 to cycle 2) and then increases (from cycle 2 to

cycle 3). Its value changes two-threefold. Such large range of changes in the energy index speaks for high sensitivity of low-velocity low-frequency waves. High sensitivity of these waves is specified by the characteristics of their propagation in conditions of the original seismic channel. The upper boundary of the channel is ground surface; the lower boundary is top of rocks with increased elastic parameters. In this case a seismic wave is generated as a result of sequential reflection of signals from the above boundaries at the angle near to 90° at which elastic waves undergo small refraction. Due to such path waves pass a distance that exceeds the sounding base (source point interval/spacing) by many times and accumulate the impact of the stress state of rock mass.

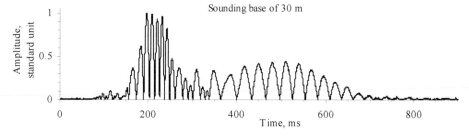

Figure 5. Low-velocity low-frequency wavetrain in the time band of 400-600 ms at seismic records obtained based on the different sounding bases.

Low-velocity low-frequency waves as a low-frequency wavetrain (11-18 Hz) are notably traced at seismic records obtained at the sounding bases of 30 and 60 m (Figure 5).

Amplitude levels of these waves in regard to the surface and refracted waves with the increase in the sounding base practically do not change. This feature shows that low-velocity low-frequency waves propagate in the seismic channel. In favor of their channel nature speaks also comparatively narrow spectrum and existence of dispersion (Figure 6).

Hence, taking into account the mode of propagation and significant changes in the energy index we can consider the low-frequency waves as the most sensitive indicator of the changes in stress state of rock mass.

It should also be stated that in spite of different nature of their generation the refracted and low-velocity low-frequency waves are characterized by correlation of energy indices changing with observation time (see Figure 2 and 4). This fact speaks for physical objectivity of the employed energy index as well as the capability of using the high-frequency components of refracted waves and low-velocity low-frequency waves for reliable estimation of the changes in the stress state of rock mass.

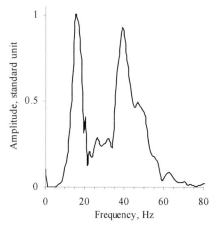 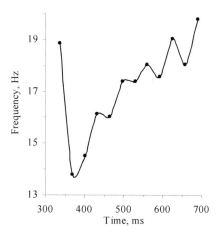

Figure 6. Seismic spectrum and frequency-wavetrain duration function for low-velocity low-frequency waves obtained at the sounding base of 30 m.

Thus, the outcomes of the survey are as follows.

1. Seismic sounding down to the overburden depth enables to detect redistribution of the stress state due to undermining of coal-rock mass and recent fault movements.

2. Due to the redistribution of the stress state the energy indicator of low-velocity low-frequency waves of channel nature can increase and decrease two-threefold and the energy indicator of high-frequency component of the refracted waves can change by a factor of four.

3. Low-velocity low-frequency waves of channel nature and high-frequency component of the refracted waves can be used for verifiable and operating control of the current changes in the stress state of coal-rock mass.

REFERENCES

Antsyferov, A. V., Tirkel, M.G., Glukhov, A.A., Trifonov, A.S. & Tumanov, V.V. 2009. *Development of the fundamentals of seismic monitoring of geomechanical state of rock mass when working coal deposits of Ukraine*. Methods and Systems of seismodeformation monitoring of induces earthquakes and rock bursts. Novosybirs'k: Mining Institute of the Siberian Branch of the RAS.

Trifonov, A.S., Kiseleev, N.N., Tumanov, V.V., Buzhdezhan, A.V. Khlyustov, N.V. & Yalputa, Ye.A. 2012. *Manifestations of human-induced faulting of rock mass in energy indices of seismic waves*. Donets'k: Transactions of UkrSRMI of the NAS of Ukraine: Collection of Scientific Papers, 11: 267-274.

Trifonov, A.S., Arkhipenko, A.I., Khlyustov, N.V. & Yalputa, Ye.A. 2010. *Seismic surveys of geomechanical state of the overburden in active development of exogenous geological processes*. Donets'k: Transactions of UkrSRMI of the NAS of Ukraine: Collection of Scientific Papers, 6: 294-300.

Antsyferov, A.V., Trifonov, A.S., Tirkel, M.G., Tumanov, V.V. 2009. *Estimation of the stress state of the undermined rock mass based on the parameters of seismic reflections*. Donetsk: Transactions of UkrSRMI of the NAS of Ukraine: Collection of Scientific Papers, 5 (Part 1): 434-440.

Trifonov, A.S., Tirkel, M.G., Tumanov V.V. & Arkhipenko A.I. 2009. *Research into the impact of stress state of the upper part of the rock mass being undermined on seismic signals parameters.* Donetsk: Transactions of UkrSRMI of the NAS of Ukraine: Collection of Scientific Papers, 4: 61-70.

Antsyferov, A. V., Tirkel, M.G., Trifonov, A.S. & Tumanov, V.V. 2008. *Seismic monitoring of coal-rock mass above production workings.* Proceedings of the Conference *Fundamental problems of formation of technogenic*

geoenvironment. Mining Institute of the RAS. Novosybirs'k.

Antsyferov, A. V., Tirkel, M.G., Trifonov, A.S. & Tumanov, V.V. 2007. *Diagnostic seismic monitoring geodynamic state of the rock mass when mining coal seams.* Proceedings of the Conference *Geodynamics and Stress State of Subsurface.* Mining Institute of the RAS. Novosybirs'k.

Specifics of percarbonic rock mass displacement in longwalls end areas and extraction workings

I. Kovalevs'ka
National Mining University, Dnipropetrovs'k, Ukraine

V. Vivcharenko
Department of Coal Industry Ministry of Energy and Coal Industry of Ukraine

V. Snigur
"DTEK Pavlogradugol", Pavlograd, Ukraine

ABSTRACT: Results of loading of support underground investigations of extracting working and longwall equipment set in end areas of longwalls were exposed. They took into account structure of percarbonic rock mass, represented by laminate massif of soft rocks.

Researches of rock pressure manifestation in extraction workings of West Donbass have history, comparable with the beginning of this region development. During last decades changed environmental conditions of seams extraction have not fundamentally transformed qualitative regularities of rock pressure rising in case of longwall approach to any fixed cross-section of extraction working and remove from it. These regularities are divided according to the most typical re-used extraction workings support periods. For each of which complex of analytical and underground investigation has been made. Eventual result – delivery of scientifically grounded recommendations of resource-saving conditions of their re-use providence.

Modern techniques of extraction workings support in West Donbass mines are characterized by common use of abutment-bolt support, essence of which includes bolts system formation in working roof, which is reinforced with rock arch, sharing part of the pressure, limiting roof subsidence and protecting frame support from extreme rock pressure of mainly vertical direction. Such combination of abutment-bolt and frame supports was analyzed in terms of extraction workings rock contour displacement measurements in proportion to stoping face progress.

Found regularities of roof rocks, walls and lying wall displacement development in extraction workings showed that:

– influence of longwall becomes obvious at distance of 20-25 m and up to 60-80 m in different mines. It is caused by different kinds of environment of mining;

– rock heaving is only 10-20% less than rock fault in the context of close meanings of rock hardness coefficients in direct roof and lying wall, and sometimes the second one is bigger by half than the roof; causes of such phenomenon are explained by "stamp effect";

– proportion of walls and roof-lying wall convergence builds up 67-105% in different mines and working areas, just as this proportion is regulated within the range of 0.20-0.39 in existing regulatory documents.

As a result, conducted experimental investigations in different mines of West Donbass discovered indispensable presence of active walls and lying wall rocks arching in extraction workings both outside and inside stopings influence zone. It also proved by numerous results of computer modeling of stress-strain state of rock massif, support and security systems. Nature of this phenomenon is following:

– partly softened rocks amount is located in working walls, inclosed between two hard bodies – reinforced rock plate in rood and coal seam;

– laminating amount of rocks in working lying wall is affected by abutment joint in walls, which is spread through hard coal seam.

The essence of the recommendation is reduced to multiplying of frame support, in accordance with known technical solution of space-flexible connections between frames and walls bolts, and to creation of united bearing system. Activation of rock pressure manifestation and appearance of its significant anomalies near working take place in front of the first longwall. Partial softening of rock layers of percarbonic mass in terms of its spatial discrete displacement in the areas of extension stresses activities and σ_y, and strong concentrations of compression stresses lead to increase of rock amount liable to active strains. And if they are spread on interacting plates system, process of partly softened near-the-contour massif walls and lying wall arching is intensified.

After longwall drivage, process of layerwise rockfall is growing rapidly, first, of direct and then of main roof with formation of classical zones in the goaf: random rockfalls, and gradual deflection of layers without discontinuity. These zones occur near side borders of the goaf, as it is shown on Figure 1. Following positions of support and security systems load should be marked here:

– extreme inequality of load spreading through working perimeter is occurred. It has a detrimental effect on its stability;

– security construction provokes formation of increased abutment load on it, and, as a result, – formation of side load on the frame;

– skewness of support load spreading is increasing, and intensive side load does not correspond to frame support design features.

Gradual stabilization of rock pressure manifestation takes place during further longwall removal. The working is maintained in such state for a long period (usually, not less than 1.5-2 years) until the time of adjacent extraction area mining. Solidification and consolidation of broken rocks with adaptation of some quaziself massif features takes place during this period. Rock pressure anomalies gradually reduce their manifestation intensity by means of stresses relaxation and strain flow. It follows, that the most dangerous area of extraction working (from the viewpoint of geostatic anomalies) is in the area of its connection with longwall: if it is possible to provide stable state of extraction working, then there is no possibility of further significant difficulties of its operating condition maintenance, taking into account some repair operations – mainly, bottom ripping.

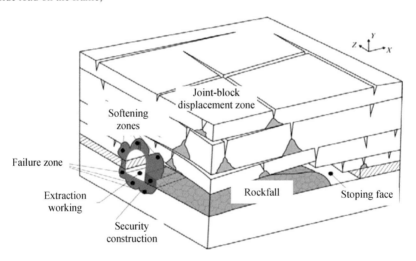

Figure 1. Spatial representation of percarbonic rock mass displacement scheme in the area of longwall and extraction working connection.

Further development of geomechanical processes in the area of extraction working concerns stoping zone of influence, which is generated by the second longwall. New significant inequality of load and its common growth trend in proportion to the second longwall movement appears. In order to choose compromise of extraction working maintenance (in terms of resource-saving) two factors must be considered:

– short length of the working maintained area behind the second longwall;

– sharp decrease of rock pressure manifestation intensity at the limited area 10-15 m length.

In order to rate influence of percarbonic rock mass movement and its structures in longwall end areas experimental investigation complex was conducted. It studies load formation processes on end sections of mechanized support depending on geomechanical and technological parameters of the stoping. It is determined, that regardless of geomechanical parameters of stoping, there is no stable regularity of load shifting on support section across longwall length, including its end areas (Figure 2). Maximum load on sections according to different coordinates X of extraction panel length occurs through longwall length Z in stochastic way. It has been repeatedly monitored both in central and end longwall areas. It is obvious, that it linked with percarbonic rock mass structure changes along the longwall length, which are extremely difficult to monitor using geological surveys. It is caused by

significant distances between holes. And mining-and-geological prognosis is based on data obtained from drilling of them. Given guess was tested with variation of other parameters, which affect formation of load on support sections. It and regular ambiguity (mainly spontaneous) of load spreading along the longwall length only approve the conclusion of massif structure dominant impact in context of other equal conditions.

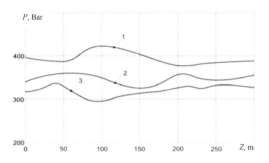

Figure 2. Change of pressure averaged values P in hydraulic props of the section along the longwall length Z caused by daily average rate V_c of stoping face advance:

$1 - V_c = 3.2$ m / day.; $2 - V_c = 7.7$ m / day.;

$3 - V_c = 11.4$ m / day.

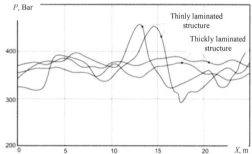

Figure 3. Change of pressure P in end areas hydraulic props during the process of main roof different structure areas extraction.

Increased attention is paid to percarbonic rock mass structure, because the process of its movement includes not only end areas of mechanized support, but also near-the-contour rocks around extraction working with its support. According to geomechanical process essence there is a united system considered. Its loading indicator is pressure P in end areas hydraulic props head ends.

Common point is that immediate mine roof amount weight, caved right after outcropping is the factor of mechanized support loading. But the main factor – subsidence of main roof rock layers, where load directly depends on hanging over the section mining bases length. This statement also concerns end sections, but mounting bases should be considered in three-dimensional representation in form of rock plates with soft jamming in coal seam depth and with bearing on side near-the-contour rocks of extraction workings. Against this background experimental investigations of rock plates formation process in longwall end areas are attractive. The objective is – further use of the results in extracting mechanism of load formation on extracting working support in the area of connection with longwall and behind it.

In order to conduct given investigations five areas with essentially different structure of percarbonic rock mass with roof height 11-13 m were selected: from mainly thinly laminated to mainly thickly laminated structures. Generally, they are represented by mudstone and silty rocks. Figure 3 diagrams show sufficient pressure stability P in hydraulic props (as section load equivalent) during the process of longwall movement (coordinate X of extraction area length, including extraction working) in terms of mainly. It shows that in some way significant mounting bases are not formed behind mechanized support sections blocking and process of rock layers fall from top to bottom is gradual and equal enough without rock pressure disturbance, which often precedes so called main roof caving. Such phenomenon is almost absent. According to this, it is logical to consider the rock plate (for every layer) in longwall end areas. It will also fall in extraction working cross-section at the border with the goaf, but without mounting bases hovering, directly in near-the-contour rocks of the working. When roof rocks has thickly laminated structure, despite low strength characteristics of mudstones and silty rocks, periodical pressure disturbance in end areas of hydraulic props takes place. It denotes mounting bases hovering with further periodical fall of them, which according to X coordinate corresponds to main roof caving spacing. It is logical to suppose, that there is also a plate with blow into the goaf for some length forming on the border of longwall and extraction working (in its cross-section). Although noted exceedence of average pressure in hydraulic props builds up 20-25%, highwall-shaped rock plates blow in terms of its roof height is able to significantly increase the load on the support of extraction working and security construction, raised behind the longwall. It reduces working berm stability and intensifies rock heaving.

As a result, experiments proved that process of load formation on the extraction working maintenance system can fundamentally differ within one extraction area; it is caused by percarbonic rock mass structure change. Marked variations of rock pressure manifestation must not only be reflected in the mechanism of load formation on the support, but also during the following geomechanical processes modeling stage, in order to predict extraction working stability and substantiate provision events of its re-using more objective and reliable.

tion working section, where main rood caving takes place. Noninteraction between parameters P and H shows that the process of load formation on the end sections is closed in some amount of main roof rocks. It does not depend on working depth, and far more connected with percarbonic rock mass structure and large variety of other parameters. On the analogy it is possible to suppose, that straining next to it unstable rocks amount, which creates load on support and security systems of extraction workings is also barely connected with mining depth along with closer relationship with other parameters, percarbonic rock mass structure in particulat. Explanation of this phenomenon lies in the context of West Donbass rock specific features, which have low strength and strain characteristics, weak connections between layers (or their almost absence), strongly marked deformation behavior; and deformable massif is extremely disposed to creation of pressure arch according to prof. M.M. Protodyakonov; it is known, that such arch volume does not depend on mining depth.

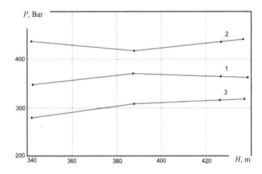

Figure 4. Change of average pressure P in hydraulic props according to the depth H of stoping with longwall position X in relation to l_n main roof caving spacing, wich is represented mainly by thickly laminated structure: $1 - X < l_n$; $2 - X = l_n$; $3 - X > l_n$.

The other experimentally proved feature of rock pressure manifestation in longwall end areas is load on mechanized support sections independence from mining depth H (Figure 4). As we can learn from the diagrams, depth change of almost 100 m (from $H = 440$ m to $H = 340$ m) has practically not affected hydraulic props sections pressure volume P. Data presented here is mainly for thickly laminated structure, when effect of main caving is manifested, in order to reflect depth effect H as significant as possible. Nevertheless, far more fluctuation P is caused by end sections location in relation to extrac-

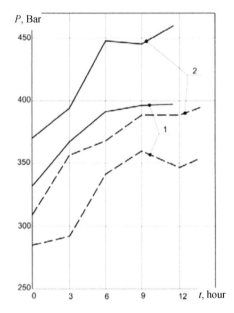

Figure 5. Pressure growth P in hydraulic props of pushed (1) to the working face and unpushed (2) end sections during t standstill: —— before main roof caving; – – – on the rest of spacing length.

Technological parameters of stoping such as daily average rate of stoping face advance V_c, standstill time t and distance from end sections to the working face δ play essential role in load formation on mechanized support (Figure 1 and 5). This regularities are limiting and widening roof rocks arch, which forms the load on end sections. Similar processes also include neighbor roof rocks over the extraction working.

CONCLUSIONS

1. Reuse efficiency of extraction workings is determined by the choose of their maintenance parameters based on objective and reliable reflection of geomechanical processes, which include "loading history" of support and security systems during the whole period of working exploitation from its construction and till its repayment after reuse. There are 4 stages of rock pressure manifestation during extraction working maintenance process with substantiation of soft rocks percarbonic rock mass movement, which is typical for West Donbass environment: outside stoping influence zone; inside abutment pressure in front of the first longwall; behind stoping face, including area of rock pressure stabilization; in the area of the second longwall stoping influence up till extraction working redemption. Such approach provides continuity of support and security systems loading mechanism reflection in different areas of extraction working and correlation of further recommendations of its resource-saving maintenance.

2. Experimental investigations in different mines of West Donbass discovered indispensable presence of active walls and lying wall rocks arching (especially when frame-bolt support is used), which we named "stamp effect". Such phenomenon is found in the load development mechanism on the support system both outside and inside stoping zone, where "stamp effect" is considered in complex with percarbonic rock mass movement process in longwalls end areas. It also affects extraction working state.

3. With security system introduction into work, high abutment pressure forms horizontally, which increase skewness of support load spreading on support system. It sharply decreases its bearing ability. Moreover, frame support itself is not meant for resistance to high wall loads, that is why it needs reinforcement of chosen direction. And, according to our opinion, in order to realize it in the most effective way it should be done by means of frame-bolt support with space-flexible connection of wall bolts and frames creation. Other direction of extraction working stability increase is the diagram of load spreading on support system by means of strain-strength characteristic of security construction regulation in accordance with rock pressure manifestation specifics.

4. Area of rock pressure manifestation is characterized by partial leveling of geostatic anomalies caused by joint action of rheology factors such as solidification and consolidation of earlier softened rocks over the goaf. It simplifies support and security systems work. That is why it is extremely important to provide stability of extraction workings in the area of their connection with longwall. It is a warranty of resource-saving perspective creation of their reuse.

Operations under combined method of mining graphite deposit

V. Buzylo, T. Savelieva, V. Serduk & T. Morozova
National Mining University, Dnipropetrovs'k, Ukraine

ABSTRACT: Transition to combined method of mining graphite deposit as the most efficient one while mining given area is considered in the paper. In these conditions combined method gives the possibility for grade homogenizing, i.e. ore separation according to grades, accumulation of various ore grades and their further measuring to form a mixture with given value of ore grade. Questions concerning graphite percentage within ore mass, extraction of sized ore, i.e. quality of ore breaking and parameters which have impact on this process, mechanical means of drilling operations and hole charging are considered. Recommendations concerning loading and delivering broken ore are given.

1 INTRODUCTION

Transition from open-pit to combined underground-open-pit method of mining Zavalievs'k graphite deposit requires study and recommendations concerning the choice of the method of ore mass winning. Conditions of deposit occurrence and the shape of ore body are complicated. That's why applying selective or bulk mining is rational only in some places (it's not efficient to use them all over the section). In these conditions combined method (combination of bulk and selective mining) is the most rational one.

Combined mining will give maximum effect while working-out the section as sized ore is possible (blending), i.e. ore separation according to grades, accumulation of various ore grades and their further measuring to form a mixture with given value of ore grade.

Ore entry into technological operations with deviation from average level leads to deviation of technological mode from optimal level and to unbalance of carrying capacity and efficiency of following operations. In case of reducing the component value productivity is decreased as the process takes place in not-optimal mode. Therefore, the basic criterion of evaluating efficiency of ore mining method is graphite percentage within ore mass which enters to dressing mill. It should not be less than 5%. The important questions are extraction of sized ore, i.e. quality of ore breaking and parameters which have impact on this process, mechanical means of drilling operations and hole charging . These problems are paid attention to in this paper.

2 PARAMETER EFFECT OF HOLE CHARGE AND DIAMETER ON BREAKING QUAILITY

Calculation of graphite percentage within ore mass was carried out according to method offered in the paper (Koval & Stebakov 1988). Two enriched formations of gneiss layers G_1 and G_2 isolated vertical ore layers K_1 and K_2 are marked in geological section.

Isolated ore layers K_1 and K_2 can be mined by pillar method with chamber location within ore body, and thin layers of formations G_1 and G_2 with thin interlayers of waste rock can be mined by pillar method with chamber location across the strike of deposit.

Calculation of graphite percentage within ore mass showed that given mining method is acceptable to these conditions. Each zone gives its own average graphite percentage within ore mass. It is required to mix ore formation G_2 with ore layers K_1 and K_2 in equal proportions to have constant graphite percentage.

Extraction of sized ore under underground mining is of great importance. Usually extraction of oversized fraction is in the wide ranges (15-40%). Ore delivery from stopes is limited by the sizes of output units and applied means of delivery, load and transportation. Volume of works concerning secondary breaking on the output horizons grows with increased number of pieces exceeding fixed standard size. It leads to reducing labor productivity and output intensity, delay of delivering, loading and transport equipment, dust and gas content in mine atmosphere.

The main reason of poor crushing of breaking ore is excessive rarefaction of hole pattern. It is connected with the fact that manufactured drilling rigs have low rate of drilling and large diameter of boring crowns. It leads to great labor intensity and high cost of one meter of the well.

There are definite patterns of location for each well diameter under which explosive energy is in full application.

Data analysis given in Table 1 showed that expediency of applying particular diameter should be finally determined by economic comparison. Furthermore, drilling productivity plays the key role. Great diameter of blast holes leads to center localization of ore mass destruction that in its turn leads to greater irregularity of massif crushing. Ore breaking by wells of small diameter leads to increasing number of wells and consequently to increasing labor intensity and charging as well as to additional requirements for quality of explosives. Besides, changing the line of minimal resistance in the range of 1.5-2.5 m doesn't greatly influence on distribution of broken ore into fractions.

Table. Parameter effect of charge location and well diameter on crushing quality.

Well diameter, mm	Line of minimal resistance, m	Distance between wells, m	Output of oversized material, %
44	0.8	0.8	8.7
65	1.0 – 1.5	1.0 – 1.5	13 – 18
80	1.5 – 1.8	1.5 – 1.8	12 – 18
105	2.0 – 2.5	2.0 – 2.5	12 – 16
130	2.5 – 3	2.5 - 3	13.5
150	3 – 3.5	3 – 3.5	7 – 12
180	3.5 – 4	3.5 – 4	11 – 20

Table shows that under well pattern with the nests equal to 20-25 diameters of wells output of oversized material doesn't depend on diameter (Tkachuk & Fedorenko 1978).

3 WELL DEPTH EFFECT ON QUALITY OF ORE CRUSHING

With well depth increase from 10 to 30 m drilling productivity is sharply reduced. It is explained by the time increase spent on tripping operations and drilling out narrowed sections of the wells. Well diameter decrease under drilling depth increase is the most noticeable in hard abrasive ores.

Figure 1 shows that output of oversized material is increased up to 20% with well depth increase up to 30%. It was stated that the well keeps straightness at the length up to 17 m. Its deviation takes place

under great depth. Under well depth of 30-35 m its end due to deviation can be declined from direction of linear part per meter. Well deviation and declining from given direction while drilling to great depth cause in homogeneity of crushing and as a consequence increased output of oversized material. Thus, wells of up to 30 m depth are the most acceptable. Drilling wells of significant depth can lead to loss of advantages received in result of rational choice of well diameter.

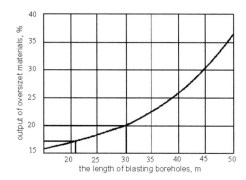

Figure 1. Output of oversized material under various depths of blast holes.

4 EFFECT OF SPECIFIC CONSUPTION OF EXPLOSIVES ON QUALITY OF ORE CRUSHING

Connection between specific consumption of explosives on breaking and quality of crushing is clearly seen within relatively monolithic rocks.

Dependence of specific consumption of explosives on ore hardness change while breaking by deep wells is shown in Figure 2.

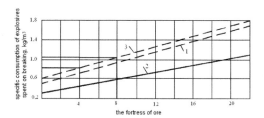

Figure 2. Change of specific consumption of explosives depending on ore hardness: 1 – well diameter 85 mm; 2 – well diameter 65 mm; 3 – well diameter 100 mm.

Ore body of Zavalievs'k deposit is not of the same hardness. Its hardness is in the range of 3 to 8.

Experimental and practical works of a large number of mining enterprises found out qualitative character of dependence of oversized material output on specific consumption of explosives on breaking (Figure 3). This dependence is given to conditions of ore mining of average hardness with insignificant fracturing. Quality of massif breaking is characterized by dependence of hyperbolical character.

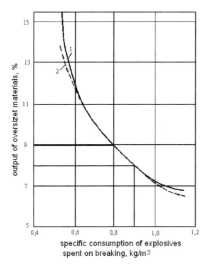

Figure 3. Change of oversized material output depending on specific consumption of explosives on breaking: 1 – according to calculation data; 2 – according to experimental data.

Data analysis given above shows that under ore hardness from 3 to 8 specific consumption of explosives is changed correspondingly from 0.65 to 0.9 kg / m^3. Furthermore, output of oversized material is in the range of 8-11%. It depends on specific consumption of explosives. The more specific consumption of explosives, the less output of oversized material.

Therefore, it can be regulated by two ways: drilling depth of blasting holes and specific consumption of explosives on breaking. Well diameter doesn't greatly influence on the output of oversized material.

Calculation of well pattern under specific consumption of explosives equal to 0.9 kg / m^3 and specific weight of 2.34 t / m^3 of ore allowed to determine well pattern taking into account the distance between the ends of adjacent wells within drill ring of 2.9 m. While loading coefficient of well charge along the length are interchanged in drill ring in sequence: 1, 2/3, 1/2, 1, 2/3, 1/2, 1, etc. Furthermore, given specific consumption of explosives according to coefficient of rock hardness corresponds to conditions.

5 MECHANICAL TOOLS OF DRILLING OPERATIONS AND WELL CHARGING

Practice shows that application of units NKR-100 with pneumatic impact tools is rather efficient for conditions of Zavalievs'k deposit. The tool has a great depth of drilling. So, well deviation will be minimal under drilling depth of up to 30 m/sec. Furthermore, drilling rate of the tool within ores of average hardness is 20-25 m / sec.

Granulated explosives are currently widely used in underground mining. These explosives are of great looseness that allows to mechanize charging efficiently. Equipment for mechanized charging depends on well direction, its depth and the type of explosives.

The unit UZDM-1 is used for mechanized preparation of igdanite explosive mixture, delivering explosives from production level to stoping face and charging well of any diameter and depth . These tools are able to charge wells by granulated explosives of factory fabrication and can produce new AS/DT explosives in the process of charging and to regulate water feed to explosive composition to dust elimination process or to charge wells by aquanite.

Plastic pipes are used as transport and charging magistral.

Charging safety concerning localization of electrostatic and electromagnetic effects is achieved by applying semiconductive pipes with definite specific volume, electric resistance, good ground of charging unit and drift watering before charging.

Well drilling takes much time in technological cycle as drilling units have a great drilling rate. It is required to select optimal parameters of well drilling with maximum productivity and minimal labor intensity.

Figure 4 shows the change of labor productivity while well drilling depending on its depth.

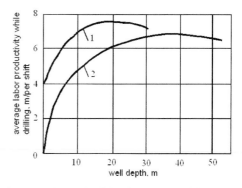

Figure 4. Labor productivity change under drilling depending on well depth: 1 – downward wells, 2 – upward wells.

Analysis shows that under drilling depth of 30-40 m downward wells and under drilling depth of 18-25 m upward wells average drilling labor productivity is maximum. Therefore, efficient depth of drilling downward wells is 30-40 m and upward wells is 18-25 m.

Figure 5 shows the change of labor productivity, drilling and charging cost under combined method of well drilling of ore massif depending on the depth of downward and upward wells.

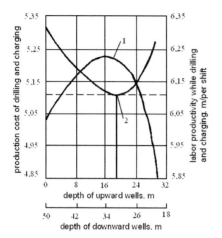

Figure 5. Change of labor productivity, drilling and charging cost under combined method depending on the depth of downward and upward wells:1 – downward wells; 2 – upward wells.

Under the height of drilling massif of 50 m rational depth of upward wells is 16-18 m concerning labor productivity and 18-20 m according to cost; downward wells is correspondingly 31-34 m concerning labor productivity and 30-32 m concerning cost.

Project is provided for ore output by natural flow to the oil ort. Loading and delivering is carried out by load-haul-dump units which are served by one worker and perform ore loading and delivering at the distance of hundreds of meters. It greatly increases labor productivity.

Load-haul-dump units carry out operations of loading, transporting and unloading loose rock mass (Baikonurov, Filimonov & Kaloshyn 1981). In comparison with other means of load and delivery they require less number of operating staff. They are more movable. It allows transporting ore from several faces during one shift. Development, final and non-productive operations take less time. Therefore, it is recommended to apply load-haul-dump units PD-12.

6 CONCLUSIONS

Analysis of the work done showed that output of oversized material can be regulated by the depth of drilling blasting holes and specific consumption of explosives spent on breaking. Furthermore, well diameter doesn't greatly influence on output of oversized material.

It was stated that under given method of mining wells of the following depth: upward – up to 20 m and downward – 30 m are more preferable.

Preliminary calculations concerning stability of chamber parameters carried out by the finite-element method were able to recommend the sizes of mining chambers.

For further design it can be possible to give following recommendations: chamber height is 48 m, chamber width is 10 m, chamber length is to the whole thickness of ore body or section of bulk mining.

Calculation of well pattern under specific consumption of explosives equal to 0.9 kg / m^3 could determine well pattern. Distance between the ends of adjacent wells within drill ring is 2.9 m. Well direction is drill ring of 90° from vertical wells to horizontal ones, well diameter is 80-105 mm, the length of upward wells is up to 20 m. Loading and delivering broken ore is recommended to carry out by load-haul-dump units PD-12.

REFERENCES

Koval, I.A. & Stebakov, B.A. 1988. *Efficiency increase of ore mining*. Moscow: Nedra: 332.
Tkachuk, K.N. & Fedorenko, P.I. 1978. *Blasting operations in mining industry*. Kiev: Higher school: 332.
Baikonurov, O.A., Filimonov, A.G. & Kaloshyn, S.G. 1981. *Complex mechanization of ore underground mining*. Moscow: Nedra: 264.

Plasma reactor for thermochemical preparation of coal-air mixture before its burning in the furnaces

A. Bulat, O. Voloshyn & O. Zhevzhik
M.S. Polyakov Institute of Geotechnical Mechanics, Dnipropetrovs'k, Ukraine

ABSTRACT: A mathematical model of the process of thermochemical preparation coal before combustion was presented; gas and particle velocities, temperature and gas composition as a result of the calculations were obtained; was founded that at invariable initial air mixture consumption increased diameter of the reactor leads to peak temperature zone shifting towards the initial section; the influence of the coal particle form coefficient and its diameter of the intensity of the air mixture burning in the reactor and the position of the ignition zone was analyzed.

1 INTRODUCTION

Effective firing of low-reactive coals in the furnaces of thermal power stations presents one of the most pressing and live challenges for the energetic complex.

Low-reaction coals of anthracite culm class with volatile content less than 4-6% are burnt in the power stations in the coal-dust flares illuminated by mazut or natural gas share of which in the heat balance can reach 30%. The process is accompanied by essential mechanical underburning and harmful emissions into the air.

The situation in the world indicates that volumes of anthracite culm coal production and usage will be not only kept at the same level but, in view of quickly exhausting oil and gas reserves, will increase. But the coal used in power complexes and heat-supplying systems is burnt in the obsolete equipment. Thus, improvement of the furnace operation through the coal prior thermochemical preparation (PTChP) is the problem No. 1.

There are several institutions, mainly in Ukraine (Korchevoy 2009) and Russia (Messerle & Ustimenko 2012), that studied processes of the coal prior thermochemical preparation and obtained good results with using natural gases or for highly-reaction coals with high volatile content which easily ignite and burn well even without prior PTChP.

The problem of productive burning of low-reaction coals of the anthracite culm class without using illuminating mazut or gas can be solved by a new plasma technology of PTChP designed by the M.S. Polyakov Institute of Geotechnical Mechanics under the National Academy of Science of Ukraine together with the Scientific and Engineering Center

"Ecology-Geos" and the Pridneprovskaya Thermal Power-Station.

Essence of the technology lies in rapid high-temperature heating of the air- mixture flow in a special aerodynamic reactor where coal-dust particles separated by their sizes are heated up to the ignition temperature. In the process of heating, the fuel is partially gasified, volatiles are released, and coal particles are ignited and burn with no added illuminating fuel (natural gas or mazut).

2 MATHEMATIC MODELING OF THERMOCHEMICAL PREPARATION PROCESS

Scheme of plasma reactor for the PTChP is shown in Figure 1. Air mixture containing air and coal particles flows to the plasma reactor 1. In the inlet into the reactor, air is heated by plasmatron 2 up to average temperature 800-1200 °C while initial temperature of the coal particles is about 240 °C. Mass consumption of the air mixture via the reactor is 5-20% of the total coal consumption via the burner. In result of the heating and chemical reactions on the surface of the coal particles and in volume of gas, a stream is formed in the outlet from the reactor which contains products of the coal dust gasifying and heated coal particles prepared for further burning in the furnace.

On the basis of mathematical description of mechanical, physical and chemical processes flowing in the reactor an approach was proposed which is described in the paper (Nigmatulin 1987).

To calculate velocity field and air mixture temperature and concentration in the reactor, a model of two-phase gas-solid particle stream was created basing on the eight chemical reactions (Vilenskiy & Khzmalyan 1977) (Table 1).

Continuity equation for each i-element of the gas component is

$$\frac{d}{dx}\left(\rho_i U_g\right)=\sum_j f_i^j,$$

where x – moving coordinate by the reactor length; ρ_i – density of i-element; U_g – velocity of gas flow; j – No. of chemical reaction from the Table 1.

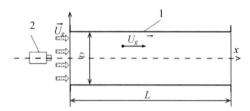

Figure 1. Scheme for calculation.

For example, for oxygen O_2:

$$\frac{d}{dx}\left(\rho_{O_2} U_g\right)=\sum_j f_{O_2}^j, \tag{1}$$

where $\sum_j f_{O_2}^j = f_{O_2}^1 + f_{O_2}^2 + f_{O_2}^6 +$

$+ f_{O_2}^7 + f_{O_2}^8 = f_{O_2}^{1,2,6,7,8}.$

Similarly are written equations for the rest components of the gas mixture participating in the reactions:

$$\frac{d}{dx}\left(\rho_{CO} U_g\right)=f_{CO}^{1,3,5,6}, \tag{2}$$

$$\frac{d}{dx}\left(\rho_{CO_2} U_g\right)=f_{CO_2}^{2,3,6,7}, \tag{3}$$

$$\frac{d}{dx}\left(\rho_{H_2} U_g\right)=f_{H_2}^{4,5,8}, \tag{4}$$

$$\frac{d}{dx}\left(\rho_{CH_4} U_g\right)=f_{CH_4}^{4,7}, \tag{5}$$

$$\frac{d}{dx}\left(\rho_{H_2O} U_g\right)=f_{H_2O}^{5,7,8}. \tag{6}$$

For nitrogen, which gets into the reactor together with air, $\sum_j f_{N_2}^j = 0$, so

$$\frac{d}{dx}\left(\rho_{N_2} U_g\right)=0. \tag{7}$$

Continuity equation for one-dimensional gas mixture stream is formulated after summing all equations for the gas components:

$$\frac{d}{dx}\left(\rho_g U_g\right)=f_g^p, \tag{8}$$

Where $\rho_g = \rho_{O_2} + \rho_{CO} + \rho_{CO_2} + \rho_{H_2} +$

$+\rho_{CH_4} + \rho_{H_2O} + \rho_{N_2}$ – gas mixture density;

$f_g^p = f_{O_2}^{1,2} + f_{CO}^{1,3,5} + f_{CO_2}^{2,3} + f_{H_2}^{4,5} + f_{CH_4}^4 + f_{H_2O}^5$ – intensity of mass exchange between gas and coal particles in the reactor

$$\frac{d}{dx}\left(\rho_g U_g^2 + P\right)= R_p^g + f_p^g U_p^g, \tag{9}$$

where P – pressure in the air-mixture stream, and $f_p^g U_p^g$ – summand which characterizes impulse exchange between gas and coal particles and is defined as

$$f_p^g U_p^g = \left(f_{CO}^{1,3,5} + f_{CO_2}^2 + f_{CH_4}^4 + f_{H_2}^5\right)U_p +$$

$$+\left(f_{O_2}^{1,2} + f_{CO_2}^3 + f_{H_2}^4 + f_{H_2O}^5\right)U_g.$$

In the Table 1, S is surface of the coal particle on which the reaction takes place.

Interfacial resistance force between gas and coal particles referred to the volume unit is defined as

$$R_p^g = C_d \frac{\pi d_p^2}{4}\rho_g \frac{\left(U_g - U_p\right)^2}{2}n,$$

where $C_d = \frac{24}{Re} + \frac{4}{\sqrt[3]{Re}}$ – coefficient of particle

aerodynamic resistance; $Re = \dfrac{\left|U_g - U_p\right|d_p}{v_g}$ – Reynolds number; d_p – coal particle diameter; v_g – gas kinetic viscosity; U_p – particle velocity.

Number of the coal particles per volume unit is

$$n = \frac{6G_{p0}}{\pi d_p^3 \overset{o}{\rho}_{p0} F U_p},$$

where G_{p0} – mass coal consumption in the inlet into reactor; $\overset{o}{\rho}_{p0}$ – real density of the coal particles; F – area of the reactor transverse section.

Table 1. Key chemical reactions in the reactor.

#	Equation for the reaction	Reaction velocity
1	$2C + O_2 = 2CO + 220$ kJ / mol	$\dfrac{d}{d\tau}\rho_{O_2} = -32967 \cdot S \cdot \rho_{O_2} \cdot e^{-14500/T_P} = f^1_{O_2}$; $\dfrac{d}{d\tau}\rho_C = \dfrac{24}{32} f^1_{O_2} = f^1_C$; $\dfrac{d}{d\tau}\rho_{CO} = -\dfrac{56}{32} f^1_{O_2} = f^1_{CO}$
2	$C + O_2 = CO_2 + 395$ kJ / mol	$\dfrac{d}{d\tau}\rho_{O_2} = -329.67 \cdot S \cdot \rho_{O_2} \cdot e^{-8500/T_P} = f^2_{O_2}$; $\dfrac{d}{d\tau}\rho_C = \dfrac{12}{32} f^2_{O_2} = f^2_C$; $\dfrac{d}{d\tau}\rho_{CO_2} = -\dfrac{44}{32} f^2_{O_2} = f^2_{CO_2}$
3	$C + CO_2 = 2CO - 175$ kJ / mol	$\dfrac{d}{d\tau}\rho_{CO_2} = -3296.703 \cdot S \cdot \rho_{CO_2} \cdot e^{-17500/T_P} = f^3_{CO_2}$; $\dfrac{d}{d\tau}\rho_C = \dfrac{12}{44} f^3_{CO_2} = f^3_C$; $\dfrac{d}{d\tau}\rho_{CO} = -\dfrac{56}{44} f^3_{CO_2} = f^3_{CO}$
4	$C + 2H_2 = CH_4 + $ kJ / mol	$\dfrac{d}{d\tau}\rho_{H_2} = -498.168 \cdot S \cdot \rho_{CH_4} \cdot e^{-13500/T_P} = f^4_{H_2}$; $\dfrac{d}{d\tau}\rho_C = \dfrac{12}{4} f^4_{H_2} = f^4_C$; $\dfrac{d}{d\tau}\rho_{CH_4} = -\dfrac{16}{4} f^4_{H_2} = f^4_{CH_4}$
5	$C + H_2O = CO + H_2 - 132$ kJ / mol	$\dfrac{d}{d\tau}\rho_{H_2O} = -1260.073 \cdot S \cdot \rho_{H_2O} \cdot e^{-16000/T_P} = f^5_{H_2O}$; $\dfrac{d}{d\tau}\rho_C = \dfrac{12}{18} f^5_{H_2O} = f^5_C$; $\dfrac{d}{d\tau}\rho_{CO} = -\dfrac{28}{18} f^5_{H_2O} = f^5_{CO}$; $\dfrac{d}{d\tau}\rho_{H_2} = -\dfrac{2}{18} f^5_{H_2O} = f^5_{H_2}$
6	$2CO + O_2 = 2CO_2 + 570$ kJ / mol	$\dfrac{d}{d\tau}\rho_{CO} = -\dfrac{1.489 \cdot 10^{12}}{T_g^{0.75}} \rho_{CO} \cdot \rho_{O_2}^{0.25} \cdot \rho_{H_2O}^{0.5} \cdot e^{-16000/T_g} = f^6_{CO}$; $\dfrac{d}{d\tau}\rho_{O_2} = \dfrac{32}{56} f^6_{CO} = f^6_{O_2}$; $\dfrac{d}{d\tau}\rho_{CO_2} = -\dfrac{88}{56} f^6_{CO} = f^6_{CO_2}$
7	$CH_4 + 2O_2 = CO_2 + 2H_2O + 800$ kJ / mol	$\dfrac{d}{d\tau}\rho_{CH_4} = -4.066 \cdot 10^{10} \rho_{CH_4} \cdot \rho_{O_2} \cdot e^{-15700/T_g} = f^7_{CH_4}$; $\dfrac{d}{d\tau}\rho_{O_2} = \dfrac{64}{16} f^7_{CH_4} = f_{7O_2}$; $\dfrac{d}{d\tau}\rho_{CO_2} = -\dfrac{44}{16} f^7_{CH_4} = f_{7CO_2}$; $\dfrac{d}{d\tau}\rho_{H_2O} = -\dfrac{36}{16} f^7_{CH_4} = f_{7H_2O}$
8	$2H_2 + O_2 = 2H_2O + 484$ kJ / mol	$\dfrac{d}{d\tau}\rho_{H_2} = -2.527 \cdot 10^8 \rho_{H_2}^{1.5} \cdot \rho_{O_2} \cdot e^{-3430/T_g} = f^8_{H_2}$; $\dfrac{d}{d\tau}\rho_{O_2} = \dfrac{32}{4} f^8_{H_2} = f^8_{O_2}$; $\dfrac{d}{d\tau}\rho_{H_2O} = -\dfrac{36}{4} f^8_{H_2} = f^8_{H_2O}$

Energy equation for the gas phase is:

$$\frac{d}{dx}\left(\rho_g U_g e_g\right) = -\frac{d}{dx}\left(P U_g\right) + E^g_p + \Sigma Q_g , \qquad (10)$$

where $e_g = u_g + \dfrac{U_g^2}{2}$ – specific gas energy; E^g_p – summand which defines intensity of energy exchange between coal particles and gas in the air mixture; ΣQ_g – total heat emission from homoge-

neous chemical reactions which is specified in accordance with the Table 1.

Internal gas energy is:

$$u_g = C_{Vg} T_g = \left(C_{Pg} - R\right) T_g ,$$

where C_{Vg}, C_{Pg} – specific heat capacities of gas at constant volume and pressure, accordingly; $R = P / \left(\rho_g T_g\right)$ – gas constant; T_g – gas temperature.

Value of E_p^g is determined by heat exchange Q_p^g between coal particles and gas and as a consequence of transformation of part of the coal particles into gaseous products of the air mixture; at the same time specific internal energy and kinetic energy transfer is:

$$E_p^g = Q_p^g + f_g^g \left(u_p^g + \frac{1}{2} U_p^{g2} \right).$$

Quantity of heat transferred with the coal particles in the gas phase due to the heat emission is:

$$Q_p^g = \alpha_p \left(T_p - T_g \right) \pi d^2 nk,$$

where T_p – particle temperature

Coefficient of the heat emission α_p is determined by the Nussel number and Katsnelson-Timofeeva formula $Nu = \frac{\alpha_p d_p}{\lambda_g} = 2 + 0.03 Re^{0.54} \times$

$\times Pr^{0.33} + 0.35 Re^{0.58} Pr^{0.35}$, which is true at $0.8 < Re^{0.54} Pr^{0.33} < 800$; $Pr = \frac{\mu_g C_{Pg}}{\lambda_g}$ – Prandtl number; μ_g – gas dynamic viscosity; λ_g – heat conduction in gas; k – geometric coefficient of the form ($k = 1.62...2.58$ for coal dust).

Equation of kinetic momentum of the particle continuum is:

$$\frac{d}{dx} \left(\rho_p U_p^2 \right) = R_g^p + f_g^p U_g^p, \qquad (11)$$

where $R_g^p = R_p^g$ – interfacial force at interaction between gas and coal particles referred to the volume unit; $f_g^p U_g^p$ – characterizes impulse exchange between gas and coal particles due to the mass exchange.

Energy equation for particle continuum is:

$$\frac{d}{dx} \left(\rho_p U_p e_p \right) = E_g^p + \Sigma Q_p, \qquad (12)$$

where ρ_p – density of the coal particles; $e_p = C_{Pp} T_p + U_p^2 / 2$ – internal energy of the particles; C_{Pp} – specific heat capacity of the gas parti-

cles; E_g^p – characterizes intensity of energy exchange between gas and coal particles; ΣQ_p – characterizes intensity of heat emission due to the homogeneous chemical reactions.

$$E_g^p = W_g^p + Q_g^p + f_g^p \left(u_g^p + \frac{1}{2} U_g^{p2} \right),$$

where $W_g^p = R_p^g \left(U_g - U_p \right)$ – work of aerodynamic resistance.

Quantity of heat lost by the particles is defined by convective heat transfer and radiation:

$$Q_g^p = \left[\alpha_p (T_p - T_g) + \sigma \varepsilon \left(T_p^4 - T_s^4 \right) \right] \pi d^2 nk,$$

where T_s – temperature of internal walls of reactor; σ – Stefan-Boltzmann constant, ε – rate of coal particle blackness.

The equations (1)-(12) in combination with the state equation for gas and approximative dependences of thermal properties of gas and coal particles on temperature form a closed system of equations for numeric calculation of air mixture parameters in the reactor for thermochemical preparation of the coal-air mixture.

Boundary conditions for calculating such parameters are conditions existing in the inlet into reactor.

3 CALCULATION RESULTS

The calculations were made for reactors with maximal length $x = 1$ m and working sector diameter $D = 160$-220 mm which use ANTHRACITE CULM coals (density $\overset{o}{\rho}_{P0} = 1630 \text{ kg} / \text{m}^3$, mass fraction of ash $A = 30\%$, carbon $C = 1$-A) with particle diameter $d_p = 20$-75 mcm, initial temperature of gas in the air mixture $T_{g0} = 800$-$1200\,°C$, coal particles $T_{p0} = 240\,°C$, initial velocity of particles in the inlet into the reactor $U_p = 0.99 U_g$, mass fraction of oxygen 23% and nitrogen 77% in the reactor. The calculation specified velocity fields U_p and U_g and temperature filed T_p, T_g and gas composition at thermochemical preparations (Figures 2-4).

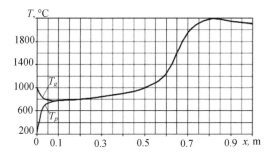

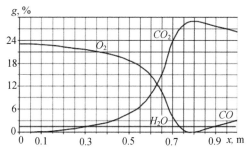

Figure 2. Gas and particle temperature in the reactor at D = 200 mm; k = 1.75; d_p = 50 mcm.

Figure 3. Gas concentration in the reactor at D = 200 mm; k = 1.75; d_p = 50 mcm.

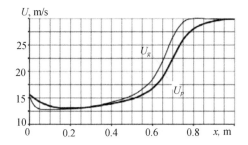

Figure 4. Velocity of the gas and particle flows at D = 200 mm; k = 1.75; d_p = 50 mcm.

4 CONCLUSIONS

At invariable initial air mixture consumption, increased diameter of the reactor leads to peak temperature zone shifting towards the initial section. For example, at D = 160 mm peak temperature is $x \approx 1$ m, and at D increased up to 22 mm it is possible to use reactor with twice less length and, consequently, reduced dimension of the whole aggregate. It can be explained by slower flow rate and, as a result, longer time period of coal particle presence in the reactor. Rate of flow in the outlet from the reactor reaches 48 m / s at D = 160 mm and 25 m / s at D = 220 mm.

At D < 160 mm, zone of intensive air-mixture firing with peak temperature moves to outside from the border of reactor. Zone of complete oxygen burning corresponds to peak temperature in the reactor.

The earlier oxygen enters into reaction the more intensive CO is accumulated in the reactor in zone where oxygen concentration is ρ_{O_2} = 0. Reduce of CO_2 concentration and drop of temperature in the zone where ρ_{O_2} = 0 is explained by endothermic reactions (Table 1).

Also, interdependence between coefficient of the particle form and intensity of the air mixture burning in the reactor was analyzed. For more round particles (k = 1.62), zone of intensive burning approaches outlet section of the reactor and corresponds to x = 650 mm. For great particles k = 2.5, peak temperature is reached at x = 420 mm. Hence, in reactor with D > 200 mm, air mixture ignites inside the reactor independently on the particle form, which is characteristic for ground coal.

When coal particle diameter increases from 20 to 75 mcm ignition zone moves towards outlet section of the reactor. When d_p = 20 mcm, peak temperature occurs at x = 0.32 m, and when d_p = 75 mcm, peak temperature occurs at $x \approx 1$ m. Coal particles with d_p > 75 mcm have no time to be heated and ignite when reactor length is 1 m.

Basing on the results of this mathematic simulation, new constructive decisions were made, and basic geometric parameters and potential operation modes were chosen for the reactor. When engineering documentation had been worked out, a pilot model of the plasma reactor for thermochemical preparation of the coal-air mixture was manufactured by the Production Incorporation "Yuzhny" Machinebuilding Plant and tested in the Pridneprovskaya Thermal Power-Station.

When 20% of the air mixture is treated in the plasma reactor, average mass temperature of the stream in the outlet from the reactor is 2-3 time higher than in the existing coal-dust vortex burners ensuring (a) complete coal burning in the furnace; (b) usage of no illuminative fuel; and (c) operation of the furnace with minimal mechanical underburning.

These results can be useful at designing plasma reactors for thermochemical preparation of coal prior to its burning in the furnaces of the thermal power stations.

REFERENCES

Korchevoy, J.P., Kukota, J.P., Dunajevskaja, N.I., Nehamin, M.M., Bondzyk, D.L. & Dedov, V.G. 2009. *Creation and preparation to experimental operation of pilot burners for boilers with pulverized high ash anthracite.* Science and innovation, 4. Vol. 5: 13-21.

Messerle, V.E. & Ustimenko, A.B. 2012. *Plasma-fuel systems for energy efficiency increase or pulverized coal thermal power plants.* Energy technologies and resource saving, 5: 42-45.

Nigmatulin, R.I. 1987. *Dynamics of multiphase media* [Dynamics of multiphase mediums]. Moscow: Nauka.

Vilenskiy, T.V. & Khzmalyan, D.M. 1977. *Dynamics of dust-type fuel burning* [Dynamics of dust fuel burning]. Moscow: Energiya.

The influence of fine particles of binding materials on the strength properties of hardening backfill

O. Kuz'menko & M. Petlyovanyy
National Mining University, Dnipropetrovs'k, Ukraine

M. Stupnik
Kryvyi Rig National University, Kryvyi Rig, Ukraine

ABSTRACT: Results of research of disperse particles influence of the domain granulated slag and dolomite on structural features and strength backfill massif while developing iron ores by the systems with a hardening backfill are given.

1 THE PROBLEM AND ITS RELATION TO THE SCIENTIFIC AND PRACTICAL TASKS

The increasing depth of underground mining of ore deposits and the increasing global demand for mineral materials can be satisfied in the application chamber mining method with a settable tab. It allows you to increase the completeness of extraction of ore and has an effective method to manage the rock pressure due to its intense manifestation, as well as stability to seismic movement blasting operations. In most cases, the strength of filling massif is proposed to increase by improving the composition of the hardening of the mixture by adding binding materials. The increasing content of expensive cement in the filling mixture do not always produce the required strength with divergent stresses that occur along the contour of the wide chamber.

The stresses in the contour wide chamber depend on its configuration and, consequently, the creating massif with constant physic-chemical features is unable to accept these crippling loads. For outcrop stability filling massif necessary to create internal structural features that must operate at different stresses.

Thus, the formation of stable structural features in hardening backfill under influence of fine particles is a new important scientific problem and actual for mining industry.

2 ANALYSIS OF RESEARCHES AND PUBLICATIONS

The main directions of improving the filling operations are mechanical activation or chemical activation of the components of filling mixture. Realizing these approaches to the formation of filling mixture are possibly with the following components: iron and steelmaking slag, fluxing limestone and crushed rocks. The operations in this direction are mainly focused on changing actual in the mixture of the components. The ability to change the strength is not considered due to the correction of infrastructural features and specific surface area of the particle binding materials.

3 GOAL SETTING

The chemical activation of binding components of filling mixture is possible by increasing the surface area of the particles in the initial stage of preparation the filling mixture. It is necessary to break the natural growths calcium oxide and silicon from other chemical entities. Thus, the conditions will be created to the formation of new structural features that will change the strength properties of the fill massif.

Investigation of the influence of the specific surface of the particles on the strength hardening backfill has been held conducted on the part of filling mixture applied at the CJSC "ZZhRK". The composition of filling mixture: granulated blast furnace slag – 18.1%, fluxing limestone – 47.5%, the crushed rock – 16.3%, water – 18.1%. Fineness slag in ball mill particles is 50-60% grade – 0.074 mm, which corresponds to a specific surface of about 2000 cm^2/g.

4 THE MAIN RESEARCH MATERIAL

In the laboratory, "ZZhRK" were tested 12 compositions filling mixtures at a quantity of granulated blast furnace slag 100, 200, 300 kg / m³. The specific surface was incremented 2000, 2800, 4300, 6600 cm²/g. In the filling mixtures crushed fluxing limestone added in an amount to 50% of quantity blast-furnace slag as a microfiller with a surface similar to the slag (Kuz'menko 2009). Granulated blast-furnace slag and fluxing limestone was crushed by the jet mill to this specific surface.

For fine grinding granulated blast furnace slag particles formed have a high reactivity due to the accumulation of surface energy and the formation of large amounts of unsaturated valence bonds.

For grinding of quartz significant amounts of bonds $Si - O$ on the surface of the grains forming ions Si^{4+}, O^{2-} (Taymasov 2003).

The main minerals of granulated blast furnace slag are calcium silicates. Contents oxides CaO and SiO_2 is more than 85% of the slag minerals with surface 2000 cm²/ g. On the surface of crushed slag particles are forming large amounts of calcium ions Ca^{2+} ions than Si^{4+}. Predominantly ionic $Ca - O$ and $Si - O$ covalent bonds are destroyed. Their energy is 1075.6 kJ / mol and 1861 kJ / mol, respectively, indicating a greater susceptibility to ion $Ca - O$ bonds to fracture. A significant part of the $Si - O$ is in a state of inertia. By its nature, the

covalent $Si - O$ bond is a strong, many silicate material characterized by high hardness. To destroy a larger amount of $Si - O$, to release the silicon ions and to use them in the hydration of binding materials is necessary to increase the degree of dispersion of the particles. This will also increase the solubility of calcium silicate particles.

Consequently, it is necessary not destroyed bonds $Si - O$ to convert from inert to active state. Thus silicon ions need to involve in the formation of hydrated calcium silicates final structure strong covalent bond.

To determine the change in the structural features filling massif used the method of infrared spectroscopy and scanning electron microscopy.

Structural features that have occurred in the hardening backfill with the increase of the specific surface particles were recorded in the infrared spectrum of the instrument SPECORD-75IR and the results are shown in Figure 1.

When comparing the spectrogram samples hardening backfill different specific surface binding material (Figure 1a, b) there has been increasing the intensity of the absorption in the interval of 900-1000 cm⁻¹, which indicates of strong increasing covalent bond $Si - O$ and weak decreasing ionic bond $Ca - O$.

(a)

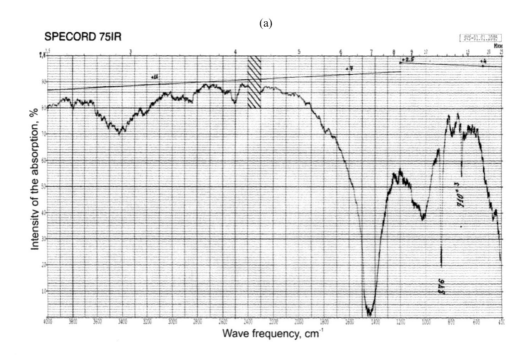

(b)

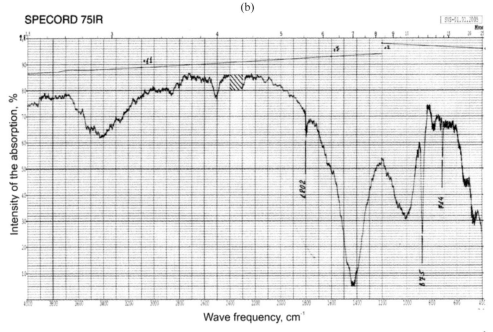

Figure 1. Infrared spectrograms of samples hardening backfill with specific surface binding materials: (a) 2800 cm^2 / g; (b) 4300 cm^2 / g.

The hardening backfill is strengthened at the expense of increasing the degree of crystallization of hydrated calcium silicate bonds. Strengthening bands of absorption in the interval 3400-3600 cm^{-1} shows the increase of calcium in the structure neocrystallisation of hydrated calcium silicates of hardening backfill.

Structural features filling mass were studied on samples using a scanning electron microscope REMMA-102-02 with a specific surface area of the slag and limestone 2000, 2800, 4300, 6600 cm^2 / g. The study of the samples hardening backfill detected a number of structural features of the neocrystallizations at different specific surface area of the slag and limestone, different form, chemical composition and the degree of influence on the strength (Figure 2).

(a)   (b)

Increased strength 1.8 – 2.5 times

(c)   (d)

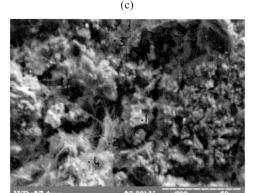   

Increased strength 3.1 – 3.5 times   Increased strength 3.1 – 4.5 times

Figure 2. Conversion structural features in hardening backfill curable depending on the specific surface of the particles of slag and limestone: (a) structure of gel ($S = 2000$ cm^2/ g); (b) large-needled structure ($S = 2800$ cm^2/ g); (c) needle-fibered structure ($S = 4300$ cm^2/ g); (d) layered structure ($S = 6600$ cm^2/ g).

Structure hardening backfill varies with increasing fineness slag and limestone, run down neocrystallizations size (Figure 2a-d). In conditions ZZHRK structure hardening backfill of period at 90 days provided hydrosilicate gel (Figure 2a). It lacks strong crystal structures which increases its destruction. The increase specific surface area contributes to the appearance of crystalline bonds in the structure (Figure 2b-d), decreasing porosity hardening backfill and increasing stability of the filling massif.

On the strength hardening backfill considerable influence exert proportion CaO / SiO_2 in the structural features. At proportion $CaO / SiO_2 \geq 1.5$ form highly basic hydrated calcium silicates and at $CaO / SiO_2 \leq 1.5$ – low basic.

The greatest strength has crystals of low-basic hydrated calcium silicate. Strength tensile filamentary structure of hydrated calcium silicates $CaO / SiO_2 \geq 1.5$ consist 1350 MPa, and highly basic $CaO / SiO_2 \geq 1.5$ – 770 MPa (Taymasov 2003). Hence, one way to creating a strength hardening backfill array is the formation of hydrated calcium silicates with basicity $CaO / SiO_2 \approx 1$.

To identify the type of basicity hydrated calcium silicates of filling massif were made measurements the content oxides CaO, SiO_2 with microscope microprobe analyzer. On Figure 2a-d test points marked in arabic numerals. The range of basicity decreases with increasing of the surface area of the particles of slag and limestone.

The structure with needle-fibered features is the most stable to stretching. Needled features distribute in artificial stone indefinitely. Needles and fibers are intersecting, creating a reinforcement hardening system and increase the adhesion between the crystals.

5 CONCLUSIONS

1. Increasing surface area of the slag and limestone in interval 2000 to 6000 cm^2/ g accompanied by changing of form the chemical features, leads to the transformation of highly basic hydrated calcium silicates to low basic and is expressed with power law. It will allow to control the structure of the hardening backfill and increasing the strength of filling massif.

2. Needle-fibered structure of the filling mass most stability to stretching, layered structure to compression. Therefore, to fill out waste area by compositions hardening backfill with the above structure is recommended in areas of concentration of tensile and compressive stresses arising with developing wide chambers that increase the stability of filling massif.

REFERENCES

Kuz'menko, A., Petlyovanyy, M., Chystyakov, E. and other. 2009. *To the question of selection filling mixture composition of extra strong.* Collection of Scientific Papers IGM. Edition 86. Vol. 1: 50-57.
Taymasov, B. 2003. *Technology of production of portlandcement.* Tutorial, SKSU: 297.

Magnetite quartzite mining is the future of Kryvyi Rig iron ore basin

M. Stupnik & V. Kalinichenko
Kryvyi Rig National University, Kryvyi Rig, Ukraine

ABSTRACT: Solution of the problem of rational use of raw materials by increasing completeness of ore extraction due to the involvement into magnetite quartzites mining occurring in the areas of acting and closed mines is given.

1 THE SCIENTIFIC AND PRACTICAL TASKS

The mining industry of Ukraine is the basis of the national economy and in the next years will maintain its leading position. Mining complex production is the foundation for steel making plants and provides a substantial part of the country's Germany.

The essential part of balance reserves of iron ore in Ukraine is concentrated in Kryvyi Rig. However, the sheer number of iron ore reserves does not guarantee stable prospects for the use of basin raw materials as industrial iron ore reserves in project ultimate pit of producing quarries and mines accounted for only 7.71 billion tons (43.1% of balance sheet reserves), of which only 0.45 billion tons are natural and rich iron ore with an iron grade of 54-62% and 7.26 billion tons of iron ore is poor or magnetite quartzite with iron magnetite grade of not more than 24-30% (Kolosov 2004).

2 PAPER ANALYSIS

A distinctive feature of the ore mines, developing their steep and sloping deposits is the increase in production costs while mining and increase of the depth of mining. Achieved by certain mines the depth of the 1400-1600 m is almost the limit for the applied technologies of ore lifting. The transition to the new, combined ore lifting diagram in the current investment environment is hardly probable.

In addition, with the increasing depth of mining the conditions of ore production and delivery, ventilation of working faces worsen, the quality of ore fragmentation reduces and losses and dilution of ore increase at output.

The main direction of the rational use of underground Kryvyi Rih raw materials many experts (Kolosov 2004; Kolosov & Dyadechkin 2005; Bliznyukov & Piven' 2002; Drobin 2002) see in the mining of magnetite quartzite, using resource-saving technologies, as at the present rates of lowering underground mining reserves of natural high-grade iron ores within the first stage of hoisting, will be worked out over the next 20-25 years (Kaplenko 2004; Conception of mine-metallurgy...).

3 THE PROBLEM

The rational use of Kryvyi Rig raw materials, according to the authors, can be achieved through scientific study and development of highly efficient and low-cost technologies for magnetite quartzite mining, located on the overlying levels of acting and closed mines. The use of these technologies will allow through reconstruction and implementation of the overlying levels to maintain the production capacity of the mines without transition to the second penetration degree.

Thus, the aim of this paper is the development of technological approaches to improve the utilization rate of Kryvyi Rig raw materials by involvement of magnetite quartzite into underground mining.

4 PRESENTATION OF THE MATERIAL AND RESULTS

The rational use of underground Kryvyi Rig raw materials having developed infrastructure, the well-run mining economy and qualified personnel determines the necessity of complex use of iron ore. Involvement into magnetite quartzite development will expand the raw material base of underground ore mines of Kryvyi Rig basin, will reduce the intensity of the development of high-grade ore, and will play an important role in solving the problem of integrated utilization of Kryvyi Rig ore deposits.

Currently, preliminary works are carried out for the resumption of magnetite quartzite underground mining in some mines, which stocks are estimated in billions of tons. Located on the overlying levels of acting mines, these stocks may be available during the reconstruction of the worked-out levels.

It should be noted that underground mining of magnetite quartzite in Kryvyi Rig on a large scale was carried out at the time of the former ore mines named after Dzerzhinsky, Pervomay and mine named after Ordzhonikidze. At some mines of the northern group of ore mines technical projects were compiled and preliminary works were started. However, with the change in the economic situation mining technology policy has been changed. Some mines carried out quartzite mining, were closed, and started preliminary works in acting mines have been suspended.

The first mine, which started magnetite quartzite mining after many years, was the mine named after Ordzhonikidze. Its current annual underground production is about one and a half million tons.

The main disadvantage of magnetite quartzite underground mining is its higher cost compared to the open pit mining. In addition, the applied technological diagram of production "room-pillar" is characterized by high losses of ore in the pillars (50% and more). The use of more advanced technological diagrams with goaf stowing was unprofitable because of the high cost of consolidating stowing.

Appeared in recent years, tendency of increase in cost of open pit quartzite mining associated with the further deepening of the pits and significant retardation of stripping, has allowed to talk about the reasonability of resumption of underground mining of magnetite quartzite. So now there is a need for new advanced technological diagrams that allow efficient underground mining of magnetite quartzite with a high completeness of extraction.

Development of magnetite quartzite located in the upper levels of mines is available at relatively low cost on hoisting, pumping, ventilation and transportation of minerals. In the case of the preparation and development of new levels considerable strength and stability of magnetite quartzite allows to pass most of the breakoff with an inexpensive lining that helps to reduce the cost of minerals production. Worked-out rooms located in the upper levels of mines, also requires low cost of transporting the stowing material to the worked-out blocks.

With a view to improving the technology of underground mining of magnetite quartzite a number of technological diagrams allowing mineral hoist with high recovery rate and low production cost of filling works has been developed (Kaplenko 2006; Kaplenko & Kalininchenko 2006).

These technological diagrams allow to save undisturbed ground surface, using a consolidating stowing with less expensive binding components. Some technological diagrams involve disposal of waste rock and rock refuse in the worked-out area of stope blocks.

The use of inert filler in the filling works along with a simultaneous decrease in the flow of binding components reduces the cost of underground mining of magnetite quartzite competitive level compared to open pit mining.

Taking into account the environmental benefits of underground mining, the possibility of disposing of waste rock surface dumps used as an inert filler, the proposed technologies not only allow rational use of Kryvyi Rig raw materials, but also to produce a highly efficient mining with environmentally clean resource-saving technologies.

The essence of the proposed technological diagrams of magnetite quartzite mining by underground methods is as follows.

Developing site (level) is divided into stope blocks of phase I and II on the principle of "room-pillar" with the block height h_{bl} (Figure 1). Mining of stope blocks deposit of phase I of width L_1 is carried out by a room-and-pillar system. After deposits mining the worked-out room space of phase I is filled with a consolidating stowing with reduced binder component consumption. After development of regulatory strength of consolidating stowing, recovery of inter-room pillars (stope blocks of phase II) with a width L_2 is started.

Developing of inter-room pillar is carried out with the development system with caving of ore and host rocks, which allows concrete block of consolidating stowing of block of phase I constantly be in a state of volume compression.

If caving overlying rocks is impossible due to stringent regulations to magnitude of host rock displacement, caved ore hosting in blocks of phase II is carried out under crushed waste rocks, which are specially served in the clearing space to create conditions of volume compression of the concrete block.

Both place driving rocks and surface dump rocks intentionally bypass to the room through rock-letting raises.

In the presence of rock refuse, a combined filling of clearing space of blocks of phase II with the mixture of waste rock and rock refuse is possible. This mixture has a higher density and a lower coefficient of shrinkage compared with waste rocks. However, in this case it is recommended to create a rock cushion on contact with a caved ore to prevent the filtration of fine rock refuse into cave ore mass.

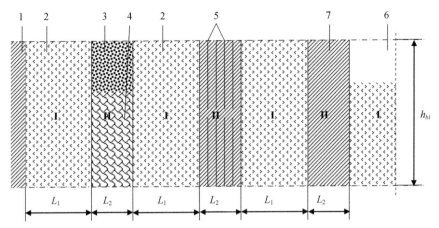

Figure 1. Technological diagram of combined mining of magnetite quartzite with room-and-pillar system with consolidating stowing (blocks of phase I) and systems with caving (blocks of phase II): 1 – rock ore; 2 – concrete block of consolidating stowing; 3 – caved waste rocks; 4 – broken ore; 5 – deep holes for drilling-off blocks of phase II; 6 – filling of blocks of phase I with consolidating stowing; 7 – inter-room pillar.

Being in a state of volume compression, concrete block of consolidating stowing with reduced binder component has a high load carrying capacity.

Thus, cost reduction of filling without violating the strength characteristics of concrete block is achieved.

There is no doubt that when developing the overlying levels with development systems with caving of ore and host rocks may be a occurrence of the geomechanical movement of overlying waste rocks. While large volumes of production by systems with caving, geomechanical violation of host rocks may cause undesired surface sagging.

In this regard, in order to reduce the harmful effects of floor (sublevel) caving systems onto the ground surface, the width of the block of phase II of the L_2 is taken as minimal to condition of necessary stability of the inter-room pillar 7 at the time of processing the neighboring stope blocks of phase I.

In developing deposit areas at great depths in the absence of severe restrictions on the amount of host rock displacement the width of blocks of phase II L_2 can be accepted under the terms of the minimum reduced costs.

Thus, the proposed technological diagram of combined mining of deposits allows to eliminate the drawbacks which are characteristic for room-and-pillar system with leaving inter-room pillars, as it allows to exclude the loss of ore in inter-room pillars. Development of the blocks of phase II by systems with caving allows to use as filling consolidating mixtures with reduced binding component, eliminating the main drawback of development systems with stowing – the high cost of consolidating filling mixtures. Being in a state of volume compression, concrete block with reduced binder component provides its high load-bearing capacity, reducing the risk of geotechnical risks in overlying rocks.

Analysis of the developed technologies has allowed to identify possible problems in combined mining of magnetite quartzite with room-and-pillar systems with a combined stowing of blocks of phase I and development systems with caving of ore and host rocks while mining of blocks of phase II.

In our opinion, the main problem with the blocks of phase II development is to reduce the loss of magnetite quartzite and clogging of ore with consolidating low-strength stowing with the hoisting of the caved magnetite quartzite at the contact with the side surface of waste pillars of blocks of phase I.

5 CONCLUSIONS

With a view to expanding the Kryvyi Rig raw materials base the resumption in some mines of underground mining of magnetite quartzite with reserves of billions of tons is appropriate. Located on the overlying levels of acting mines, these reserves may be available during the minimum reconstruction of worked-out levels.

REFERENCES

Kolosov, A. 2004. *Condition and prospects of mining industry development of Ukraine*. Ore deposits development. Kryvyi Rig: Publisher KTU, 85: 37-41.
Kolosov, V. & Dyadechkin, N. 2005. *Condition and pros-*

pects of raw base development of Ukrainian mine-metallurgy complex. Mining magazine, 5: 10-13.

Bliznyukov, V. & Piven', V. 2002. *About feasibility of board iron content changing in ore on mining enrichment plant Kryvbas deposit*. Metallurgy and ore mining industry, 1: 68-71.

Drobin, G., Yatsenko, L., Rimaruk, B. & Kornienko, V. 2002. *Ways of efficiency increasing of magnetite quartzite underground extraction*. Metallurgy and ore mining industry, 3. 73-75.

Kaplenko, Y., Fed'ko, M., Bezverhyi S. & Kuznetsov V. 2004. *Search of ways of efficiency increasing of mag-netite quartzite underground extraction*. Ore deposits development. Kryvyi Rig: Publisher KTU, 85: 42-45.

Conception of mine-metallurgy complex development of Ukraine to 2010. Voice of Ukraine, 206-207: 10-12.

Kaplenko, M. & Kalininchenko, V. 2006. *Perspectives of efficiency increasing of technological schemes of magnetite quartzite underground extraction*. Messenger of Kryvorizs'kyi technical university, 4 (14): 22-25.

Kaplenko, M. & Kalininchenko, V. 2006. *Increasing of ore extraction indices during magnetite quartzite underground extraction*. Messenger of Kryvorizs'kyi technical university, 13: 25-28.

Microalloyed steels for mining supports

M. Rotkegel & S. Prusek
Central Mining Institute, Katowice, Poland

R. Kuziak
Institute of Ferrous Metallurgy, Poland

M. Grodzicki
Huta Łabędy SA, Poland

ABSTRACT: The desire to reduce the cost of coal mining leads among other things to look for new, more cost-effective ways to protect mine roadways. GIG studies show that the mine supports are more than 60% by weight and material costs. Therefore, the work focused on the search for more efficient mine supports should primarily include high-performance. Such approaches also tend increasingly difficult geological and mining conditions. This direction was adopted during the project number 6ZR8 2008 C/07012 performed between 2010 and 2012 by Huta Łabędy SA, Institute of Ferrous Metallurgy and the Central Mining Institute. The work was conducted multi-threaded. One task was to develop a new type of steel with high mechanical characteristics, as well as the production technology of the steel sections. The second task was to design a high performance mine supports with some modifications achieved the mine supports are characterized by a higher load capacity than similar work ŁP supports, widely used in Polish mines. This effect has been achieved, by varying the curvature of the curves cooperating frictional coupling.

This paper presents the progress and results of research on a new steel grade and course design and research of new mine roadway supports.

1 INTRODUCTION

Steel mining supports is a basic way of securing roadways in Polish coal mines. The mining supports are a fundamental element made of profiled sections hot-rolled steel. Measures aimed at reducing the cost of mining supports should refer to the steel arches. The main way to be reinforced steel arches spacing. However, this can be done only in the case of increasing their performance or more efficient to use them. These issues have been addressed through targeted project No. 6ZR8 2008 C/07012 implemented in 2010 ÷ 2012 by Huta Łabędy SA Institute of Ferrous Metallurgy and the Central Mining Institute. The work was conducted multiple threads. One task was to develop a new type of steel with high mechanical characteristics, and manufacturing technology of the steel sections The second task was to design a high performance steel arches, optimally utilizing the strength of the individual elements.

2 CHARACTERIZATION OF A NEW TYPE OF STEEL

The development of the new steel grade for the V-shapes was aimed at combining high strength with good ductility with the enhanced corrosion resistance. This was achieved by designing a proper steel chemical composition along with the careful control of steel melting and casting practice, as well as rolling conditions. The main issue in the steelmaking and casting practice was the control of the volume fraction and morphology of non-metallic inclusions, and prevention against cracking occurrence at the ingot surface and edges. As the starting point, the strength versus ductility properties of the steel were balanced by defining a proper base chemical composition (C, Mn and Si content) and by using microalloying additions of V and Ti added to steel in precisely controlled quantities. A precise control of temperature during rolling and the presence of TiN resulted in obtaining fine-grained ferritic-pearlitic microstructure, containing around 50% of both phases. Additional strengthening was achieved as a result of vanadium carbo-nitride, $V(C, N)$, precipitation. On the contrary, enhanced corrosion of the steel was achieved trough adding alloying elements of Cr, Cu and Ni in contents eliminating the risk of achieving the negative effect of these elements on ductility. By this way, the following mechanical properties were achieved in V-type shapes:

- $R_e - min$ 550 MPa;
- $R_m - min$ 730 MPa;
- $A_5 - min$ 18%;
- KCU2A $- min$ 50 J / cm^2.

An example of the typical chemical composition of industrial heat of steel grade S550W is given in Table 1. The ingots of this heat were rolled into V-32 shapes and the results of mechanical properties measurement are given in Table 2.

Table 1. An example of chemical composition of industrial heat of steel S550W.

C	Mn	S_i	P	S	Cr	Ni	Cu	Mo	Al	Ca	Sn
0.25	1.35	0.43	0.017	0.012	0.28	0.08	0.25	0.02	0.001	0.001	0.022

Ti	V	As	Pb	Nb	W	N
0.004	0.073	0.008	0.003	0.002	0.01	0.009

Table 2. Mechanical properties of V-32 shapes rolled out of heat having composition given in Table 1.

#	R_e, MPa	R_m, MPa	R_e / R_m	A_5, %	Z, %	KCU2A, J / cm^2
1	585	761	0.77	28.0	67	97
2	581	764	0.76	25.4	66	69
3	584	757	0.77	25.6	68	67
Average	**583**	**761**	**0.77**	**26.3**	**67**	**78**

Table 2 shows that the mechanical properties of V-32 shape are much better than the expected ones. Specifically the elongation to fracture and notch toughness are much higher than demanded. This is due to the uniform and fine grained microstructure and the limited amount of non-metallic inclusions (Figure 1 and 2).

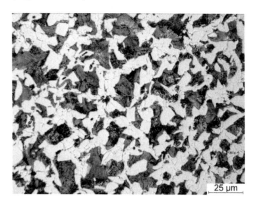

Figure 1. Ferrite-pearlite microstructure of the V-32 shape. The sample was cut from the shape's flange. Light optical microscopy – Nital etching.

The corrosion test performed according to special methodology has shown that the resistance against coal mines salty environment of the investigated steel was around twice that of the conventional

$C - Mn$ steels without any alloying additions.

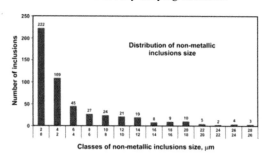

Figure 2. Distribution of non-metallic inclusion size in the sample shown in Figure 1.

3 CHARACTERISTICS MINING SUPPORTS OF A NEW TYPE OF STEEL

An important step was the development of the work under construction mining supports of high performance made of a new type of steel. The bench studies suggest that only a small part of the strength of mining supports elements (45÷55%) is used during the excavation work. Therefore, for optimal use of heavy-duty mining supports elements also need to increase the working capacity. The Figure 3 presented the idea of increasing the parameters mining supports – characteristics stiffened and susceptible. While the "increase performance" mining supports rigid stems from the use of more durable compo-nents, so "increase performance" mining support susceptible associated with an appropriate choice of

friction joints and appropriate shaping of the con-nected parts.

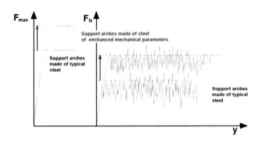

Figure 3. Raising the characteristics mining supports: F_{max} – maximum capacity mining supports (strained); F_{N2} – working load mining supports (susceptible); y – lower mining supports.

Figure 4. Increasing the number of contact surfaces in the joint friction by eliminating the "lens".

Increasing the parameters mining supports sus-ceptible make better use of the strength of arches. However, the excessive growth leads to stiffening mining supports. Then, their capacity and function-ality exhausted even at low deformation. Balance becomes like the appropriate formation of friction joints and proper selection of the clamps to obtain the optimal operating mining supports.

The research, analysis and experience, it is obvi-ous that the joint friction carry heavier loads when using third clamps in the joint and to some extent – increasing torque on bolts nuts. However, this in-crease in capacity is caused even higher for the elimination of "lens" – the gap between the flanges cooperating arches, resulting from their different curvatures. It excludes contact and as a result of friction between the mating arches in the area be-yond the ends of the friction joints. The idea of these changes is shown in Figure 4. Consequently, the mining supports of ŁPw were designed. The mining supports of these uses co-equal curvature

arches.

Newly designed mining supports are made of sec-tions V29, V32 and V36, and include sizes 7-19. In terms of size and contour are compatible with min-ing supports series ŁP according to PN-G-15000/02 (PN-G-15000/02:1993). Figure 5 shows the outline mining supports and the range.

Due to the safe use of the newly designed mining supports important issue is to determine the actual parameters. This is done by testing bench (Pacześ-niowski 1997), numerical analysis by finite element method (COSMOS/M 1999) and test mining. The project carried out a number of bench testing mining supports ŁPw different sizes. During the study min-ing supports loaded in line with the PN-G-15000-05 to give both the deformation characteristics, as well as working in Figures 6 and 7 Averaging the results of the studies, the capacity mining supports. The Figure 8 collected load specified for the mining supports ŁPw10/V36/4/A.

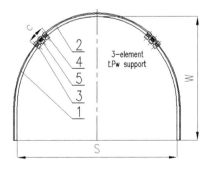

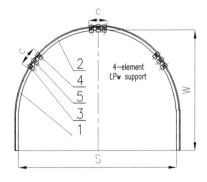

	Support arch markings	F	S	W	C			Weight of support without clamps		
					V29	V32	V36	V29	V32	V36
		m^2		mm				kg		
3 elem.	LPw7/-/A	11.0	$4200^{\pm84}$	$3100^{\pm62}$	$500^{\pm25}$	$500^{\pm25}$	$500^{\pm25}$	277.6	313.6	346.7
	LPw8/-/A	13.0	$4700^{\pm94}$	$3300^{\pm66}$	$500^{\pm25}$	$500^{\pm25}$	$500^{\pm25}$	297.3	335.8	371.2
	LPw9/-/A	14.6	$5000^{\pm100}$	$3500^{\pm70}$	$550^{\pm28}$	$550^{\pm28}$	$550^{\pm28}$	316.2	357.0	394.8
	LPw10/-/A	17.4	$5500^{\pm110}$	$3800^{\pm76}$	$550^{\pm28}$	$600^{\pm30}$	$600^{\pm30}$	341.2	388.7	429.9
3 elem.	LPw7/-/4/A	11.2	$4200^{\pm94}$	$3100^{\pm66}$	$500^{\pm28}$	$500^{\pm25}$	$500^{\pm25}$	297.0	330.6	365.7
	LPw8/-/4/A	13.1	$4700^{\pm94}$	$3300^{\pm66}$	$550^{\pm28}$	$500^{\pm25}$	$500^{\pm25}$	316.4	352.5	389.8
	LPw9/-/4/A	14.8	$5000^{\pm100}$	$3500^{\pm70}$	$600^{\pm30}$	$550^{\pm28}$	$550^{\pm28}$	336.9	375.6	415.4
	LPw10/-/4/A	17.6	$5500^{\pm110}$	$3800^{\pm76}$	$600^{\pm30}$	$600^{\pm30}$	$600^{\pm30}$	362.0	408.3	451.6
	LPw11/-/4/A	19.8	$5800^{\pm116}$	$4025^{\pm81}$	$600^{\pm30}$	$600^{\pm30}$	$600^{\pm30}$	380.2	428.9	474.3
	LPw12/-/4/A	21.8	$6100^{\pm122}$	$4225^{\pm85}$	$600^{\pm30}$	$600^{\pm30}$	$600^{\pm30}$	396.2	446.8	494.2
	LPw13/-/4/A	23.9	$6400^{\pm128}$	$4425^{\pm89}$	$600^{\pm30}$	$600^{\pm30}$	$600^{\pm30}$	412.7	465.5	514.8
	LPw14/-/4/A	26.2	$6700^{\pm134}$	$4550^{\pm91}$	$600^{\pm30}$	$600^{\pm30}$	$600^{\pm30}$	429.2	484.1	535.3
	LPw15/-/4/A	28.4	$7000^{\pm140}$	$4700^{\pm94}$	$600^{\pm30}$	$600^{\pm30}$	$600^{\pm30}$	444.0	500.8	553.8
	LPw16/-/4/A	30.8	$7200^{\pm144}$	$4900^{\pm98}$	$600^{\pm30}$	$600^{\pm30}$	$600^{\pm30}$	460.0	518.7	573.7
	LPw17/-/4/A	33.2	$7500^{\pm150}$	$5110^{\pm102}$	$600^{\pm30}$	$600^{\pm30}$	$600^{\pm30}$	476.0	536.7	593.6
	LPw18/-/4/A	35.8	$7800^{\pm156}$	$5325^{\pm107}$	$600^{\pm30}$	$600^{\pm30}$	$600^{\pm30}$	491.9	554.7	613.4
	LPw19/-/4/A	38.3	$8000^{\pm160}$	$5465^{\pm109}$	$600^{\pm30}$	$600^{\pm30}$	$600^{\pm30}$	506.7	571.4	631.9

Figure 5. Enclosure mining supports of ŁPw: 1 – arch; 2 – arch; 3 – bottom clamp; 4 – top clamp; 5 – middle clamp.

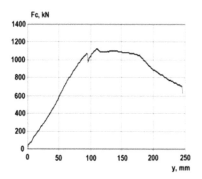

Figure 6. Example of a mining supports ŁPw10/V36/4/A deformation (Pacześniowski 2010).

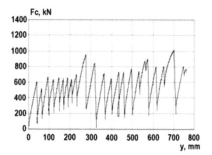

Figure 7. Example of a mining supports susceptible ŁPw10/V36/4/A (2 clamps in the joint, bolts tightening torque $Md = 450$ N·m) (Pacześniowski 2011).

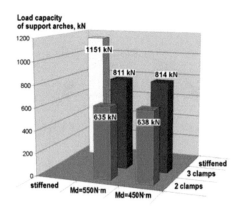

Figure 8. Averaged results mining supports ŁPw10/V36/4/A.

The research bench allowed the calibration of numerical models mining supports and highly accurate computer simulation. These studies were carried out by finite element method (FEM) (Cook 2002) using the COSMOS/M (COSMOS/M 1999). Models reflect the size mining supports and parameters are supported for cross-section arches and charged in line with the PN-G-15000-05 (PN-G-15000/05:1992), as well as in the research bench. As a result of the calculations is obtained primarily reduced stress distribution, which defines the load indices mining supports. In the Figure 9 is an example of the stress distribution in the model reduced mining supports. However, the Figure 10 – load indices obtained mining supports ŁPw.

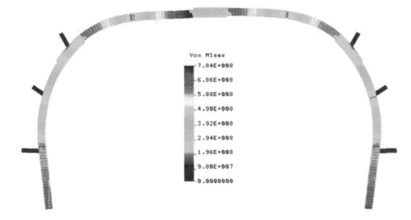

Figure 9. Sample of the stress distribution in the model mining supports.

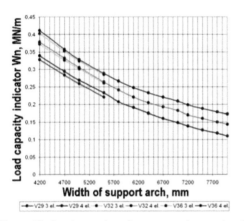

Figure 10. Load capacity of support arches made of S550W steel for different size of the section V.

Positive results of test bench and numerical analysis was the basis for accession to the test mining. These tests were carried out in the mines "Bobrek-Centrum", "Bogdanka" and "Jas-Mos". Tests were carried out on different structures made of steel mining supports S550W. Tested were newly designed mining supports of ŁPw (new design), supports of ŁPSC (for LW "Bodganka") and the supports of the SPŁ. In this way it was possible to obtain a broader information on the work of the various enclosures made of a new type of steel.

As a result of this step was verified positively newly designed mining supports (ŁPw), and previously used (ŁPSC and SPŁ) made of steel S550W. During the study confirmed a high load mining supports made of steel with improved mechanical properties of second generation. Thus, they can be successfully used to protect roadways in difficult geological and mining – at great depths, in environments with legacy operating or geological disturbances. In Figure 11 shows housing and research stations installed in the mine.

Figure 11. ŁPw mining supports installed in the riser 3 research on board the 504 in coal mine "Bobrek-Centrum" (Kuziak 2012).

4 CONCLUSIONS

The present target project, a new steel mining supports and structure (ŁPw) high performance. These parameters result from the use of steel S550W, with improved mechanical properties and change the shape of the mining supports. This modification consists in the unification of cooperating curves of arches. This results in a better performance in terms of mining supports susceptible and higher working capacity. Mining supports with a new steel grade successfully completed the entire test cycle, including test bench and mine in selected mines.

Therefore it can be concluded that the designed mining supports can be successfully used in underground mines, particularly to protect the excavations carried out in difficult geological and mining conditions that require high performance supports.

REFERENCES

COSMOS/M – User's Guide, Structural Research & Analysis Corp. Los Angeles. 1999. USA.

Kuziak, R., Żak, A., Woźniak, D., Rotkegel, M., Grodzicki, M. & Nawrot J. 2012. *Odrzwia obudowy chodnikowej ze stali II generacji.* Gliwice. Prace Instytutu Metalurgii Żelaza, 4.

Pacześniowski, K. 1997. *Wpływ wybranych czynników mechanicznych i geometrycznych na nośność łukowych odrzwi ŁP.* Prace Naukowe Katowice: GIG, 825.

Cook, R.D., Malkus, D.S., Plesha, M.E. & Witt, R.J. 2002. *Concepts and applications of finite element analysis.* John Wiley & Sons, Inc. USA.

Pacześniowski, K. + zespół. 2010. *Sprawozdanie z badań nr BL-2/10-267 – Stanowiskowe badania odrzwi obudowy typu ŁP10/V36/4/A (gat. stali S480W) ze strzemionami SDw32/34/36 (jarzma – gat. stali S480W).* Katowice: GIG.

Pacześniowski, K. + zespół. 2011. Sprawozdanie z badań nr BL-2/11-173 – Stanowiskowe badania odrzwi obudowy typu ŁPw/4/A ze strzemionami SDw. Katowice: GIG.

PN-G-15000/02:1993. *Obudowa chodników odrzwiami podatnymi z kształtowników korytkowych. Odrzwia łukowe podatne ŁP, z kształtowników typu V, typoszereg A. Wymiary.*

PN-G-15000/05:1992. *Obudowa chodników odrzwiami podatnymi z kształtowników korytkowych. Odrzwia łukowe otwarte. Badania stanowiskowe.*

Induction heating in electrotechnology of machine parts dismantling

V. Driban
UkrSRMI of the NAS of Ukraine, Donets'k, Ukraine

A. Novikov & I. Shestopalov
Donets'k National Technical University, Donets'k, Ukraine

ABSTRACT: Energy approach for the analysis of geomechanical processes in the working area of the roadway-enclosing strata and justification of rigid support parameters (including frame-anchor systems) are proposed. The outcomes of underground instrumental observations in vertical and horizontal mine workings are described that confirm competence of such approach.

1 INTRODUCTION

It is known that the main task of geomechanics is numerical estimation of different manifestations of rock pressure developed in strata during mining. Whereas many researchers all over the world undertook efforts in solving the problem, it has not been solved as yet. This is proved by not only large amounts of mine workings, which are in improper conditions, but also by high cost and manual labor content for their maintenance.

Current governing documents that regulate procedure for computation of roadway supports (Guidelines for efficient location… 1986; Location, protection and maintenance… 1998 & Industry-specific standard СОУ 10.1.05411357.010:2008) stipulate the following sequence of computations: properties of enclosing rock mass-displacement of rocks into mine working (U) – support load (P) – parameters of the support. One of the weakest links in this chain of decision-making is transition from properties of enclosing rocks to computations of displacements and loads (P – U dependence). This is due to the circumstance that rock displacements shall be determined either for specific model of rock mass or shall be defined for the whole life time of mine working, starting from condition of fabricating support therein with minimal resistance. Then, according to the data obtained, empirical curve P – U shall be constructed.

The results of computations for the first case are frequently not correct due to a number of computation errors capable to change the outcome by an order of magnitude (factor of rock loosening within the mass, angle of internal friction, module of dip of a deformation curve etc.). Not always correct selec-

tion of the model of rock mass deformation is also the reason of incorrect computation results.

The second case involves serious difficulties connected with the necessity of obtaining reliable P – U relation: dependence between real relationship and reference resistance-free displacements is determined; the fact that the value of support resistance changes the nature of deformation of edge rock mass is not taken into account.

We propose alternative approach for the solution of this problem, which enables not only to overcome the above difficulties, but also to take a fresh look at the nature of geomechanical processes that occur in the working zone and to employ it to control stability of enclosing strata. The main point of the approach is that when rigid (concrete, reinforced concrete, composite anchor-based systems etc.) support is involved in work, the support – strata system becomes immovable that is energy conservation law being effectively kept here. The system energy balance, taking into account rock collapse of the edge zone of mine working, enables us to obtain the necessary parameters of the support.

2 DESCRIPTION OF THE RESULTS

Let us consider the process of changing conditions of the support – strata system when driving a working and fabricating its support. When driving a roadway failure of rocks due to concentration of stresses occur in the working zone and consequently a zone of inelastic deformations occur, which is developed while working face movement away from the section under consideration. At sufficient distance from the face this zone becomes stationary because the impact of the face tends to zero. Thus, during advancement of the face energy dissipation out of rock ring that coves the zone of inelastic deformations takes place.

In case of fabricating in a roadway a rigid support (composite anchor-system-based supports can be considered as such) this support plays a role of energy receiver, which compensates energy losses. Changes in the energy state and in the elastic part of rock mass (off the zone) due to the redistribution of stresses in development of inelastic regions also take place.

In our further considerations we shall use the following designations: r_1, r_2 – roadway inside and outside radii respectively; r_3 – radius of the zone of inelastic deformations; E_k, E_m – modules of elasticity of support and strata respectively; v_k, v_m – Poisson's ratios; W_k – energy of rock ring (r_2, r_3); W_p – energy of rigid support; W_y – energy of elastic part of rocks (off the zone inelastic deformations); 1 – index related to energetics of rock mass with development of inelastic deformations; 2 – index which designates reference energy state in case of maintenance of elasticity of the whole strata and maintenance of the same deformation characteristics; ξ – relative part of the energy lost before fabricating a support.

Changes in the energy of rock mass can be written as:

$$\Delta W = W_k^{(2)} - W_k^{(1)} + W_y^{(2)} - W_y^{(1)}. \tag{1}$$

For the moment when a rigid permanent support will be involved in work part of the energy ξW dissipates. Then, assuming that support – strata system is conservative, we can write:

$$W_p = W^p - \xi \Delta W, \tag{2}$$

where W^p – energy of rock mass with resistance P.

This means that stress-deformation state of support is directly related to the value of dissipation energy before support has been erected and physical-mechanical characteristics of strata that enables to do without P – U diagram.

To determine the W value we use the known elastic energy formula:

$$W = \frac{1}{2} \int_\Omega \sum_{ij} \sigma_{ij} \varepsilon_{ij} d\Omega, \tag{3}$$

where Ω – region of rock mass.

Then energy of the rigid support ring will be equal to:

$$W_p = \frac{1}{2} \int_0^{2\pi} \int_{r_1}^{2r} r \sum \varepsilon_{ij} \sigma_{ij} dr dy =$$

$$= \frac{\pi P^2 r_2^2}{E_k \left(r_2^2 - r_1^2 \right)} \cdot \left[(1 - v_k) r_2^2 + (1 + v_k) \cdot r_1^2 \right]. \tag{4}$$

For calculation of W_p we shall use Lame's equation:

$$\sigma_{r(\Theta)} = \frac{P_2 r_2^2 - P_1 r_1^2}{r_2^2 - r_1^2} \mp \frac{(P_2 - P_1) \cdot r_1^2 r_2^2}{\left(r_2^2 - r_1^2 \right) \cdot r_1^2}, \tag{5}$$

where P_2 and P_1 – outside and inside loads on the ring respectively; r – current radius.

In the case under consideration $P_2 = P$; $P_1 = 0$.

As is the case above we shall determine energy of rock ring (r_2, r_3) in elastic state:

$$W_k^{(2)} = \frac{\pi \left(r_3^2 - r_2^2 \right)}{E_m} \times$$

$$\times \left[(1 - v_m) \sigma^2 + (1 + v_m)(\sigma - P)^2 \right] \cdot \frac{r_2^2}{r_3^2}, \tag{6}$$

where σ – remote stress in axial-symmetric problem. In our case $\sigma = \gamma H$.

Energy of the elastic part of rock mass with $r > r_3$ in the send state can be easily found using the known solution for distribution of stresses in the plate with round opening:

$$\sigma_{r(\Theta)} = \sigma \pm \frac{\sigma - Q}{r^2}, \tag{7}$$

where Q – internal resistance.

As the outside radius of a roadway is equal to r_2, we have:

$$Q = \sigma_r \big|_{r=r_3} = \sigma \left(1 - \frac{r_2^2}{r_3^2} \right) + P \frac{r_2^2}{r_3^2}. \tag{8}$$

Then

$$W_m^{(2)} = \frac{\pi (\sigma - P)^2}{E_m} (1 + v_m) \frac{r_4^2}{r_3^2} +$$

$$+ \frac{1 - v_m}{E_m} \int_\Omega \sigma^2 d\Omega. \tag{9}$$

In order to estimate energy state of rock mass with development of the zone of inelastic deformations we use the concept of the edge stratas the elastic-plastic environment wherein Coulomb-Mohr condition is performed:

$$(\sigma_r - \sigma_\Theta)^2 + 4\tau_{r\Theta}^2 = \sin^2\rho(\sigma_r +_\Theta + 2Kctg\rho)^2 \quad . (10)$$

It is to be noted that we use Coulomb-Mohr condition to simplify our computations and it won't make a great impact on the final result because the most significant parameters to determine energy state is cohesion of rocks and stress level at the elastic-plastic boundary.

Radius of the zone of inelastic deformations is equal to (Protocenya 1981):

$$r_3 = \left[\frac{\sigma + Kctg\rho}{P + Kctg\rho}(1 - \sin\rho) \right]^{1/\alpha} . \quad (11)$$

Distribution of stresses in the plastic region:

$$\begin{cases} \sigma_r = (P + Kctg\rho)r^\alpha - Kctg\rho \\ \sigma_\Theta = (P + Kctg\rho)(\alpha + 1)r^\alpha - Kctg\rho \end{cases}, \quad (12)$$

wherefrom we find:

$$\sigma_r|_{r=r_3} = \sigma(1 - \sin\rho) - K\cos\rho. \quad (13)$$

Thus, taking into account (7), (9) and (13) we obtain:

$$W_y^{(1)} = \frac{\pi(1 + v_m)}{E_m}(\sigma \cdot \sin\rho + K\cos\rho)^2 r_3^2 +$$

$$+ \frac{1 - v_m}{E_m}\int_\Omega \sigma^2 d\Omega. \quad (14)$$

For the solution of equation (2) it is necessary to obtain the value of $W_k^{(1)}$, that is the energy of rock ring of the zone of inelastic deformations. In the process of generation of this zone rocks are being broken by a system of fractures at separate blocks (pieces). Though the zone around a roadway derives non-reversible (inelastic) deformations, each of the blocks is a solid body (rock samples selected from the zone of inelastic deformation behave like that). Plastic deformation is implemented by means of movements along the slip line (systems of fractures). Such approach enables us to estimate energetics for generating plastic zone as a solid body, stress state of which is equal to stresses in the zone of inelastic deformations that is to apply (12).

Thus,

$$W_k^{(1)} = \frac{1}{2}\int_\Omega \sum_{ij} \sigma_{ij}^{\pi\pi} \cdot \varepsilon_{ij} d\Omega. \quad (15)$$

When transformations are effected we obtain:

$$W_k^{(1)} = \frac{\pi}{2E_m}\left\{ (P + Kctg\rho)^2\left[\alpha^2 + 2(\alpha + 1)(1 - v_m)\right]\frac{r^{2\alpha+2}}{2r_2^{2\alpha}(\alpha + 1)}\bigg|_{r_2}^{r_3} - \right.$$

$$\left. - 2(1 - v_m)(\alpha + 2)(P + Kctg\rho)Kctg\rho\frac{r^{\alpha+2}}{(\alpha + 2)r_2^\alpha}\bigg|_{r_2}^{r_3} + 2(Kctg\rho)^2(1 - v_m)\bigg|_{r_2}^{r_3} \right\}, \quad (16)$$

or

$$W_k^{(1)} = \frac{\pi}{E_m}\left[\frac{\alpha^2 + 2(\alpha + 1)(1 - v_m)}{\alpha + 1}\left(\left(\frac{\alpha\sigma + R_c}{\alpha + 2}\right)^2 r_3^2 - \frac{R_c}{\alpha^2}r_2^2\right) - \right.$$

$$\left. - \frac{(1 - v_m)R_c}{\alpha}\left(\frac{(\alpha\sigma + R_c)^2}{\alpha + 2}r_3^2 - \frac{R_c}{\alpha}r_2^2\right) + (1 - v_m)\frac{R_c^2}{\alpha^2}(r_3^2 - r_2^2) \right], \quad (17)$$

where R_c – uniaxial compression strength of rocks.

Then equation (2) will take the form:

$$\frac{P^2 r_2^2}{E_k(r_2^2 - r_1^2)}\left[(1 - v_k)r_2^2 + (1 + v_k)r_1^2\right] = \frac{r_2^2}{E_m}\left\{ 2v_m(1 - \xi)\sigma^2 + \right.$$

$$\left. + \left[(1 - v_m)\sigma^2 - \left(\frac{\alpha\sigma + R_c}{\alpha + 2}\right)^2\right]\frac{(r_{3p}^2 - \xi r_{30}^2)}{r_2^2} + P^2(1 + v_m) - 2(1 + v_m)\sigma P \right\}. \quad (18)$$

where p and 0 – indices are referred respectively to the modes after and before fabricating the support.

Equation (18) involves ξ value – energy portion lost before support has been erected, which depends on the trail value of support being erected from the face and roadway radius. To determine it we use energy approach. In a driven mine working (without a face) elastic displacements of the walls can be calculated using equation:

$$f(\infty) = \frac{1+v_m}{E_m}\sigma_2 . \qquad (19)$$

According to theorem of Castigliano internal energy of elastic body equals half of the work of external force for movements. Then energy shortage with deformation of roadway walls will be determined from the expression:

$$2\pi_2 \int_0^\infty (f(\infty) - f(\ell))\sigma d\ell \qquad (20)$$

and shall be compensated by the work involved in deformation of the face:

$$\int_0^{2\pi} \int_0^{r_2} \rho\psi(s)\sigma ds , \qquad (21)$$

where $\psi(s)$ – face displacements.

Designating $f_1(\ell) = f(\ell)/f(\infty)$ we can write the equation:

$$\int_\ell^\infty (1 - f_1(z))dz = \int_0^{1-f_1(\ell)} \rho\psi(s)ds . \qquad (22)$$

Differentiating by ℓ (22) and changing $U = f(\ell)$, we obtain:

$$1 - U = U \cdot U_\ell' \psi(U), \qquad (23)$$

or

$$\ell(r) = \int_0^r \frac{U\psi(U)}{1-U}dU , \qquad (24)$$

where $\ell(r)$ – function inverse to $f_1(\ell)$.

To calculate equation (24) it is necessary to know the value (r), which we can find using Hankel transformation:

$$\psi(r) = \frac{\sigma_2(1-v_m^2)}{E_m} \cdot \begin{vmatrix} \dfrac{2}{\pi}E\left(\dfrac{2}{r_2}\right) r < r_2 \\[2ex] \dfrac{1}{2}\cdot\dfrac{r_2}{r}\cdot F\left(\dfrac{1}{2};\dfrac{1}{2};2;\dfrac{R_0^2}{r^2}\right) r > r_2 \end{vmatrix} , \qquad (25)$$

where E – complete elliptic integral of the second order; F – Gaussian hypergeometric function.

Thus, computing equation (24) with the condition (25) we obtain deformation curve in question for shaft walls. From here according to theorem of Castigliano it follows that ξ equals $f_1(\ell)$.

Figure 1 shows energy dissipation curve before support has been erected for different Poisson's ratios (v) and process variable of the trail of support being erected from the face (ℓ/r).

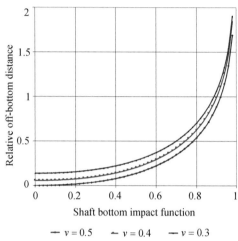

Figure 1. Shaft bottom impact function.

From the plots shown Figure 1 we can see that with $v = 0.3$ for the case when the rigid support has been involved in work with trail from the face by the radius of the roadway ($\ell = r$) there occurs dissipation of up to 88% of the elastic energy (dissipation of 100% of the stored energy will occur with trail from the face by $2.3\,r$).

In order to verify convergence of the obtained theoretical solution and data of in-situ measurements we get to know the outcomes of the observations in mines for displacement of the outlines of different-purpose mine workings in the impact zone of the heading face.

For example, outline measuring system with the inside diameter of 6.5 m was laid in the shaft #3 (Kochegarka Mine). The measuring system included four benchmarks placed at the distance of 0.5 m from the face was laid on the strike and transverse to the strike of the seam (Figure 2).

Shaft support is 0.5 m thickness reinforced concrete. The height of the entry for covering with concrete is 3.0 m. The station was placed down to the depth of 1.167 m at the junction of sandstone and clay shale layers. Four series of measurements were made as far as the face left measuring station as many as 4.4 m. Results of the measurements to-gether with the estimated data according to the proposed model are shown in Figure 3.

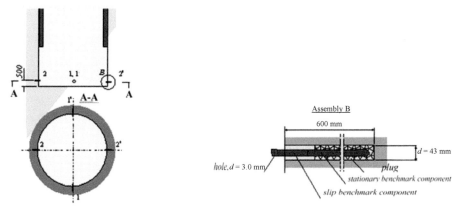

Figure 2. Layout of the observation station in the shaft #3, Kochegarka Mine.

During computations we made calibration by target values of the measured displacements of shaft walls. We can see from the plots that discrepancy of the estimated and real data does not exceed 12%. To our opinion this speaks for satisfactory convergence of the computational method to obtain ξ ratio.

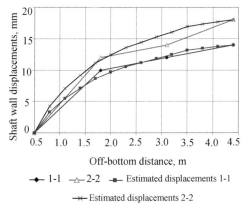

Figure 3. Results of the measurements and in-situ observations for wall displacements of the shaft #3 during road-heading.

At present one of the most promising directions in increasing stability of horizontal and sloping roadways is reinforcement of conventional frame support with anchors that enables employing natural rock strength involving rock mass in the joint work with the support. Frame-anchor systems used to support mine workings are rigid bearing structures in the form of rock-anchor case. Its internal surface and working space of a roadway are divided by frame support. Experience of employing composite support (Novikov 2012) shows that efficiency of its use depends on the implementation of deformation processes in rock mass enclosing roadway at the moment of execution of works for support reinforcement. Let us analyze in-situ observations made by us in the roadways where control of the strata stability was made using frame-anchor systems with different time and space parameters for fabricating a strengthening anchor support employing the proposed energy approach.

Underground survey was conducted in two development roadways of the Dobropolskaya Mine: belt heading of north longwall #5 and belt heading of south longwall #5, seam m_4^0. Integrated stations were laid in roadheads, strengthening of the main support (frames were been erected in the breast) with anchors was made with different time lag when rocks have been removed. Test measuring stations were being laid at the areas of roadways only with frame support.

Because at the Dobropolskaya Mine observations were being conducted in the same mining-geological conditions and the results obtained were similar, we shall consider as an example data of the observations at the measuring stations laid south belt heading #5, seam m_4^0. The test area in a roadway occupied the first five stakes. Extract from subsurface map with indication of locations of measuring stations is shown in Figure 4 and 5 shows scheme of measuring station.

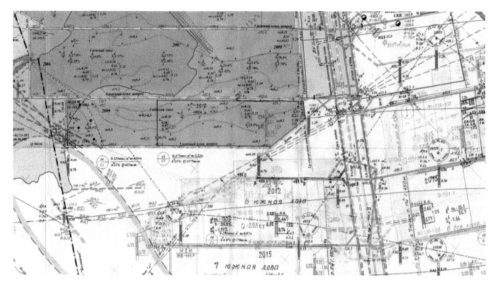

Figure 4. Extract from the subsurface map in the seam m_4^0.

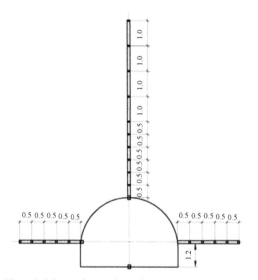

Figure 5. Scheme of measuring station.

A roadway was driven in seam in rocks with uni-axial compression strength in the range of 15-45 MPa with double-sided breaking. Maximal height of underbreaking was 1.7 m. Average seam thickness was 1.2 m. Bedding angle was 10°. Depth of driving a roadway was 700 m. Height of the arch section in drivage was 3.44 m; design length was 1.275 m. Drivage was carried out by combined machine (roadway cutter loader) 1П110. Average driving speed was 110 m per month. Part of the roadway (the first five stakes) was supported with metal frame yielding support (2 frames per meter). The rest of the roadway was anchored with composite frame-anchor support (1.25 frames per meter). Reinforcing anchor support in the place of laying integrated measuring stations was being laid with different trail from the face (0 m, 3 m, 5 m, and 10 m) that corresponds to the time lag between installing anchor support and removal of rock in the face: 0; 0.68; 1.14 and 2.27 days. Anchors were being embedded in the roof among the frames under metal W-shape bar the length of which was 3.2 m. By a template there were made holes in the bar (in the line with the distance among the anchors) through which anchor bolts were being drilled. With the help of backplates the bar was being pressed to top covers. The distance between the rows of anchors lengthwise of the drift was 0.8 m; the distance in the row was 1.0 m. Four anchors were being embedded in the roof. The distance from the outermost anchor in the roof to the roadway wall was 0.9 m. The length of anchor bolts embedded among the support frames was 2.4 m and wood ties were being used.

Observation data are shown in the form of the curves of deep benchmark displacements and changes in fragmentation index (k_p) at the areas of the hole between benchmarks versus the time of installing anchor support after removal of rock in the face for 5th, 20th, 50th, 80th and 220th days of observations (Figure 6-11).

Figure 6 shows the curve of changes in the grain size of the zone of broken rock (ZBR) in time at the test measuring station. We see that by the time of installing reinforcing anchor support with 0, 3, 5 and 10 m trail from the face there was formed ZBR of respectively 0, 0.28, 0.44 and 1.3 m around a roadway (predicted ξ values were respectively 0.14-0.2; 0.91; 0.97 and 1.0).

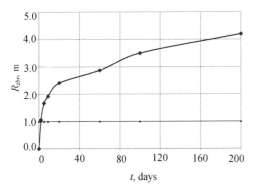

Figure 6. Changes in the grain size of the zone of broken rock (ZBR) in time at the test measuring station.

Analysis of the curves in Figure 7 shows that on the 5th day of observations at the test measuring station displacements of top covers were 32 mm. From the curve of changes in k_p we see that the area of 0-0.45 m is damaged with the maximal $k_p = 1.040$.

Frontal damage was moving from the outline into the depth of strata. With distance from this area displacement of the outline damped. ZBR in the roof exceeded 7.0 m. Anchor-tied shell (borehole area of 0-2.1 m) moved as a single block without any damage (maximal k_p value did not exceed 1.003). We note that 2.1-3.0 m area conditions are close to breaking and there irreversible deformations will occur in the future.

Similar conclusions can be made for the analysis of the rest stations where reinforcing of frame anchor support was made with a trail of 3, 5 and 10 m. Top covers displacements at the stations were respectively 22, 26 and 28 mm ($\xi = 0.91$; 0.97 and 1.0 respectively). At the borehole areas between the outline and benchmark #2 (0-0.5 m) there was intensive loosening of the edge strata. The more is the gap in the trail between removal of rock and subsequent anchoring, the higher is loosening of the edge area: $k_p^{max} = 1.008$; 1.015 and 1.028 respectively. It is obvious that later on this area will be damaged.

(a)

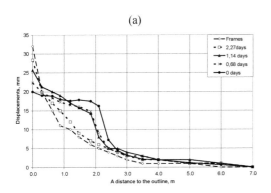

(b)

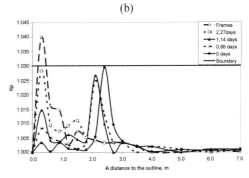

Figure 7. Displacements of deep benchmarks in the top covers (a) and changes in k_p (b) on the 5th day of observations.

On the 20th day of observations (Figure 5) at the test measuring station and stations with composite support top covers displacements were 116, 62, 71, 82 и 107 mm. The size of ZBR formed around a roadway was 1.35 m, 0.45 m, 0.7 m, 0.85 m and 0.75 m respectively, while the size of the zone of irreversible deformations (ZID) in the roof remained unchanged. From the analysis of in-situ measurements we make the following conclusions related to the deformation features of the roadway-enclosing strata:

a) further increase in ZBR from the roadway outline into the depth of strata up to 1.35 m was observed, $k_p^{max} = 1.065$;

b) at the area where anchor support was erected immediately after rocks have been removed, anchors broke rocks at once (2.25-2.70 m), $k_p^{max} = 1.046$; the anchor-tied shell saved its integrity: $k_p^{max} = 1.030$;

c) at the measuring station laid in the place where anchor support was erected with the lag time of 0.64 days (3 m) we observed heavy deformation in the edge part of strata (0-0.4 m), $k_p^{max} = 1.032$. However, the main rock failure occurred outside the anchor-tied shell (2.25-2.65 m), $k_p^{max} = 1.051$;

d) similar results with increase in the broken condition of enclosing strata were obtained also at two

other measuring stations where the gap was 1.14 and 2.27 days respectively (trail of 5 and 10 m).

(a)

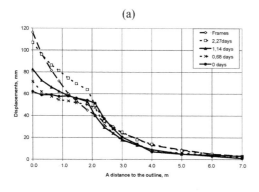

(b)

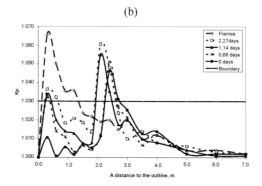

Figure 8. Displacements of deep benchmarks in the top covers (a) and changes in k_p (b) on the 20th day of observations.

For example, front of heavy deformation has moved from the roadway outline up to 0.45 m and 0.65 m with $k_p^{max} = 1.034$ and 1.036. Outside the anchor-reinforced shell $k_p^{max} = 1.055$ and 1.060 respectively.

On the 50th day of observations (Figure 9a) top covers displacements were 179, 122, 131, 145 and 166 mm. The size of ZBR formed around a roadway was 2.65, 1.55, 1.3, 1.05 and 1.1 m respectively.

We point out the following deformation features of the roadway-enclosing strata:

a) at the test area further increase in ZBR from the roadway outline into the depth of strata (broken

area of 0-2.65 m), $k_p^{max} = 1.098$;

b) at the area where anchor support was erected immediately after rocks have been removed, an area of 2.15-3.70 m has been broken, $k_p^{max} = 1.074$; the anchor-tied shell saved its integrity.

c) at the measuring station laid in the place where anchor support was being erected with the lag time of 0.64 days (3 m) we observed partial collapse of the edge part of strata (0-0.5 m) with $k_p^{max} = 1.035$. The main rock failure occurred outside the anchor-tied shell (2.25-3.35 m) with $k_p^{max} = 1.078$;

(a)

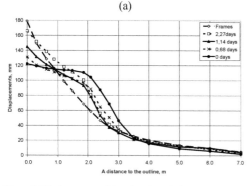

(b)

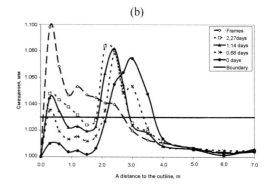

Figure 9. Displacements of deep benchmarks in the top covers (a) and changes in k_p (b) on the 50th day of observations.

d) similar results were obtained at two other measuring stations. The only difference was increase in the size of the broken edge part of strata and broken con-

dition of rock mass (0-0.7 and 0-1.25 m) with $k_p^{max} = 1.045$ and 1.048 respectively. Outside the anchor-reinforced shell $k_p^{max} = 1.082$ and 1.084 respectively.

On the 80th day of observations (Figure 10) at the test measuring station and stations with composite support top covers displacements were 237, 145, 161, 175 and 221 mm respectively. The size of ZID of all of the stations exceeded 7.0 m. The size of ZBR formed around a roadway within test measuring stations was 2.9, 1.3, 1.7, 2.0 and 3.9 m.

Data related to the deformation process are as follows:

a) at the test area further increase in ZBR from the roadway outline into the depth of strata (0-2.9 m), $k_p^{max} = 1.121$ was observed;

b) at the area where anchor support was erected immediately after rocks have been removed, an area of 2.05-3.35 m has been broken, $k_p^{max} = 1.087$; the rock-anchor shell saved its integrity;

c) at the measuring stations laid in places where anchor support was being erected with the trail of 3 and 5 m we observed partial collapse of the edge part of strata (0-0.6 m and 0-0.75 m) with $k_p^{max} = 1.037$ and 1.050 respectively. The main rock failure occurred outside the anchor-tied shell (2.15-3.85 m and 1.85-3.85 m) with $k_p^{max} = 1.092$ and 1.100 respectively;

d) with the trail of 10 m the edge anchor-reinforced strata were completely broken; top covers displacement was small (up to 7%). The main failure is confined to the area of strata located outside the rock-anchor shell. For example, if in the range of 0-1.8 m $k_p^{max} \leq 1.057$, then at the area of 1.8-4.0 m $k_p^{max} = 1.113$.

(a)   (b)

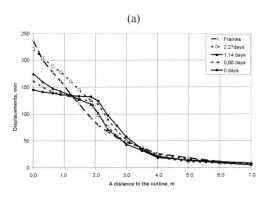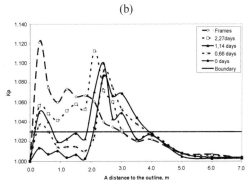

Figure 10. Displacements of deep benchmarks in the top covers (a) and changes in k_p (b) on the 80th day of observations.

By the end of the observations (Figure 11) top covers displacements were 307, 204, 229, 241 and 280 mm. The size of ZID at all of the stations exceeded 7.0 m. The size of ZBR formed around a roadway within test measuring stations was 3.45, 2.3, 2.0, 2.05 and 3.8 m, respectively.

The resulting outcomes of the observations are the following:

a) at the test area further increase in ZBR from the roadway outline into the depth of strata (0-3.45 m), $k_p^{max} = 1.136$ was observed.

b) at the area where anchor support was erected immediately after rocks have been removed ($\xi = 0.14$-0.2), only part of the edge strata located behind the anchors (1.95-4.25 m, $k_p^{max} = 1.112$) has been broken; area of 2.05-3.35 m has been broken,

$k_p^{max} = 1.087$; the rock-anchor shell saved its integrity;

c) at the areas where anchor support was erected with the trail of 3.0 and 5.0 m ($\xi = 0.91$ and 0.97) partial collapse of the edge strata (0-0.85 and 0-1.05 m) with $k_p^{max} = 1.043$ and 1.053 respectively occurred. The main failure was outside the rock-anchor shell (1.9-3.9 m) with $k_p^{max} = 1.119$ and 1.122 respectively.

d) with the trail of 10 m ($\xi = 1.0$) the edge strata were completely broken. However, the main rock failure has occurred not at the roadway outline but outside the anchor-reinforced region. In the range of 0-1.8 m $k_p^{max} \leq 1.070$, and outside that region (1.8-

4.0 m) $k_p^{max} = 1.131$.

We make the following conclusions.

Maximal technical effect (decrease in top covers displacements) is achieved in case where anchor reinforcing of frame support was made immediately after rocks have been removed from the face ($\xi = 0.14\text{-}0.2$). Anchor-tied shell is not broken and

bears loads from enclosing strata playing the role of the support. Failure front is being transferred beyond the shell. In this case top covers displacements decrease approximately by one third in comparison with the roadway anchored only with frame support. Broken condition of enclosing strata is also reduced by 50%.

(a)

(b)

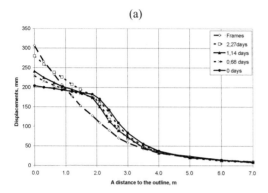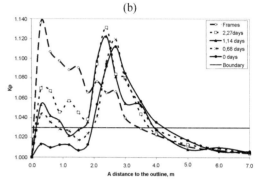

Figure 11. Displacements of deep benchmarks in the top covers (a) and changes in k_p (b) by the end of observations (on the 220th day).

With reinforcement of frame support with anchors in case when ZBR formed by the time of anchoring does not exceed half of the embedment depth that corresponds to 5.0 m ($\xi = 0.97$), the following deformation features of enclosing strata are determined. Immediately after frames are erected in the face (after rocks have been removed) there starts development of ZBR from the outline into the depth of strata. Due to the early embedment of anchors it is capable to reduce this process and then to stop it. Rock-anchor shell saves its integrity. Top covers displacements decrease by 20%.

With larger lag time between rock removal and anchoring ($\xi = 1.0$) it fails to stop completely development of ZBR from the roadway outline into the depth of strata. Anchor-tied shell ruptures completely, and displacements of the outline of roadway with composite frame-anchor support approach to displacements of the outline of roadway supported in similar conditions only with framed support structures.

3 CONCLUSIONS

The outcomes of underground survey related to deformation of strata enclosing roadways with rigid reinforced concrete and composite frame-anchor

support depending on the time of involvement of the support into work are as follows:

1. Use of energy approach in designing rigid support structures in roadways enables to determine support engineering and duty parameters, which provide stability of supported roadways, easily and to a high degree of accuracy

2. By fabricating reinforced concrete support in the shafts with trail from the face by 3.5 m ($\xi = 0.9$) condition of support stability is provided with the dissipation of 90 % of stored elastic energy from strata bottom zone.

3. The most efficient way of using frame-anchor systems to control stability of enclosing strata is to erect reinforcing anchor support with trail from the face by no more than $(0.16\text{-}0.23)\,r$. Prevention of energy dissipation from the area of rocks being anchored ($\xi \leq 0.2\text{-}0.4$) allows not only to reduce density of framing but also to decrease top covers displacements up to 66% in comparison with roadway areas supported only with frame support.

4. Frame-anchor support enables to decrease broken condition of enclosing strata. Loosening of rocks is reduced maximally to 0.024, and its average value is decreased by 0.015 in comparison with roadways with frame support.

5. With due use of frame-anchor support structures anchor-stitched area (0-2.2 m) was not break-

ing (within its limits maximal value of k_p was 1.013 that does not exceed limiting values). Strata collapse occurs outside its limits in the range of 1.94-4.25 m.

6. By using frame-anchor support systems (anchor reinforcement is fabricated with trail by 5 m from the face, $\xi = 0.97$) with partial dissipation of stored elastic energy from anchor-tied part of strata it became possible to retard and then to stop development of deformation processes started after removal of rocks from the face into the depth of strata. The size of failed edge part of strata did not exceed half of the anchoring depth (0-1.05 m). The main failure due to anchor fabrication occurred outside the anchor-tied shell. The size of ZBR decreased by 41%, average value of k_p decreased by 0.011 and maximal – by 0.017 in comparison with roadway with frame support only.

7. Use of frame-anchor support systems with trail by 5-10 m from the face (when complete dissipation of stored elastic energy from anchor-tied part of strata occurred, $\xi = 1.0$) does not allow to stop completely development of ZBR started from the outline. Anchor-tied shell ruptures completely. However, due to embedded anchors the degree of broken condition of edge strata at the area of 0-2.1 m is much less than in the roadways with frame support only. For example, at the test area k_p was 1.094, and at the area with composite support where anchors were embedded with trail by 10 m k_p was 1.058.

8. Use of frame-anchor support systems with trail by more than 10 m (with $\xi = 1.0$) from the face is inappropriate. In this case due to collapse of the edge strata down to a large depth, before support will be involved in work, displacements of top covers approach to displacements in roadways with frame support only, being maintained in similar conditions.

REFERENCES

Guidelines for efficient location, protection and maintenance of mine workings at coal mines of the USSR. 1986. VNIMI: 222.

Location, protection and maintenance of mine workings with coal seam mining. 1998. Methodological instructive regulations КД 12.01.01.201-98: Approved by the Ministry of Coal Industry of Ukraine on 25.06.98. Donets'k: UkrNIMI: 154.

Industry-specific standard СОУ 10.1.05411357.010:2008. System for providing reliable and safe functioning of mine workings with anchor support. General technical specifications: 89.

Protocenya, A.G., Stavrogin, A.N., Chernikov, A.K. & Tarasov, V.G. 1981. *Towards defining equations of condition with deformation of rocks in superlimiting region.* Physical-Technical Problems of Useful Mineral Development, 3: 33-43.

Novikov, A.O., Petrenko, Yu.A., Shestopalov, I.N. & Reznik, A.V. 2012. *Justification of the time limit for employing additional measures directed towards increase in mine working stability.* Transactions of the Donets Natioanl Technical University. Mining-Geologic Series. Editorial board: Bashkov E. O. (Chairman) et al. Issue 16 (206). Donets'k: DonNTU: 179-184.

© 2013 Taylor & Francis Group, London, ISBN: 978-1-138-00108-4

Studies of stationary supporting zone sizes varied in the course of mining operations in deep horizons

O. Voloshyn & O. Ryabtsev

M.S. Polyakov Institute of Geotechnical Mechanics, Dnipropetrovs'k, Ukraine

ABSTRACT: This article presents theoretical findings on changing with time of stationary supporting zone length towards deep into the rock mass. The result obtained are compared with results of instrumental and visual monitoring of seams k_5^1 and k_6 the Krasnyy Partyzan Mine, where depth of mining exceeds 1000 m.

Having analyzed a big massive of up-to-date knowledge of and experimental findings on mechanism of the rock contiguity we can conclude that in the course of mining sedimentary rocks they approach each other in the form of consecutive bending of strata which are mildly trapped into the tunnel contour. Rock deformation spread into the rocks outside the tunnel walls. In the thickness over the coal stratum, the rock contiguity is accompanied by rock stratifying and strata moving relatively to each other. Curves of these rocks subsiding show inflection points where the bend curvature changes its sign; these points are the boundaries of supporting zone above the goafs. The closer are the inflection points to the surface the more intensively they move towards the tunnel axis.

With increasing of the tunnel span length (as advancing away from the face entry), sagging of the rocks over the goaf grows, and cavities in the stratified rocks can disappear. A part of undermined rocks falls on to the bottom of the seam. Normal loads are also distributed unevenly: they reach maximum, then decrease while approaching the inflection points and again increase over the tunnel center.

As mineral are extracted, and face drives, stress field round the stope changes. Massive area within which all these changes occur is named a stope zone of influence. As opposed to the preparatory roadways, zone of influence round the working excavations covers essentially greater areas of rocks. Very often the processes cover the whole thickness of overlying rocks, sometimes reaching the day surface. The deformation processes also involve great areas of rocks from the side of the face bottom.

Taking into account continuity of the stope driving, it is a usual practice in mining theory to separate temporary (dynamic) or operational support pressure occurred near the moving boundaries of the stope. As contrasted to this, a zone where stresses are concentrated near the fixed boundary of stope is named a residual pressure zone or steady-state pressure zone.

Due to certain difficulties in determining stress-strain state of the rocks round the stope when methods of mathematical simulation were not widely used, parameters of supporting pressure area were calculated with ignored components of static stress fields in the rocks round the goafs. However, the most reliable results can be ensured not by calculated parameters of supporting pressure zone but by the parameters measured in-situ. In particular, numerous instrumental studies show that usually peak stress occurs in supporting pressure zone at a distance equal to 2-5 seam thicknesses and most often is $(3\text{-}7)\,\gamma H$.

Parameters of supporting pressure zone are specified by various factors. In the first turn, it is parameters of initial stress field, sizes and configuration of the goafs, physical, mechanical, deformational and strength features of the rock-hosting massive and, of course, by method of the seam mining. Figure 1 shows example of dependences between supporting pressure area width, stope depth and seam thickness in the coal deposits.

Today, one of the pressing challenges for the researchers is to determine changing with time of stationary supporting zone length. Especially it is the strongest business case for mines which extract coal reserves at the depth more than 1000 m as majority of existing practices were designed for depth less than 1000 m. Knowing of stationary supporting zone parameters is very important because they make possible to specify parameters of protective pillars and to avoid negative impact of mining operations on the main tunnel, spine road, pit-bottom road and even on the shafts. Besides, knowing of stationary supporting zone length allows filling tunnels outside of abutment pressure zone, which is, undoubtedly, a stationary supporting zone, and, hereby, improving mining conditions and reducing operational costs of the tunnel supports.

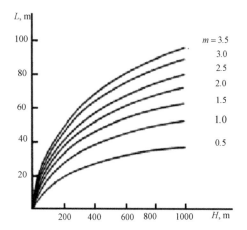

Figure 1. Dependences between supporting pressure area width L, stope depth H and seam thickness m.

The Krasnyy Partyzan Mine of the Limited Liability Company "The State Fuel-Energy Complex "Sverdlovantratsyt" long ago crossed depth mark of 1000 m in the coal seams k_2^1, k_5^1 and k_6, and today the mine performs mining operations at the depth of 1200-1300 m. While mining the longwall #71 – Vostochnaya (East), seam k_5^1, the mine faced the necessity to protect bypass round the belt boundary incline and belt boundary incline itself against mining-induced negative affects in the longwall. The Mine's previous experience in mining seam k_5^1 at the depth more than 1000 m shows that existing methodological recommendations on specifying stationary supporting zone length do not work precisely enough, and very often the protected zones are found under the negative affect of the mining operations forcing the mines to involve additional material and human resources in order to maintain the longwalls in active state. In the case of the Krasnyy Partyzan mine, it concerns tunnels that should be exploited for not less than 10-15 years more, hence, it is essential to know changes with time in stationary supporting zone length from the longwall #71 – Vostochnaya deep into the rock mass with view to choose substantiated sizes for protective pillars in order to protect the belt boundary incline.

Changes with time of stationary supporting zone length were studied with the help of one of the latest innovative complex methodological approaches – a program and technological complex the "Technology for Planning Strategic Development of Mining Operations" designed by the M.S. Polyakov Institute of Geotechnical Mechanics under National Academy of Science of Ukraine together with the Scientific and Engineering Center (SEC) "Ecology-Geos" (Bulat 2011; 2012).

Initial data for simulating changes in the stationary supporting zone sizes are as follows: predicted mining and geological conditions for mining the longwall #71 – Vostochnaya, seam k_5^1; mining and technical parameters of the longwall mining; stratigraphic column of exploratory hole #И 3251; file of this hole; and duplicated extraction from the mining operation plan of the Krasnyy Partyzan mine for the seam k_5^1.

Numerous studies held in the mine shows that active geomechanical processes occurred at mining panels start to damp in 90-120 days and continue to damp for 800-900 days, i.e. till this time stationary supporting zones will have been completely formed in its final dimension. With this view in mind, we can conclude that each particular case needs individual approach.

Our findings state that at the rate of the longwall #71 – Vostochnaya, seam k_5^1, advancing 2.5-3 m per day active geomechanical processes, while a stationary supporting zone is being created, damp when the longwall has been advanced to 300 m, that equals, by time, to 100-120 days. In order to demonstrate completeness of our studies, intervals of 50, 100 and 200 m were chosen as intermediate values for the longwall advancing, which corresponded by time to 17-20 days, 33-40 days and 65-80 days, respectively.

Figure 2 shows regularities of changes in stationary supporting zone parameters depending on distance of the longwall advancing away from the section under the consideration.

Figure 3 shows dependence of changes in stationary supporting zone length in the longwall #71 – Vostochnaya, seam k_5^1, on distance of the longwall advancing away from the section under the consideration.

As findings presented in Figure 3 show, the stationary supporting zone sizes varied between 89 m (at the longwall advancing to 50 m, or in 17-20 days) and 105 m (at the longwall advancing to 300 m, or in 100-120 days).

72

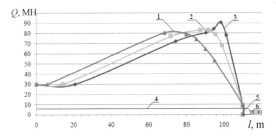

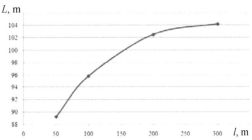

Figure 2. Regularities of changes of normal loads Q in stationary supporting zone along its length l depending on distance of the longwall advancing: 1 – advancing to 300 m; 2 – to 100 m; 3 – to 50 m; 4 – belt boundary incline; 5 – belt entry of the longwall #71 – Vostochnaya; 6 – goaf of the longwall #71 – Vostochnaya.

Figure 3. Dependence of changes of stationary supporting zone length L on distance of the longwall advancing l.

Our findings presented in Figure 2 and 3 made possible to conclude that in order to ensure reliable protection of the belt boundary incline against mining-induced negative affects in the longwall #71 – Vostochnaya, seam k_5^1, size of the protective pillar should be not less than 105 m to each side from the protected tunnel, i.e. 210 m – along the panel in the longwall and 105 m – across the longwall. Based on these findings and own experience, the mine adopted, agreed with appropriate organizations and implemented these recommendations. Instrumental monitoring held at the longwall mining within the pillar sizes showed that there were no negative impact on the belt boundary incline (vertical convergence never exceeded 60 mm at the entry to the tunnel, and no rock movements were detected at a distance of 20-25 m along the tunnel road). It is a good evidence of high precision of the results obtained with the help of the Technology (Bulat 2011; 2012) and practical recommendations worked out on its basis.

In the course of mining longwall #365, seam k_6, in the Krasnyi Partyzan mine, intensive contiguity between the rocks were detected in the contour of air roadway – a main tunnel that supplied fresh air and delivered materials, equipment and people for the whole wing of the mine. The panel was stowed in the longwall in such a way that after it was extracted a stationary supporting zone should be formed which would not impact on the tunnel. According to the existing methodological recommendations on specifying parameters for abutment pressure zone after the longwall, width of the abutment pressure zone in the longwall #365, seam k_6, was 60 m. The air roadway was located at a distance of 60-140 m from the boundaries of winning operations and such distance should exclude any mining-

induced negative impact on the longwall. Nevertheless, intensive movements were detected in the longwall roof, floor and walls, the most intensive of them were fixed at a distance of 60-80 m from the tunnel towards longwall boundaries. It was significant that the largest shifts were observed from the side of the extracted pillar.

Repair and maintenance managed to provide only temporary improvement of the tunnel conditions as in the course of time the movements recommenced again indicating that geotechnical state of the rocks round the tunnel had changed towards increased stresses, which always negatively impact on the tunnel supports.

In order to avoid further material and technical expenses for supporting this tunnel, stress-strain state of ambient rocks was studied with taking into account mining operations in the longwall #365, seam k_6. In particular, a model was created to simulate parameter changes in the forming stationary supporting zone of the longwall #365.

Our studies showed that in conditions under the consideration damping of process of the stationary supporting zone forming would end in 800-900 days after the longwall was mined out. Till that date changing of the zone parameters – zone sizes and values of stresses and loads – would be in the process.

Figure 4 shows regularities of loads Q changing in the longwall #365, seam k_6, depending on time periods.

As can be seen from Figure 4, at initial stage of the stationary supporting zone forming in the longwall #365, seam k_6, its length is about 57 m which corresponds to the value 60 m calculated for the mine by the existing normative method. However, as our studies show, parameters of the zone will change with the time: maximal load will decrease by its value and will advance further deep into the rocks away from the mining operation boundaries; length of the zone will increase and by the moment of geomechanical processes being completely damped (according to our studies – in 800 days) its length will have been about 127 m. Width of the stationary supporting zone shown by our simulation model is more than twice greater than the value calculated by the normative method.

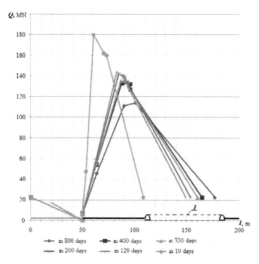

Figure 4. Regularities of normal loads Q changes in stationary supporting zone by its length l depending on time periods: 1 – air roadway.

Dependence of changes of the stationary supporting zone length from the moment when winning operations are completely finished in the longwall #365, seam k_6, are shown in Figure 5.

Reliability of our findings were verified by comparing calculated values of the tunnel height with results of underground surveying. Results of the comparison are presented in Figure 6.

Comparison between values of the air roadway height predicted by the simulation model and factual values obtained by the underground surveying showed average 87 and 93% matching of the results for June and August, respectively, indicating good reliability of the results obtained with the help of the Technology (Bulat 2011; 2012).

Our studies much contributed into understanding essence of the geomechanical state changes taken place in the rocks round the air roadway which led to such negative affects for the tunnel supporting. It was determined that key reason of the situation was essential increase with time of the stationary supporting zone length away from the longwall #365, seam k_6. The result was that the tunnel appeared in abutment pressure zone with all the ensuing consequences. Basing on the results of our studies, practical experience and results of futile efforts undertaken by the Mine to repair the tunnel it was concluded that its further exploitation was not reasonable due to high costs of maintenance under the permanent negative impact of the abutment pressure zone. It was also decided to mine a new tunnel with functions of the air roadway and to maintain the existing tunnel at minimal operating costs until the new one starts working.

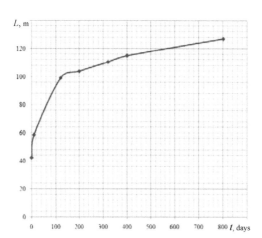

Figure 5. Dependence of changes of the stationary supporting zone length L on time period t.

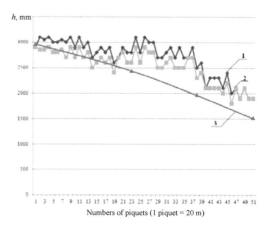

Figure 6. Comparison between calculated values of the tunnel height h with results of underground surveying: 1 – underground surveying in June; 2 – underground surveying in august; 3 – calculated values.

According to preliminary estimate of the mine's specialists, economic effect of such decision is about UAH 100 mln. The estimate is based on comparison between costs of the new tunnel mining in amount of about USD 3-4 mln. and operating costs of maintaining the existing one for 20 years when this tunnel shall operate in technological mode: these costs were estimated in amount more than USD 15 mln.

Combination of various studies held in the Krasnyy Partyzan mine has produced new knowledge for the conditions under the consideration concerning regularities of stationary supporting zone formation after the longwall is mined out, and these regularities include the following:

– at mining panel along the strike, stationary supporting zone is formed from the side of the coal fall area, and when mining operations are finished in the longwall face this formation continues to develop both in length and by stress and normal loads during up to 900 days. By the end of this period, the stationary supporting zone will have reached its final sizes, and no further changes will occur either by time or by length;

– at the conditions under the consideration, length of the stationary supporting zone from the side of the coal fall area can reach 130 m, i.e. more than twice exceeding its value calculated by the normative methods;

– maximal loads are concentrated in the stationary supporting zone at a distance of 30-60 m from the boundary of mining operations in the longwall and their value is $(5\text{-}8)\,\gamma H$.

Thus, having relatively precise method and instrument for predicting regularities of stationary supporting zone formation it is possible, and in case of complicated mining, geological and technical conditions and at mining at depth more than 1000 m, it is necessary to apply them for planning strategic development of mining operations in order to avoid negative impact of mining operations and drivage on the tunnels of various purposes.

REFERENCES

Bulat, A.F., Voloshyn, O.I., Ryabtsev, O.V. & Koval, O.I. 2011. *Technology for Planning Strategic Development of Mining Operations.* Coal, 2: 22-25.

Bulat, A.F., Voloshyn, O.I., Smirnov, A.V., Ryabtsev, O.V. & Koval, O.I. 2012. Improvement of Coal Mine Productivity on the Basis of Absolutely New Approach to Strategic Development of Mining Operations Proceedings of the international scientific-practical conference. Kemerovo: SB RAS. Kemerovo scientific center of SB RAS, institution of coal of SB RAS. Kuznetsky state technical University. Institute of coal chemistry and chemical materials science of SB RAS. LLC "Expo-Siberia": 174-177.

Influence mechanism of rock mass structure forming a stress on a face support

G. Symanovych & M. Demydov
National Mining University, Dnipropetrovs'k, Ukraine

V. Chervatuk
"Fuel-energy company of Donetsk" Donets'k, Ukraine

ABSTRACT: The mechanism that forms a stress on a face support under conditions of flat seams in a layered weak rock massif is examined and the value of influence of rock mass structure on it is researched.

There is one i-th layer of strata that has a thickness m_i on scheme of deformation (Figure 1a) and pressure that pushes on it separates into three components. It is a weight q_i one layer, a pressure $q_{i+1}(x)$ that acts on neighbors overlying layers, a pressure $q_{i-1}(x)$ that acts as a reaction from neighbors underlying layers. Points 1 and 2 are describing a border of disappearance a contact between adjacent layers because of their separation. The size of opening cavities $y_i(x)$ and their length on a plane of stratification of adjacent layers are reducing the distance from the face. Therefore point 1 is shifting to the right (towards the goaf) in relation to the point 2.

As it is known (Pisarenko 1979), the deformation of any beams and plates (including a rock), that have a height of section m_i that many times less than length of a span is characterizing above all a flexure moment $M_i(x)$. This flexure moment causes a stress that is exceeding normal severing pressure and one to two orders. Therefore on the scheme of rock layer the distribution diagram of a flexure moment is shown (Figure 1b), that is the main feature of condition and stability of i-th layer. On the site of stratification and loss of contact with adjacent layers (to point 1) a bending of i-th layer is determining only its own weight q_i and flexure moment $M_i(x)$ is increasing by classical parabolic law. There is an additional pressure $q_{i+1}(x)$ to overlying rock layer that is increasing a gradient of growth of flexure moment when it approaches to anrigid clamping i-th layer on the site between points 1 and 2. The

reaction $q_{i-1}(x)$ of underlying layers that increases a stress $q_{i+1}(x)$ by the condition of equilibrium is beginning to affect at i-th layer on the left a point 2. Thus a growth $M_i(x)$ is slowing in a point 3 and it is reaching a maximum M_{max}. Pressures $q_{i+1}(x)$, q_i and $q_{i-1}(x)$ are counterpoised by each other and flexure moment asymptotically approaches to zero when it moves deep into the massif. The maximum of a flexure moment M_{max} (point 3) are characterizing the most probable section of break of i-th layer when its right part completely falls on underlying layers and face support at finally. Thus a volume of rock of i-layer, that forms a pressure on a support, will be defined by a position of point 3 (to coordinate X). The mechanism of influence of grouped parameters to coordinate X of point 3 should be researched

The first group of parameters – is a structure of rock of the roof and thickness m_i separate layers. The adhesive power c_i of a plane of stratification is an analytically negligible in West Donbas.

An influence of thickness of i-th layer is that its bending $y_i(x)$ is inversely proportional to moment of inertia and maximum of the horizontal pressure $(\sigma_x)_{max}$ (that determines a cave-in rock console after bending) is inversely proportional to moment of resistance of a cross section of rock layer (Pisarenko 1979).

At first, let's describe a connection of bending the rock layer with its thickness m_i. Moment of inertia I_i is directly proportional to thickness rock layer to the third power m_i^3. Thus if it will increase its thickness bending $y_i(x)$ decreases at hyperbolic law and it means (other things being equal) a reduction the height $y_{i+1}(x)$ of cavity on contact with overlying rocks and increasing the height y_{i-1} of cavity on contact with underlying rocks.

There is point 1 when it loses of contact with overlying rocks (Figure 1a) and moves to the right – thus it grows pressure $q_{i+1}(x)$. There is point 2 when it

loses of contact with underlying rocks and moves to the left – reaction $q_{i-1}(x)$ decreases to coordinate X of the rock console of i-th layer. When a pressure increases from above and a reaction decreases from below in anrigid clamping i-th rock console leads to increase flexure moment M_i at the site $1^I - 2^{II}$ (Figure 2a, dotted line) and its maximum M_{max}^I in point 3^I.

(a)   (b)

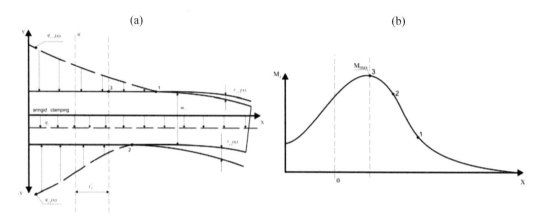

Figure 1. Scheme of a pressure (a) on separate layer of the roof under face support and distribution diagram (b) of flexure moment.

At the same time if the flexure moment M_i is increasing when thickness of the i-th layer becomes larger and moment of resistance is increasing too W_i of cross section layer thereby it is becomes stable. Thus there are two mutually opposing trends of estimate coordinate $\left(\frac{l_2^i}{2}\right)^y$ of the most probable caving of rock console and its lowering on underlying layers at the face support. Let us analyze these trends. The first regularity $M_i(m_i)$ shows:

– at the first, growing of pressure $q_{i+1}(x)$ from above and decreasing of reaction $q_{i-1}(x)$ from below are inversely proportional of heights $y_{i\pm1}(x)$ of cavities when i-th layer is bending. These cavities are inversely proportional of the i-th layer's thickness at third power and moment of inertia is directly proportional to m_i^3. Thus a pressure $q_{i+1}(x)$ is growing and reaction $q_{i-1}(x)$ is decreasing by hyperbolic law (m_i^3);

– secondly, a flexure moment $M_i(x)$ is directly proportional to pressures $q_{i+1}(x)$, q_i and $q_{i-1}(x)$ in rock layer. An increase of arm support equivalent pressure $q_{i+1}(x)$ and decrease of arm support equivalent reaction $q_{i-1}(x)$ must be kept in mind with regard to flexure moment has (Fenner 1969) a parabolic (quadratic) law;

– at the third, there are two polynomial laws (cubic on the deflection layer and quadratic on arm support equivalent pressure). Thus there is an emergency growth a flexure moment $M_i(x)$ if thickness m_i of i-th layer will increase. It describes on Figure 2a with dotted line.

The second rule $W_i(m_i)$ is characterized by growing of resistance moment W_i section and it has only parabolic law on thickness m_i of i-th layer. If we compare two functions $M_i(m_i)$ and $W_i(m_i)$ the first will increase more intensively than the second one. Thus their ratio $\frac{M_i}{W_i}(m_i)$ will increase. On the other hand the ratio $\dfrac{M_i}{W_i}$ characterizes a stress σ_x of bending rock layer and maximum of them was limited by strength limit of rock σ_{pr} in pressure's area.

Thus there is a maximum of flexure moment M'_{rf} at which the strength limit of rock and rock console falls down and lays on underlying layers. Since increased thickness m_i of i-th layer a flexure moment increases intensively that limiting condition of console (Figure 2a point 4^I) occurs before the flexure moment will be the maximum M_{max} in point 3^I and the distance $\left(l_2^i\right)^I$ from anrigid clamping will increase so length of console will grow. The in-

crease of distance $\left(l_2^i\right)^I$ with growing a i-th layer's thickness results (scheme on Figure 1(Kovalevs'ka 2012) to decrease volume rocks that create a pressure on face support. On the other hand when m_i will increase the i-th layer's weight, after it lays on underlying layers, that participates in formation of pressure on face support. These two trends show ambiguous relationship of stress $Q_i(m_i)$ with i-th layer's thickness that describes qualitative of full line on Figure 2b. When i-th layer has low thickness the stress from own weight isn't very big and when $m_i = 0$ total stress from weight of i-th numbers of layers of the roof corresponds to the weight Q_{i-1} that is generated by the underlying layers.

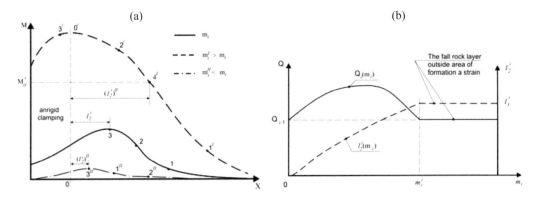

Figure 2. Qualitative laws influence of thickness m_i on distribution diagram (a) of flexure moment M_i in the i-th layer of the roof and parameters (b) of stress Q_i (——) and length l_2^i (- - -) of rock console.

Thus low thickness m_i layer has low rigidity that its increased deflection moves the point 2 to the right while increasing the reaction $q_{i-1}(x)$ and arm support of its application relatively of anrigid clamping (Figure 1a). The low thickness m_i layer and increase of reaction $q_{i-1}(x)$ provide a low gradient of growth a flexure moment and its maximum is near a coordinate of anrigid clamping in point 3^{II} (Figure 2a) and length of rock console will not very big. Therefore at the beginning of growth m_i a stress Q_i will be increased by the law what close to linear because i-th layer's weight increases linearly with the parameter m_i (Figure 2b). The further growth m_i leads to increase a length l_2^i of hanging

console and gradient of growth a stress Q_i slows. At depending on the height of i-th layer from mechanized support unit a slowdown of growth a function $Q_i(m_i)$ can decrease and growth of rock consoles $l_2^i(m_i)$ that describe as the dotted line will reduce the volume of rock i-th layer more than the increase of this volume that associated with the growth m_i. The increase up of growth m_i to a certain limited value m_i^I leads to the fact that hanging console l_2^i reaches to the limit l_3^i of formation a stress on a face support from mine goaf (Figure 1) Kovalevs'ka 2012). The hanging console lays on fall rocks in mine goaf and doesn't load a face support. Thus when $m_i \geq m_i^I$ (Figure 2b) face support is loaded by only weight Q_i of underlying rocks.

Thus a mechanism of influence of thickness m_i separate layer on the loading on a face support is examined. Their interaction with each other is a complex process in the description of that we need use a computer simulation of displacement coal overlaying strata. Therefore we can suggest the mechanism of influence of structure coal overlaying strata on the status of a separate layer of the roof. Let's describe a version of thin-layered rocks structure of the roof, which lay over described rock layer on a constant structure underlying layers. Overlying low layers $m_{i+\kappa}$ ($\kappa = 1, 2, 3 \ldots n$) have low moment of inertia $I_{i+\kappa}$ and moment of resistance $W_{i+\kappa}$ of cross section and after that it have a strong bends and discontinuity and lay on described rock layer that have thickness m_i. The stress $q_{i+\kappa}(x)$ including by the contact $i+1$ increases simultaneously with the movement to the right of the losing contact's point 1^I (Figure 1a). The rock console m_i has high flexure moment M_i (Figure 2a), that is many times greater than bearing capacity that is caused by two factors:

– on the one hand there is coactions of increasing stress (linear relationship with flexure moment M_i) and increase of arm support equivalent (quadratic relationship include stress distribution along the length of the console);

– on the other hand a resistance moment W_i of cross section is a constant.

In total there is an exponential function of growth M_i with more than two exponent when W_i is a constant. This growth moves a point 4^I fall of console to the right (Figure 1a) and leads to growth of a face support's stress. But the process of fall and lowering of layer m_i don't finish (Figure 3a): a fault of layer in point 4^I reduces the flexure moment (solid line) because it disappears a part of stress $q_{i+1}(x)$, which is to the right of the point of the point 4^I and it decreases equivalent span length of stress from i-th layer's rock console that is on the left of abscissa of the point 4^I. Nonetheless a value of stress $q_{i+1}(x)$ on area $\left(l_2^i \right)^I$ there is such that flexure moment M_i on rock console length $\left(l_2^i \right)^I$ exceeds the maximum possible value M_{rf}^I and it breaks down and lays on underlying rocks in point 4^{III}. The result is a stable rock console that

has a length $\left(l_2^i \right)^{III}$ and it is many times smaller than the original length $\left(l_2^i \right)^I$. Thus arc contour of stress formation on face support (Figure 1 (Kovalevs'ka 2012) come close to the limit of anrigid clamping of layer m_i. Therefore a thin-layered structure the upper part of the roof (with regard to the layer m_i) increases volume of rocks that form a stress on face support. When thin-layered structure is not only above but below of layer m_i the last part of the roof that has increased deflection and it decreases a value of reaction $q_{i-1}(x)$ simultaneously with the movement of point 2 to the right (Figure 1a). Thus flexure moment at layer m_i increases even more and it proposes a next mechanism of its strain (Figure 3b). The intensive growing flexure moment M_i when it approaches to the anrigid clamping increases a value M_{rf}^I and the console destroys in point 4^I. At the remainder of its length $\left(l_2^i \right)^I$ flexure moment increases from zero (point 1^{III}) to maximum (point 3^{III}) and when the stress reaches the limit $M_{rf}^{III} = M_{rf}^I$ in the point 4^{III} and the next layer is collapsed with the formation of a truncated console $\left(l_2^i \right)^{III}$. But reaction $q_{i-1}(x)$ of underlying thin-layered rocks is not big because of its increasing deflection and flexure moment M_i increases again and it exceeds a value limit $M_{rf}^{III} = M_{rf}^I$ in point 4^{IV}. There is formed a low console that has a length $\left(l_2^i \right)^{IV}$. Thus thin-layered rock structure of the roof with its increased bending results to significant stress on i-th layer with its gradual destruction and lowering. During this it forms low length rock consoles beyond the limit of anrigid clamping and arc contour of stress formation on the face support (Figure 1 (Kovalevs'ka 2012) has a more upright position that increases a volume of incompetent rocks above face and stress on a face support. In this situation, the stress limit is only possible with a very strong i-th layer (Figure 2b) that has high rigid (when it is whole) and it is loaded by overlying layers in its turn it loads falling rocks in goaf on one side and it loads a coalface area ahead face on other side. This structure of coal overlaying strata almost does not exist in West Donbas therefore it is enough to forecast an increase of volume of incompetent rocks when arc contour of stress formation on the face support moves to wall face in thin-layered mass. If there is a middle-layered structure the arc contour of stress formation moves to the goaf (Figure 1 (Kovalevs'ka 2012) and the volume of incompetent rocks above coal face decreases.

The mechanism of influence of coal overlaying strata structure and thickness m_i of its separate layer on development of the stress on face support is presented in our opinion.

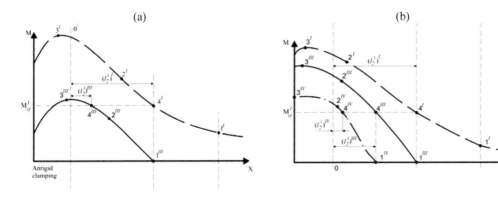

Figure 3. Qualitative distribution diagram of flexure moment M_i when thin-layered rocks of the roof lay: (a) below the i-th layer; (b) below and under the i-th layer.

REFERENCES

Pisarenko, G. 1979. *The resistance of materials.* Kyiv: Height school: 696.
Fenner, R. 1961. *The research of rock pressure.* Problems in the theory of rock pressure. Moscow: State Mining Technical Edition: 5-58.

Kovalevs'ka, I. 2012. *The features of displacement coal overlaying strata when mining coal seams is mined on the Western Donbass.* Materials of research and practice conference "School of underground mining". Dnipropetrovs'k: NMU: 243-252.

Prospects for the bioindication methods implementation in the environmental management system of industrial enterprises

A. Gorova & A. Pavlychenko
National Mining University, Dnipropetrovs'k, Ukraine

T. Kholodenko
State Enterprise Research-Industrial Complex "Pavlograd Chemical Plant", Ukraine

ABSTRACT: The possibilities of bioindication techniques using in environmental management system of industrial enterprises are considered. The necessity of implementing of bioindication methods in the environmental management system of State Enterprise Research-Industrial Complex "Pavlograd Chemical Plant" is validated.

The economic development of Ukraine was accompanied for a long time by unbalanced exploitation of natural resources, which has led to the depletion of resources, rising costs for protection of population and territories, as well as restoration of natural balance. Complicated environmental situation that emerged in the industrial regions of Ukraine requires integrated ecologization of production systems (Mishenin 1999). In toda''s economic environment the special attention is paid to the development of environmental management for the purpose of continuous improvement of environmental activity of industry and accounting requirements of the international community's attention (Kovalenko 2012).

The problem of reducing pollution caused by industry and its negative consequences requires the use of an effective system of environmental control (Nyzkodubova 2005).

With the progressing influence of anthropogenic factors on the abiotic and biotic components of the biosphere it is quite important to develop highly sensitive express methods for evaluation and control of environmental objects and human health. However, the use of traditional physical and chemical methods makes it impossible to determine the entire effect of the totality of pollutants that can be achieved by using bioindication methods at the different levels of organization (Gorova 2007).

Use of bioindication methods is appropriate to receive the information about recent, short-term and long-term effects of pollutants over a certain period. Analysis of response of some bioindication test systems at different levels of organization (molecular, cellular, organismal) to the impact of contaminants in man-overloaded areas allows to set appropriate reactions that are integrated in time and space.

Improving the system of ecological control of environmental objects through the use of express bioindication methods brings us to adequately assessment of environmental situation in the territories of industry functioning. This in turn will allow making periodic informing and maintaining an open dialogue with all parties interested in the activities of industrial enterprises in the field of environmental security.

Therefore the aim of the work is to study the possibility of bioindication methods using in the environmental management system of industrial enterprises.

To achieve this goal the following tasks were formulated:

– to study the ecological state of the industrial enterprises using highly sensitive bioindication methods;
– to conduct a survey of the employees health using biophysical and cytogenetic methods;
– to assess the state of environmental objects at the territory of industrial enterprise;
– to develop recommendations on improving the ecological state of environmental objects in the zone of the enterprise;
– to develop recommendations to improve the organism state of employees using physiologically active adaptogens and homeopathic remedies.

Two test polygons with six monitoring points were identified to conduct monitoring studies at the territory of State Enterprise Research-Industrial

Complex "Pavlograd Chemical Plant" (SE RIC "PCP"). Sampling of soil and plants were carried out according to the "envelope" rule at the territory of each monitoring point. The group of 100 employees working in different departments of plant with different conditions has been selected and justified also.

To assess the ecological status of environmental objects the following studies were carried out (Gorova 2007 & Mincer 2006):

– Assessment of general toxicity of air through bioassays "Pollen Plant Sterility".

– Assessment of toxicity and mutagenicity of soils using cytogenetic tests "Aberrant Chromosomes Frequency" and "Mitotic Index" in *Allium cepa L.* cells.

– Conducting express diagnostics of organism state of employees using micronucleus cytogenetic test Kirlian-graphy method.

Complex assessment the effect of anthropogenic pollution will become a necessary complement to existing tool control. Summarizing of results of analytical monitoring as well as bioindication and Kirlian-graphy methods will promote adequate ecological zoning of the man-loaded area, including identification of areas with the greatest ecological hazard.

Methods for determining levels of toxic and mutagenic activity of the environmental objects are based on establishing the difference between the values of cytogenetic indicators (level of pollen sterility of indicator plants, mitotic index and the frequency of aberrant chromosomes in root meristem) in bioindicators that are analyzed (experiment), and similar values in clean conditions. The toxicity criterion is the inhibition percentage of bioindicators growth and mitotic index in meristematic cells in experiments compared with the control for 48 hours in *Allium cepa L.*

Mutagenicity criterion is the increasing frequency of sterile pollen cells and cells with aberrant chromosomes in root meristem of indicator plants. It is proposed to carry out the study by the structural scheme of cytogenetic environmental monitoring that allows assessing the state of natural objects by their toxic and mutagenic background which is necessary for determining the overall environmental and genetic risk for humans and biota.

As a result of complex research the following results were obtained:

– the complex bioindication assessment of the ecological state of the plant using cytogenetic tests "Sterility pollen", "Aberrant Chromosomes Frequency" and "Mitotic Index" was conducted for the first time;

– ecological zoning of the company territory was carried out;

– the health of employees of different departments using cytogenetic and biophysical methods was assessed;

– the list of bioindication test-features was defined for further implementation of effective monitoring system within the enterprise as well as its introducing to other areas;

– the expedience of the bioindication methods use for assessing the ecological status of environmental objects on the territory of enterprise was validated;

– recommendations on the biological indication methods use in the environmental management system of the company were formed;

– highly sensitive bioindication methods were introduced to the environmental management system of SE RIC "PCP".

The results of integrated ecological assessment of environmental objects at the plant territory are the theoretical and practical basis for solution of such practical tasks as planning of ecologically grounded level of anthropogenic load, schemes design of development and deployment of technological areas as well as the improvement of environmental protection measures.

An information database of the ecological state of the environment and health of plant workers is constantly used to develop nature protection measures aimed to improve environmental safety level of Pavlograd chemical plant.

REFERENCES

Mishenin, E.V. 1999. *The economic mechanism of ecologization of production.* Sumy: Mriya-1: 138.

Kovalenko, G.M. 2012. *Monitoring of ecological and economic aspects of the industrial enterprise impact on environment.* Journal of East Ukrainian National University named after Volodymyr Dahl, 10: 57-62.

Nyzkodubova, K.V. 2005. *Methodological approaches to quality assessment of the effectiveness of the environmental management of industrial enterprises.* Municipal Economy, 5: 350-353.

Gorova, A.I., Ryzhenko, S.A. & Skvortsova, T.V. 2007. *Guidelines 2.2.12-141-2007. Survey and Zoning by the Degree of Influence of Anthropogenic Factors on the State of Environmental Objects with the Application of Integrated Cytogenetic Assessment Methods.* Kyiv: Polimed: 35.

Mincer, O.L., Gorova, A.I. & Pesotskaya, L.A. 2006. *The use of Kirlian-graphy for the express assessment of the functional state of the human body for industrial enterprises.* Kyiv: Guidelines, approved by the Ministry of Health of Ukraine, 15.

Hydrogeodynamics of the contact surface "lining-saturated rocks" in opening mine working

I. Sadovenko & V. Tymoshchuk
National Mining University, Dnipropetrovs'k, Ukraine

ABSTRACT: Questions of hydrogeomechanical interaction in the "lining of opening mine working – saturated rock" are considered. Analytical solution for determination of the weighting unloading of concrete lining with stable water-bearing rocks presence is given. The dependence of lining horizontal displacements and relative deformations on the hydraulic pressure was studied by numerical simulation. Limit values of hydraulic loads on the contour of lining and water-saturated rocks are defined for the given range of values of the physical and mechanical properties of rocks and lining.

The mechanism of the load formation on the lining in disconnected water-saturated rocks quite clearly understood by researchers, supported by experimental measurements and shaft operating experience. Such a state is not observed at presence of stable water-bearing rocks (sandstone, limestone, etc.). For example, the formulas determine the load on the supports by the arithmetic sum of the pressure of the rock mass and the hydraulic component. It should be noted that the separation of two components is theoretically justified because of complexity of accounting formed contact and subsequent lining interaction with rocks.

Most of the known solutions of calculating loads and displacements of the filter lining based on the analysis of the effect of volume hydraulic force in the infinitely small element of the rock and lining (Kozel 1976 & Rumin 1987). The difference in stress on element surfaces adopted equivalent to hydraulic gradient changes by linear filtration law, e.g. in the form of formula Dupuis. Noting excessive formalization of individual decisions should be pointed out that there are no fundamentally erroneous statements in this approach. However, the validity of its use for the whole scheme "lining-rock" is not sufficient for a number of reasons.

Firstly, the calculated value of residual head out lining, which is estimated from a comparison of flow rate in the aquifer and the lining on Dupuis (Kozel 1976) does not account for transition from the pressure to the free-flow filtration modes on contact "lining-rock".

Secondly, in the case of an impermeable lining in contact with the water-bearing rocks, calculating formulas (Kozel 1976 & Olovyannyi 1976) which determine the hydraulic component of lining load, is

not converted to a clear expression

$$p_h = \gamma_v H_r n_p,\qquad(1)$$

where p_h – hydraulic component of load; γ_v – the volumetric weight of water; H_r – residual head out lining; n_p – porosity of the water-bearing rocks;

Thirdly, the accounting weighing discharge as a factor that increases the stability of unfixed rock, has no real physical meaning, because in this case there is an additional burden of the skeleton of water-bearing rocks due to the hydraulic depression.

Known theses of the weighing theory define a partial unloading of the skeleton of water-bearing rocks under pressure filtration around the shaft. The dynamics of this process can be divided into two phases:

– formation of a hydraulic depression when shaft sinking and lining in the water-bearing rocks with a water head reduction in the value of natural pressure;

– smoothing the hydraulic depression during the level recovery outside constructed lining.

The second phase occurs when the contact of lining and rocks is formed. Here it is possible to evaluate the weighing unloading by lining resistance difference, when the rock mass elastically transfers the load at the first and second phases. Relations of the components of stress σ_z, σ_θ and displacement u_r caused by the hydraulic gradient to the shaft are true for rock mass (Figure 1), which is bounded by the radius of hydraulic depression R_d and an outer radius of the concrete lining ring r_c (Kassirova 1969)

$$\sigma_z = \frac{E_n}{1-v_n}A - \frac{E_n}{(1+v_n)r^2}B -$$

$$-\frac{1}{2(1-v_n)}\left(D\ln r + D\frac{1-2v_n}{2} + C\right); \qquad (2)$$

$$\sigma_\theta = \frac{E_n}{1-v_n}A + \frac{E_n}{(1+v_n)r^2}B -$$

$$-\frac{1}{2(1-v_n)}\left(D\ln r - D\frac{1-2v_n}{2} + C\right); \qquad (3)$$

$$u_r = Ar + \frac{B}{2} - \frac{(1-2v_n)r}{4(1-v_n)G_n}\left(D\ln r - \frac{D}{2} + C\right), \qquad (4)$$

where r – current coordinate between R_d and r_c; E_n, G_n, v_n – accordingly deformation modulus, shear modulus and Poisson's ratio for the water-bearing rocks; A, B – constants determined from the boundary conditions; C, D – constants determined by the change in pressure H around the shaft by the law

$$H = C + D\ln r, \qquad (5)$$

which is the general solution of the differential equation of the form of flat radial filtration

$$\frac{d}{dr}\left(r\frac{dH}{dr}\right) = 0. \qquad (6)$$

The boundary conditions for determining the constants C and D in the problem which is considered with notations

$H = H_e$ at $r = R_d$; $H = H_r$ at $r = r_c$.

Then

$$D = \frac{H_e - H_r}{\ln\frac{R_d}{r_c}}, \qquad (7)$$

$$C = -\frac{(H_e - H_r)\ln r_c}{\ln\frac{R_d}{r_c}} + H_r, \qquad (8)$$

where H_e – head of the aquifer in natural conditions; H_r – external head, reduced by filtration through a lining (Figure 1).

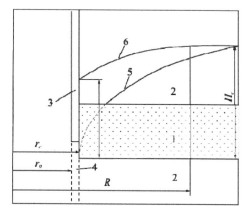

Figure 1. The scheme of weighting hydrodynamic unload within the contact surface "lining of opening working – saturated rock": 1 – water-bearing rock; 2 – water-resistant rock; 3, 4 – lining position during the first and second phases of the formation of a hydraulic depression; 5, 6 – position of depression curves in the first and second phases.

Hydrostatic weighing on the circle R_6 reaches its maximum, and

$$\sigma_z = \sigma_\theta = -\gamma_v H_e, \qquad (9)$$

where γ_v – volumetric weight of water.

It is possible to obtain the expression for calculating the weighting unloading of lining, equating the radial displacement of the loaded concrete ring with radii r_c and r_0 (Figure 1) by Lame, and the elastic displacement of rock, with a joint solution of equations (2-5), (7-9)

$$P_p = \frac{\frac{\gamma_v H_r}{2E_n(1-v_n)}\left[\frac{(1-v_n)r_c}{1-\frac{r_c^2}{R_d^2}} + \frac{1+v_n}{r_c\left(\frac{1}{r_c^2} - \frac{1}{R_d^2}\right)}\right] + \frac{\gamma_v H_r r_c}{2E_n}\left[\frac{1-2v_n}{2\ln\frac{R_d}{r_c}} - 1 + \frac{(1-2v_n)(1+v_n)}{2(1-v_n)}\left(2 + \frac{1}{\ln\frac{R_d}{r_c}}\right)\right]}{r_c\left\{\frac{(1+v_\kappa)\left[\frac{r_c^2}{r_0^2}(1-2v_\kappa)+1\right]}{E\left(\frac{r_c^2}{r_0^2}-1\right)} - \frac{1}{E_n}\left[1-v_n - \frac{1-v_n}{1-\frac{r_c^2}{R_d^2}} - \frac{1+v_n}{r\left(\frac{1}{r_c^2} - \frac{1}{R_d^2}\right)}\right]\right\}}, \qquad (10)$$

86

where E_κ, v_κ – deformation modulus and Poisson's ratio of concrete lining at the end of the first phase of the formation of a hydraulic depression.

Changing the filtration mode during the transition from the aquifer into permeable shaft lining can be accounted for by equating water flow rate on the contact rock and lining surface:

$$H_r = \frac{K_w H_e}{K_w + \dfrac{K_l r_c \ln\left(\dfrac{R_d}{r_c}\right)}{r_c - r_0}}, \qquad (11)$$

where K_w and K_l – accordingly, the conductivity of water-bearing rocks and lining.

Taking into account that the elastic geomechanical perturbation caused by the sinking of the shaft, fades at a distance of $\sim 10\,r_c$, it can be shown that the volume of suspended hydrostatically rocks increases in the specified zone with less R_d.

Thus, the cross-sectional areas between depression curve and the sole of the aquifer (Figure 1) are determined by integrating the product of equation right-hand side (8) and differential dr within from r_c to $10\,r_c$ for the first ($H_r = 0$) and second phases accordingly

$$S_1 = \frac{r_c H_e}{\ln\left(\dfrac{R_d}{r_c}\right)}\left[10(\ln 10 r_c - 1) - 11 \ln r_c)\right], \qquad (12)$$

$$S_2 = \frac{r_c(H_e - H_r)}{\ln\left(\dfrac{R_d}{r_c}\right)}\left[10(\ln 10 r_c - 1) - \right.$$

$$\left. - 10 \ln r_c + 1)\right] + 9 r_c H_r . \qquad (13)$$

It is easy to show that $S_1 \rightarrow S_2 \rightarrow 0$ at $R_d \rightarrow \infty$ other things being equal, i.e. the growth of cross-section area of hydrostatically suspended rocks in the second phase will be less than for large radii of hydraulic depression.

Values of the weighting unload of concrete filtering lining obtained by numerical analysis of (10) $0.054 \cdot 10^5 - 6.12 \cdot 10^5$ Pa within reasonably practicable range of parameters: $H_r = 0{-}50$ m; $R_d = 25{-}300$ m; $r_c = 4.5$ m; $r_0 = 4.0$ m; $v_n = 0.1{-}0.4$; $v_\kappa = 0.1{-}0.5$; $E_n = 50 \cdot 10^8 - 600 \cdot 10^8$ Pa; $E_\kappa = 10 \cdot 10^8 - 300 \cdot 10^8$ Pa.

Equation (10) is simplified, considering the importance of influential parameters in mentioned range

$$P_p \cong \frac{\dfrac{\gamma_v H_r r_c}{E_n}\left(\dfrac{0.375}{\ln\left(\dfrac{R_d}{r_c}\right)} + 1.2\right)}{1.2 r_c\left[\dfrac{0.6\dfrac{r_c^2}{r_0^2} + 1}{E_\kappa\left(\dfrac{r_c^2}{r_0^2} - 1\right)} + \dfrac{r_c a_p}{E_n}\right]}, \qquad (14)$$

where a_r – ratio equal to one and has a dimension of length in the degree of (-1).

Obtained solution, except physical description of the shaft lining interaction with stable water-bearing rocks, explains the known facts of discrepancies of calculated loads and real loads on the lining (Sadovenko 1986), and also it has an important aspect of the application.

The values of weighting unloading of concrete lining $0.054 \cdot 10^5 - 6.12 \cdot 10^5$ Pa may be close to the tensile strength of contact "concrete-rock". It is necessary to choose a concrete lining with a lower deformation modulus at the design confirmation of this approximation. At the same time, shaft construction technology should be more rigid, so that hydrodynamic regime reliably ensured formation of the first phase of depression during the lining mounting and its development of strength (consistent scheme of shaft deeping).

Under these conditions, the construction of waterproof layers when fixing sustainable water-bearing rocks may leads to a sharp water head increase outside of lining and weighing unload lining. Rock and lining strength at theirs' contact is inevitably lost in this case, so maybe its rapid violation that occurred in practice. It is based on the variant review of complex action of geomechanical and hydraulic components at the contact of the system "rock contour – lining of shaft". Finite element axially symmetric (radial) model of the rock mass with the placed therein mine opening (Figure 2) is accepted as a base. The values of physical and mechanical properties of rocks and concrete lining, which are listed above, were taken as input parameters for numerical solutions.

Interaction in modeled system "mine opening-concrete lining-rock mass" is controlled by using the contact surface which provides the possibility of autonomic displacement of calculation points along the lining plane under hydrostatic load. The hydrostatic load is controlled by hydraulic pressure value outside of lining provided that the gradual transition from the first phase (formation of a hydraulic depression while "deeping and casing" in water-bearing rocks, water head decline by the value of natural pressure) to the second one (smoothing the hydraulic depression when recovering level behind a constructed lining).

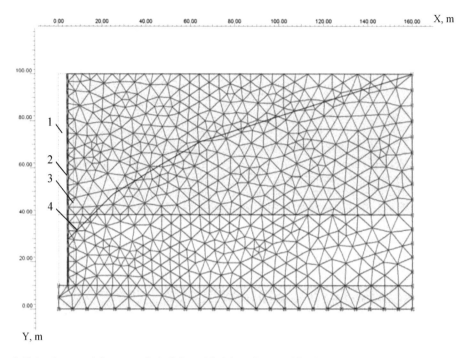

Figure 2. Finite element axially symmetric (radial) model of the rock mass with mine opening: 1 – mine opening; 2 – lining of mine opening; 3 – contact surface; 4 – hydraulic depression curve.

Obtained numerical solutions show that constructed lining deforms during the transition from the phase when a hydraulic depression forms to phase of its smoothing when water level restores. This leads to destruction of lining when a certain pressure threshold at contact "rock-concrete lining" reached (Figure 3).

Relative volume strains in rock mass and concrete lining shown in Figure correspond to the characteristic value of their physical and mechanical properties: $E_\kappa = 1.0 \cdot 10^6$ kPa, $\nu_\kappa = 0.2$, $\gamma_k = 23.0$ kN / m³, $C = 2.0 \cdot 10^4$ kPa, $\varphi_k = 25°$, $E_n = 1.0 \cdot 10^7$, $\nu_n = 0.2$, $\gamma_p = 22.0$ kN / m³, $C_n = 2.5 \cdot 10^3$ kPa, $\varphi_n = 30°$. The concrete lining destroys when reaching a critical level of load, which in given conditions corresponds to the magnitude of hydraulic pressure on the lining contour about 15.0 m. The resulting solution corresponds to the variant of interaction lining and rocks, which is characterized by low-strain stiffness of the contact surface – within it the modulus of deformation is $0.1\,E_n$.

It should be noted that the development of behind limit strain on contour "rock-lining of opening working" under these conditions is characterized by damaging strains in the bottom of lining within its 5-meter interval. Moreover maximum deformation corresponds to the middle of the interval.

Graphs in Figure 4 illustrate the dependence of the horizontal displacements in the internal lining contour, as well as relative deformations within loaded contours on the value of hydraulic load when groundwater pressure is from 0 to 30 m on the outer contour of lining.

The sharp increase in displacements and strains in the graphs corresponds to the appearance of inelastic deformation zones in lining, which are accompanied by lining destruction under hydraulic load. As in the case shown in Figure 3, the maximum allowable stresses in lining form in the lower part of opening working, where hydraulic pressure values are maximum. Moreover the interval of destructive strain coincides with the radius of shaft when the size of its, deformation and strength properties of the lining material are given.

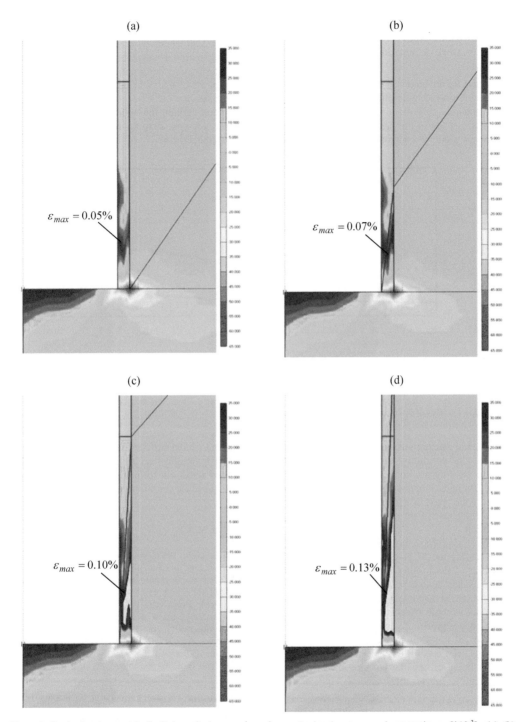

Figure 3. Strain development in the lining of mine opening when water level restores, volume strain ε, [*10^{-1}]: (a), (b), (c) and (d) – achieved strain level ε_{max} when water head is 0.0, 5.0, 10.0, 15.0 m at the outer contour of lining relatively to the mine face.

A somewhat different situation is observed when deformation properties of the lining material and rocks are similar within hydraulically loaded contour. Thus, when deformation modulus of lining and water-saturated rocks are $E_n = E_\kappa = 1.0 \cdot 10^6$ kPa destructive deformations in the concrete lining body are missing during hydraulic head is reduced from 0.0 to 25.0 m

In this case, as in the previous variant, the stiffness in the plane of contact surface corresponds to the deformation modulus $0.1\ E_n$.

When system "lining-rock" reaches the contact strength, equivalent to deformation characteristics of lining material and rock around the mine working, destructive deformations are missing during all the phase of hydraulic depression smoothing outside the lining.

For this variant, the hydraulic load on the lining contour corresponds to the reduction of water Table to 90.0 m height.

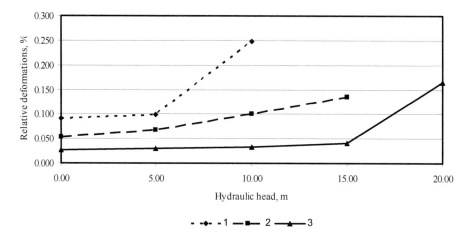

Figure 4. Diagrams of dependence volume deformations of concrete lining on hydraulic load on contour "rocks-lining" for deformation modulus of lining material: $1 - 0.5 \cdot 10^5$ kPa; $2 - 1.0 \cdot 10^6$ kPa; $3 - 5.0 \cdot 10^6$ kPa.

Analytical calculations and numerical simulations show that the destructive deformations in the lining of mine working are connected with the failure of contact strength of rock and lining during the transition from the phase of a hydraulic depression formation to the phase of its smoothing with water head restoration.

In these conditions, hydrogeodynamics of "lining-water-bearing rocks" is defined by the deformation properties of shaft lining and water-saturated rocks, as well as their contact resistance, which depends on the construction technology of lining.

Contact strength of hydraulically loaded contour is lost at a significantly reduced the deformation properties of lining in comparison to the surrounding rocks. The established mechanism of hydraulic load on lining of opening working has two important scientific and practical aspects. First, it allows reasonably approach to assessment the geomechanical stability of lining under the action of a hydraulic component on the basis of study of hydrogeodynamics of contact surface in the "concrete lining-water-saturated rocks".

Second, the definition of maximum permissible level of hydraulic load for a filter lining in water-saturated rocks allows performing parameterization of it for specific conditions of shaft sinking.

REFERENCES

Kassirova, N.A. 1969. *Study of the influence of steady-state seepage on stress state of the circular tunnel lining and the surrounding mass.* News of Vedeneev's VNIMI, 89: 55-62.
Kozel, A.M., Borisovets, V.A. & Repko, A.A. 1976. *The overburden pressure and ways to maintain a vertical shaft.* Moscow: Nedra: 294.
Olovyannyi, A.G. 1976. *The calculation of shaft lining with the hydrodynamic pressure of filtered water.* Stability and fastening the mine workings. Interuniversity collection, 3: 88-94.
Rumin, A.N. 1987. *On calculation of concrete lining of shafts with the pressure of groundwater.* Mining construction, 9: 30-31.
Sadovenko, I.O. 1986. *Calculated dependences for the definition of lining shaft loading.* Mining construction, 11: 15-17.

Conditions for minerals extraction from underground mines in the border areas against the seismic and rockburst hazards

A. Zorychta
AGH University of Science and Technology, Cracow, Poland

P. Wojtas & A. Mirek
Institute of Innovative Technologies EMAG, Katowice, Poland

P. Litwa
State Mining Authority

ABSTRACT: The seismic and rockburst hazards in Polish underground mining are the background to highlight the geomechanical conditions for underground mining in the border areas as well as movements (of parts or mining fields) of the integrated mines. Brief characteristics of problems encountered upon such operations are followed by respective conclusions.

1 INTRODUCTION

Induced seismicity in underground mining (Dubiński 1994) as well as the associated phenomena of rockbursts (tremors effecting in minor or extensive damage of the underground workings and possibly in accidents) appear as the major, dynamic manifestations of geomechanical conditions of underground mining operations. Such seismicity is generated (Gzik at al. 2012; Kłeczek & Zorychta 1991; Zorychta at al. 2000) at favourable properties of strata forming the deposit, the roof and the floor having certain geomechanical parameters, the depth, occurrence of continuous and discontinuous tectonic disturbances, the geometrical pattern of operations performed so far, etc. Time changes of the geomechanical parameters of rock (Mutke at al. 2001) during mining operations, including the widely understood seismic activity (seismoacoustics, seismology) are investigated with the use of different methods, including the geophysical ones, in order to assess the strata conditions (Lurka at al. 1997; Mirek 2001; Mirek & Oset 2013) in the light of their susceptibility to generate high energy seismic tremors (in Poland the criterion accepted is the energy $E \geq 1 \cdot 10^5$ J) threatening the workings' safety or security of the engineering structures on the surface.

2 SEISMIC AND ROCKBURTS HAZARDS IN POLISH MINING INDUSTRY

Underground mining of minerals (Mirek 2002) always appears as deep interference into the original strata balance which consequently puts the mining operations under conditions of diverse natural hazards, including the seismic and rockburst ones. In Poland in 2012 they occurred (Stan bezpieczeństwa w górnictwie 1980-2012) in 22 out of 31 of the operating hard coal mines and in all three copper ore mines. Such hazards (seismic and rockburst) are dynamic manifestations of the geomechanical processes taking place in strata interfered by the mining operations and appearing as ones most difficult to forecast and to fight effectively.

Recently and basically due to mining operation entering the mine border areas (e.g. as deposits in the central fields are being exhausted and mine plants integrated) growing tendency to strong seismic shocks may be observed in such areas where minerals are mined out of the border pillars of mines or mining areas. Such areas usually coincide with major tectonic dislocations complicating even more the geomechanical pattern of the area.

Apart from hazards imposed on the underground workings by seismicity induced by mining operations, it brings also devastating effects to the surface facilities, starting with minor faults of the engineering structures (cracking of wall plaster, concrete levelling and chimneys or loosening of ceramic building elements, etc.) to reach major damages (very seldom) of larger fragments of the building structures, e.g. of the gable walls, with falling down roofing tiles, cornices, etc.

The range of the seismic and rockburst hazard in Polish mining industry is illustrated by the statistics of high energy tremors and rockbursts as well as accident caused by such phenomena, against the production figures for coal and ore mining (Stan bezpieczeństwa w górnictwie 1980-2012). The figures for the period 1980-2012 are presented in Table 1 and 2 and illustrated by Figure 1 and 2. Table 3 points to the selected (greatest) events associated with the rockburst hazards in Polish hard coal and copper mines, throughout the period.

Table 1. Production, high energy tremors, rockbursts and accidents in hard coal mines throughout 1980-2012.

Year	Production [Mt]	Production from seams with rockburst hazard (RH)				Tremors $\geq 1 \cdot 10^5$ J (after the Central Mining Institute)		Number of rockbursts	Accidents caused by rockbursts	
		Class I-III RH [Mt]	%	Class III RH [Mt]	%	number	Σ E [GJ]		fatal	total
1980	192.8	57.1	29.6			3.432	13.58	21	7	59
1981	162.7	49.4	30.3			2.336	11.03	29	4	73
1982	188.9	55.5	29.4			2.545	4.39	20	29	105
1983	190.5	55.2	28.9			2.749	11.05	14	4	46
1984	191.0	54.3	28.3			2.970	14.59	16	20	66
1985	191.1	51.9	27.1			2.480	14.04	16	9	54
1986	191.3	51.6	27.0			2.606	9.66	27	22	83
1987	192.7	50.9	26.4			2.260	6.33	11	7	51
1988	192.7	49.5	25.7			1.599	2.05	13	3	48
1989	177.7	49.5	27.9			1.076	2.44	16	7	77
1990	147.4	42.2	28.6			1.038	2.09	16	6	36
1991	140.1	41.9	29.9			863	1.25	9	7	27
1992	131.3	41.8	31.8			833	6.00	10	9	45
1993	130.2	42.6	32.7			932	12.60	16	11	37
1994	132.7	43.0	32.4			750	1.49	12	4	47
1995	135.3	45.4	33.6			465	1.94	7	7	39
1996	136.2	44.2	32.5			564	1.07	2	3	21
1997	137.1	46.2	37.7			547	0.87	2	-	6
1998	115.9	41.9	36.2			663	0.68	5	2	17
1999	110.4	39.4	35.7			1.135	1.59	2	-	3
2000	102.5	37.2	36.3			1.088	2.12	2	-	-
2001	102.6	37.4	36.5			1.137	1.85	4	2	21
2002	102.1	41.8	40.9			1,324	1.96	4	3	20
2003	100.5	42.3	42.1			1.524	2.82	4	2	18
2004	99.5	39.2	39.4			974	1.30	3	-	11
2005	97.0	41.6	42.9	13.3	13.7	1.451	1.79	3	1	13
2006	94.4	42.1	44.6	15.9	16.8	1.170	2.06	4	4	20
2007	87.5	40.5	46.3	13.1	15.0	885	2.21	3	-	10
2008	83.6	41.9	50.1	14.3	17.1	883	2.00	5	-	26
2009	77.4	34.3	44.4	12.1	15.6	741	2.25	1	-	5
2010	76.1	35.8	47.1	13.4	17.6	1.203	4.44	2	2	15
2011	75.5	34.2	45.3	11.8	15.6	1.044	1.86	4	1	7
2012	78.6	37.6	47.8	12.7	16.2	1.069	1.78	1	1	3

The events specified in Table 3 are the major accidents and disasters in underground mines caused by rockbursts, in the years 1980-2012.

Analysis of the statistics (Stan bezpieczeństwa w górnictwie 1980-2012) of tremors and rockburst in hard coal mines throughout the past years points to the number of events reduced to a few cases per year, as compared to figures dating back to the eighties of the previous century. Similarly, the number of accidents due to such rockbursts was lower too.

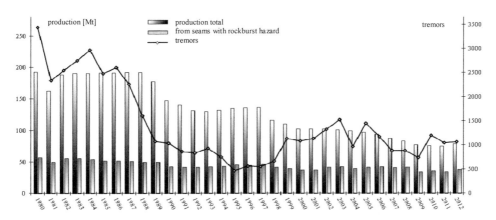

Figure 1. High energy tremors against the production total (including output from seams with the rockburst hazard) in coal mines, throughout 1980-2012.

Table 2. Statistics of production, high energy tremors, rockbursts and accidents in copper mines, throughout 1980-2012.

Year	Production [Mt]	Production from seams with rockburst hazard		Tremors ≥ 1·105 J		Number of rockbursts	Accidents caused by rockbursts	
		[Mt]	%	number	$\sum E$ [GJ]		fatal	total
1980	26.6	26.6	100	206	n/a	6	3	4
1981	22.8	22.8	100	168	n/a	7	6	26
1982	27.0	27.0	100	221	n/a	9	3	9
1983	29.0	29.0	100	217	n/a	5	2	7
1984	29.4	29.4	100	286	n/a	5	1	3
1985	29.4	29.4	100	325	1.73	2	1	9
1986	29.6	29.6	100	446	1.72	4	-	10
1987	29.8	29.8	100	484	1.72	5	7	27
1988	30.0	30.0	100	482	1.75	1	1	2
1989	26.5	26.5	100	407	2.82	4	3	8
1990	24.4	24.4	100	447	1.29	2	2	12
1991	23.7	23.7	100	359	0.92	2	2	4
1992	24.1	24.1	100	499	1.22	-	-	-
1993	27.1	27.1	100	492	3.05	4	1	7
1994	26.1	26.1	100	433	2.84	2	5	6
1995	26.5	26.5	100	389	1.87	4	2	13
1996	27.4	27.4	100	644	1.82	4	3	12
1997	24.0	24.0	100	567	2.76	-	-	-
1998	26.8	26.8	100	443	2.80	2	3	9
1999	27.0	27.0	100	414	3.96	3	2	14
2000	28.0	28.0	100	514	7.11	4	2	4
2001	30.9	30.9	100	729	6.22	5	-	3
2002	29.7	29.7	100	694	7.36	8	3	15
2003	30.0	30.0	100	570	3.39	9	5	28
2004	31.8	31.8	100	621	6.56	8	1	15
2005	32.0	32.0·	100	786	4.02	3	1	22
2006	32.9	32.9	100	872	5.65	2	-	5
2007	31.8	31.8	100	1011	1.97	3	4	10
2008	30.9	30.9	100	785	1.28	2	1	18
2009	31.2	31.2	100	474	1.82	4	1	11
2010	30.8	30.8	100	581	2.59	8	6	46
2011	31.2	31.2	100	581	1.44	1	-	5
2012	31.7	31.7	100	525	0.94	1	-	3

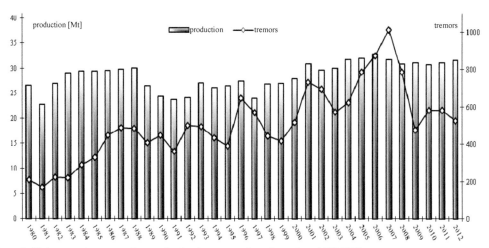

Figure 2. High energy tremors against production in copper mines, throughout 1980-2012.

Table 3. Selected group accidents due to rockburst in underground mines, throughout 1980-2012.

#	Mine	Date	Time	Energy [J]	Accidents			
					fatal	grievous	minor	total
1	POWSTAŃCÓW ŚLĄSKICH	08.11.1984	11.56 pm	$1.0 \cdot 10^7$	8	4	–	12
2	SIEMIANOWICE	22.05.1985	7.52 pm	$6.0 \cdot 10^7$	6	–	1	7
3	BOBREK	25.06.1986	6.42 pm	$7.0 \cdot 10^6$	9	–	1	10
4	ŚLĄSK	13.03.1987	10.00 am	$2.0 \cdot 10^6$	4	2	1	7
5	LUBIN	20.06.1987	2.17 am	$3.2 \cdot 10^7$	4	1	10	15
6	HALEMBA	07.03.1991	11.38 am	$1.0 \cdot 10^7$	5	-	–	5
7	PORĄBKA KLIMONTÓW	05.06.1992	5.47 pm	$1.0 \cdot 10^7$	4	2	5	11
8	MIECHOWICE	17.09.1993	1.34 am	$3.0 \cdot 10^5$	6	–	2	8
9	RUDNA	14.04.1994	8.10 pm	$1.0 \cdot 10^8$	3	–	–	3
10	NOWY WIREK	11.09.1995	1.57 pm	$5.0 \cdot 10^7$	5	–	4	9
11	ZABRZE-BIELSZOWICE	12.12.1996	4.57 am	$5.0 \cdot 10^7$	5	–	6	11
12	RUDNA	21.08.1998	1.29 am	$4.0 \cdot 10^6$	3	–	1	4
13	RUDNA	29.01.1999	2.12 am	$2.5 \cdot 10^8$	2	1	3	6
14	WESOŁA	09.08.2002	7.34 pm	$3.0 \cdot 10^7$	2	-	9	11
15	LUBIN	04.08.2003	10.17 am	$1.9 \cdot 10^8$	3	2	5	10
16	RUDNA	05.08.2005	6.33 pm	$6.4 \cdot 10^7$	1	3	13	17
17	POKÓJ	27.07.2006	2.08 am	$9.0 \cdot 10^7$	4	–	6	10
18	RUDNA	13.12.2007	3.52 pm	$5.1 \cdot 10^7$	2	–	5	7
19	POLKOWICE-SIEROSZOWICE	29.10.2008	3.30 am	$6.1 \cdot 10^7$	–	–	13	13
20	HALEMBA-WIREK	21.11.2008	10.05 pm	$1.0 \cdot 10^7$	–	–	19	19
21	RYDUŁTOWY-ANNA	21.10.2010	10.28 pm	$7.3 \cdot 10^5$	1	1	6	8
22	RUDNA	30.12.2010	9.56 am	$1.5 \cdot 10^8$	3	-	13	16
23	JAS-MOS	28.01.2011	5.17 pm	$2.3 \cdot 10^6$	1	-	3	4

Also in copper mines no growth in the number of rockbursts was observed, despite substantial expansion of the mining operations. Such tendency seems to be associated mainly with the reduced coal production (from about 200 Mt in the eighties, down to 78.6 Mt in 2012), more efficient distressing of rockbursting seams, abandoning parts of seams or beds showing the highest hazard, improved methods of assessing the hazard as well as better preventive measures. At the same time the tremors (in particular high energy ones) became more frequent in the border areas of mines or the mining areas within the mine limits. Obviously, the areas where extraction has been completed are characterized by numerous, usually irregular mining edges, frequently throughout several coal seams and often (e.g. in ore mines) in the vicinity of diverse continuous and discontinuous geological disturbances – folds or faults which oftentimes mark the borders of the mining areas or the mining fields. However, such borders – which is true especially in case of the adjacent mines – are artificial limits coinciding with the administrative division of the coal deposit into the mining areas.

In areas showing natural borders (marked by tectonic dislocation of high thrust) it is necessary to preserve safety pillars. Among the effects of mining operations is the increased seismic activity (Figure 3) due to superposition of impact manifested by zones of increased stress (effort), both in the safety pillar and in the overlying tremor prone strata. Generally, the occurrence of a tremor is the consequence of the process of cracking of the tremor prone layer when the critical effort is reached.

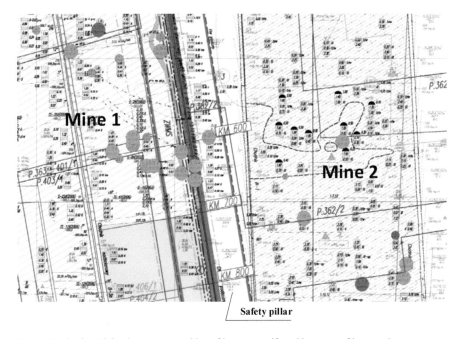

Figure 3. Increased seismic activity due to superposition of impact manifested by zones of increased stress.

What changes in the stress (effort) status of the tremor prone layer would be imposed by the ongoing extraction of the underlying seam was defined assuming mining operations at both sides of the safety pillar and occurrence of a tremor prone layer (Figure 4) overlying the batch of the seams extracted. Cracking of such layer (and consequently a tremor) may produce the rockburst prone seismic energy of $\geq 1 \cdot 10^5$ J. It was also assumed that in both mines, both sides of the pillar were mined out in the overlying seam while now operations are carried out only at one side of the pillar.

Taking into account that the effort (decisive in tremor occurrence) is proportional to energy density of elastic strain, changes in of the following indices of elastic strain energy density were analysed:

$$k_{A_c} = \frac{A_{cw}}{A_{cp}} \; ; \; k_{A_c} = \frac{A_{cw}|_2}{A_{cw}|_1} ,$$

where A_{cp} – value of the elastic strain energy density for primary stress; A_{cw} – value of the elastic strain energy density for secondary stress; $A_{cw}|_1$ – value of the elastic strain energy density for secondary stress – Option 1; $A_{cw}|_2$ – value of total elastic strain energy density for secondary stress – Option 2.

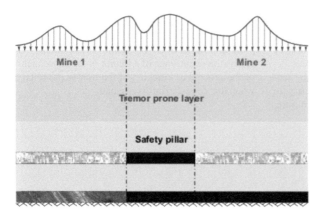

Figure 4. Safety pillar and occurrence of a tremor prone layer.

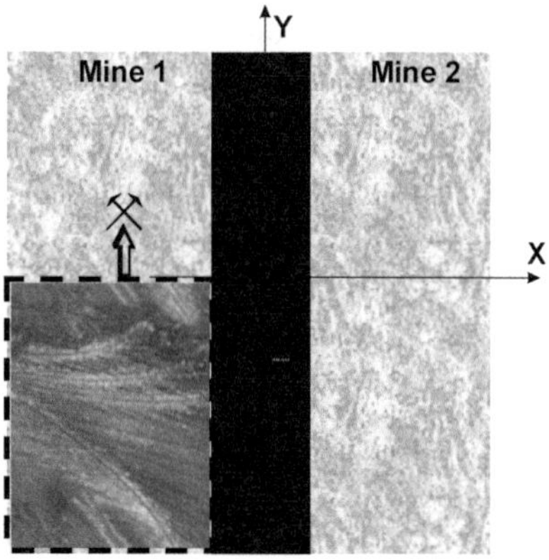

Figure 5. Cross-section of safety pillar.

For comparison of the changes, two options were taken into consideration:

Option I – describes changes in the stress conditions in the tremor prone layer after extraction (in both mines) of the overlying seam.

Option II – describes changes in the stress condi-tions in the tremor prone layer during extraction of the underlying seam.

Respective changes in energy density indices k_{A_c} for such mining situations are illustrated in Figures 6÷9.

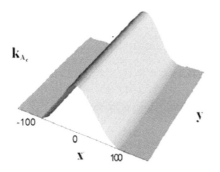

Figure 6. Distribution of k_{A_c} index in the tremor prone layer for Option I.

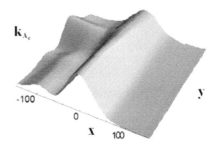

Figure 7. Distribution of k_{A_c} index in the tremor prone layer for Option II.

Referring to the results of the numerical calcula-tions presented, one should emphasize two issues:

– preserving the safety pillar effects in some un-favorable changes (significant growth of value of the elastic strain energy density and consequently greater effort) in the stress conditions in the overly-ing tremor prone strata;

– the changes in the value of k_{A_c} index, indi-rectly pointing also to the range of effort in the tremor prone layer, occur not only above the pillar but also at both sides, while areas of the peak values are those above the pillar axis (Figure 8).

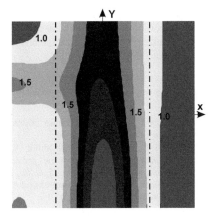

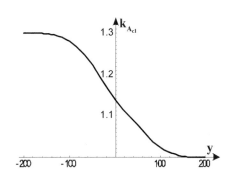

Figure 8. Distribution of k_{A_c} values in the tremor prone layer for Option II.

Figure 9. Distribution of $k_{A_{c1}}$ changes in the tremor prone layer (over the axis of the pillar $x = 0$).

Results of the calculations above confirm that mine border areas (mining fields) with usually geomechanically complex pattern of strata interfered by mining operations, may produce real high seismic and rockburst hazard. Such areas are in most cases outside the mines and their seismological monitoring networks which brings certain errors to evaluation of parameters of high energy seismic phenomena (foci outside the network) (Lurka at al. 1997), and (for the same reasons) loss of part of information on low energy seismic phenomena, which are often unregistered due the detection threshold of apparatus recording the seismic tremors. An additional disadvantage of the seismological systems used in mine geophysical stations is their vertical expansion which makes it considerably difficult to define " z " component of the tremor foci. Examples of seismic networks of the adjacent mines are illustrated in Figure 10. Example of spacing of the seismic sensors is illustrated in Figure 11.

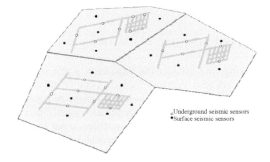

Figure 10. Example of spacing of the seismic sensors in networks of the adjacent mines.

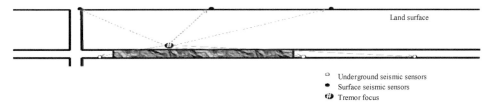

Figure 11. Example of spatial location of seismic sensors in underground mine.

Figures 10 and 11 present the aforementioned drawbacks of the networks used to record and interpret the seismic data. Figure 10 shows areas off the local (mine) monitoring networks, difficult to observe, while Figure 11 presents the flattening effect of the underground seismological network (continuous line) and the option to make it spatial with the use of surface seismic stands (dotted line) set to evaluate the impact of tremors upon the engineering structures – measurement of land vibrations (velocity or acceleration) on the surface caused by seismicity induced by the mining operations. More and more observations point to such organization of the seismological tests where the area of the measurements taken and of the geophysical tests performed exceeds the area of a single mine in order to improve accuracy of determining parameters of the strongest tremors and to collect information on some minor phenomena. A possible solution (Mirek & Oset 2013) is to establish strong, well equipped geophysical centres staffed with experienced engineering personnel. Their range would cover a few (possibly a dozen or so) mining plants operating within a given geological unit.

Another factor (Mirek & Oset 2013) hampering collection of satisfying geophysical data is substantial diversification of apparatus and software throughout individual mines; these include both, format of the recorded data as well as software applied to process information which makes it difficult if not impossible to compare the results and measurement data. As a result, evaluation of the hazard level (in particular in mine or mining field borders) may miss some geomechanical parameters of strata, essential for such evaluation.

3 CONCLUSIONS

Taking into account the expanding scope of geophysical tests demanded upon operation of underground and opencast mines and recognizing potential opportunities in the manner and organization of such investigations, it seems justified to put forward the following conclusions (proposals) regarding the discussed issues:

1. Despite the developing evaluation methods as well as respective preventive measures (active and passive prevention), the seismic and rockburst hazards effecting from the ongoing geomechanical processed still remain on a relatively high level as difficult to forecast and to fight effectively.

2. It seems needed for organization of the geophysical service to improve quality of the test results through urgently demanded geophysical centres which would comprise a larger number of mining plants operating within a geological (tectonic) unit, ensuring therefore proper evaluation of the geomechanical phenomena throughout areas larger than a single mine.

3. In order to improve the quality of interpretation of strata tremors induced by mining operations, efforts should be taken to unify (or at least to ensure common availability) formats of the recorded parameters of the seismic phenomena, ensuring comparability of the interpretation data.

4. Proper design of mining operations, including active prevention, in mine border areas should more extensively make use of analytical methods to assess the strata effort.

REFERENCES

Dubiński, J. 1994. *Związki przyczynowe wstrząsów i tąpań.* Przegląd Górniczy, 2. Katowice.

Gzik, K., Laskowski, M., Świder, M. & Mirek, A. 2012. *Wpływ uwarunkowań tektonicznych na prowadzenie robót eksploatacyjnych w polu G-7/5 O/ZG Rudna.* Zagrożenia i Technologie. Praca zbiorowa pod redakcją J. Kabiesza. Katowice: Wydawnictwo GIG.

Kłeczek, Z. & Zorychta, A. 1991. *Geomechaniczne warunki powstawania wstrząsów górniczych.* Materiały III Krajowej Konferencji Naukowo-Technicznej nt.: Zastosowanie metod geofizycznych w górnictwie kopalin stałych, T. I. Kraków: AGH.

Lurka, A., Mutke, G. & Mirek A. 1997. *Lokalizacja ognisk wstrząsów z uwzględnieniem zjawiska wieloznaczności.* Bezpieczeństwo Pracy i Ochrona Środowiska w Górnictwie – Miesięcznik WUG, Nr 9(37). Katowice.

Mirek, A. 2001. *Geofizyka – pomoc czy uciążliwość przy prowadzeniu robót górniczych w warunkach zagrożenia tąpaniami.* Prace naukowe Głównego Instytutu Górnictwa. VIII Międzynarodowa Konferencja Naukowo-Techniczna Tąpania 2001 nt.: „Miary ocen stanu zagrożenia tąpaniami i skuteczności profilaktyki". Katowice.

Mirek, A. 2002. *Wpływ systemu z szerokim otwarciem na kształtowanie się zagrożenia wysokoenergetycznym wstrząsami w kopalniach LGOM.* (Praca doktorska). Archiwum Biblioteki Głównej. Kraków: AGH.

Mirek, A. & Oset, K. 2013. *Geofizyczne badania stanu górotworu w świetle nowych uregulowań prawnych dotyczących geofizyków górniczych.* CUPRUM – Czasopismo Naukowo-techniczne Górnictwa Rud, Nr 1(66). Wrocław.

Mirek, A. & Wowczuk, G. 2007. *Prawne uwarunkowania bezpieczeństwa eksploatacji filarów oporowych w polskich kopalniach rud miedzi.* Mechanizacja i Automatyzacja Górnictwa – Czasopismo Naukowo-Techniczne, Nr 9(440). Katowice.

Mutke, G., Lurka, A., Mirek, A., Bargieł, K. & Wróbel, J. 2001. *Temporal changes in seismicity and passive tomography images: a case study of Rudna copper ore mine – Poland.* Rockburst and seismicity in mine, Symposium Series S27. The South African Insitute of Mining and Metallurgy. Johannesburg.

Stan bezpieczeństwa w górnictwie. Praca zbiorowa. za lata 1980-2012. Katowice: WUG.

Zorychta, A., Burtan, Z., Chlebowsk,i D. & Mirek, A. 2000. *Geomechaniczne warunki powstawania wysokoenergetycznych wstrząsów górniczych i ich oddziaływanie na zagrożenie tąpaniami.* Wrocław: Kwartalnik Cuprum, Nr 2.

Mining of Mineral Deposits – Pivnyak, Bondarenko, Kovalevs'ka & Illiashov (eds)
© 2013 Taylor & Francis Group, London, ISBN: 978-1-138-00108-4

Formation principles of the scientific system of ecological management at the industrial enterprises

A. Bardas & O. Parshak
National Mining University, Dnipropetrovs'k, Ukraine

ABSTRACT: Analyzed the principles of formation of system of ecological management and reviewed the organizational structure. Systematized methods of development and the main stages of the implementation of the ecological management at the industrial enterprises.

1 INTRODUCTION

The word "management" comes from English and means "administrating", and in the modern sense arises in the beginning of XX century in the USA, when the American researcher Frederick Winslow Taylor suggested the scientific principles of the management processes organization. Since that time the understanding of the nature management has changed significantly and has become more complex, there were different kinds of management, including ecological.

2 ANALYSIS OF RECENT RESEARCH AND PUBLICATIONS

Theoretical and methodological basis of the research of the environmental management system are covered in many works of Ukrainian and foreign scientists. The works of N. Pakhomova, K. Richter, A. Sadekov, O. Balatsky, T. Galushkina and other scientists, devoted to the study of environmental management as the type of management of modern metallurgical enterprise, have the great importance, as well as the works of V. Palamarchuk, H. Cherevok, M. Yatskiv, I. Yaremchuk and etc., that reveal the economic mechanism of nature management.

The studies of Olexander Amosha, Gunter Isfort and Stephan Schmidheiny, consider the problems of the interaction of the metallurgical enterprises and the environment, have the special significance, as well as the research of G.A. Chernichenko, B.V. Burkynsky, M.G. Chumachenko, O.P. Veklich, B.M. Danylyshyn, devoted to the question of economic and environmental resistance of the society, the economy and the regions.

The most of experts consider management as a type of control that the most relevant to the needs and conditions of market relations. I agree, because the existence of other types of control is perfectly natural. This typological diversity of management is determined by many factors, among the most important are the following: the level of the management object development, the factors of external socio-economic environment, the role of the human factor in the management, the degree of professionalism, the socio-organizational traditions.

3 PRINCIPLES OF FORMATION OF THE ENVIRONMENTAL MANAGEMENT SYSTEM AND ORGANIZATIONAL STRUCTURE

Principles of environmental management can and should act only in the system, in interdependence. Because each of them is a Supplement and the specification of the other. The basic principles of environmental management are presented in Figure 1 (Fostolovich 2008).

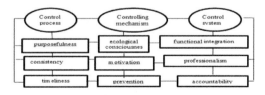

Figure 1. The principles of ecological management.

From the very beginning of the environmental management was formed as an ecologically grounded system of market-based environmental management, which is to become the new conceptual approach to the production activities. The term "environmental management" is seen mainly in the following values:

a) as the activities of public environmental organisations;

b) as the set of measures, methods and means of control of environmental activities of the enterprise (Nikolaenko 1981 & Myagchenko 1997);

c) as the control system of separate natural complex, which is subject to special protection (Popova 2001 & Moiseev 2000);

CLASSIFICATION FEATURES		TOOLS OF ENVIRONMENTAL MANAGEMENT
On the substantive nature of the impact	1. Normative regulation, payment for resources and environmental pollution, penalties; 2. Activities: privileges, advantages, the tax system; 3. Economic action, subsidies, credits, loans at reduced interest, reduction of and exemption from taxes, the payment of environmental funds.	
In view of effect	1. Administrative-and-control instruments: environmental and natural resource law, environmental monitoring, environmental standards and regulations, licensing of economic activity, environmental certification, environmental expert examination of projects, the environmental targets of the programme, the ecological audit; 2. Market-oriented instruments: natural resource payments and payments for pollution, market prices for natural resources, the mechanism of purchase and sale of rights to pollution, the system of Deposit-refund, the intervention with the purpose of adjustment of market prices, the direct market negotiations; 3. Financial instruments: i.e., instruments for financing environmental, what credit facilities, loans, subsidies, accelerated depreciation of environmental equipment, environmental in resource taxes, the system of insurance of environmental risks; 4. Instruments of moral-ethical influence and persuasion: ecological education, environmental marketing, social work, public pressure, negotiation processes, voluntary agreements.	
On the economic role		
On the Implementation mechanism		
On the nature of the impact on the economic interests of the subject	1. The prices of resources,- the prices for raw materials, materials, energy, payments for the right of use of natural resources; 2. Economic benefits-of additional income; 3. The redistribution of payment, the procedure for the withdrawal of income, rates seizure of proceeds, the order of transmission of the collected funds of economic entities, rates of payments to the recipients.	
On the original principle of impact on target groups of economic entities	1. Administrative redistribution of funds: a fine, a subsidy; 2. Financial transfers: taxes, payments, loans, payment; 3. Free market mechanisms of redistribution of resources: trade of permits for the emission of pollutants; 4. Promotion on the market: the awarding of special signs, free advertising.	
	1. Fines, sanctions; 2. Such tax breaks, accelerated depreciation of environmental assets.	
	1. Based on the principle of «polluter pays»; 2. Based on the principle of «user pays»; 3. Based on the principle of «all of society pays».	
The criteria for calculating rates of environmental and economic instruments	1. Fiscal indicators, which reflect the economic condition of the economic entities (in particular, the solvency of the enterprises) 2. Ecological-economic assessment can be calculated on the basis of: expenses on reproduction of natural factors; benefits (profits, income), obtained thanks to the use of natural factors; economic loss from the degradation of the quality of natural factors (direct losses, expenses on compensation for losses, loss of profits, etc.); economic effects from the improvement of the quality of natural factors.	
The character and the direction of the instruments	1. Non-fiscal methods (introduction of the mandatory environmental conditions, changing legal conditions taking into account the environmental factor, co-operation activities on a broad basis, ensuring the benefits of resource users, who use environmentally friendly products and production methods); 2. Methods associated with government revenue (licensing the use of natural resources, payments for pollution of the natural environment); 3. Methods associated with public expenditure targeted environmental investments, Central government funding, ecological-directed policy of employment of the population, support of the environmental priorities of R&d, state financing of institutions of environmental protection).	

Figure 2. Classification features and tools of environmental management.

d) as part of the overall management system that includes organizational structure, planning activities, responsibility, experience, methods – what is the way of knowledge, research and practical implementation of the (experimental method, the comparative method of study, methods of impact), the method is like a totality of methods and techniques appropriate for any work, the process and the resources for the formation, implementation, analysis and mainstreaming of environmental policy of a particular organization (Andreytsev 2002).

Methodology in the literal sense is the doctrine of the methods of cognition. At that, there are two main concepts of this definition: one considers the "methodology" and "method" synonyms, and therefore, the dialectical method of a unified methodology of the Sciences of nature, society, man; the second defines the methodology of a system that implements three functions: creation of new knowledge, structuring of this knowledge in the form of new concepts, categories, laws, hypotheses, theoretical ideas, theories; organization of new knowledge in practical social activity. The second is the approach used in the study of theoretical bases of ecological management (Semenov 2004). Classification attributes and characteristics of the groups of environmental management tools are shown in Figure 2 (Tendyuk 2010).

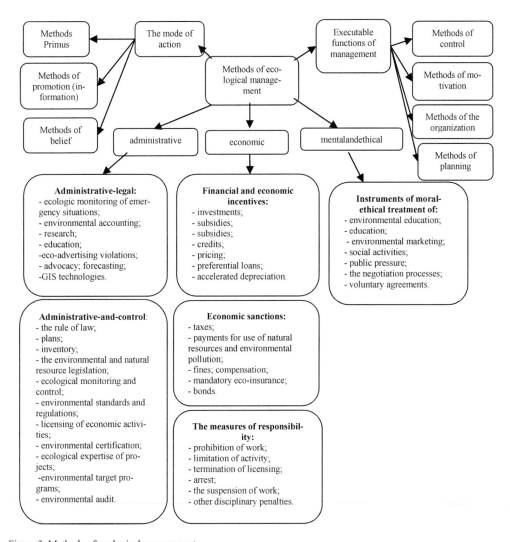

Figure 3. Methods of ecological management.

Fundamentally different are the methods of environmental management and in its content. So, the first group of methods (Figure 3) designed to provide direct impact on the environmental performance of the activities, is based on the use of administrative levers of influence. The second group through the use of economic instruments, as a rule, indirectly influences the activity of economic entities, creating the economic conditions of the achievement of environmental efficiency. The third group of morally-ethical tools of environmental management, which directly or indirectly contribute to the achievement of the environmental effectiveness by the use of various levers of persuasion and pressure.

We believe that the concept of environmental management should be applied at the level of separate industrial enterprises, since the classic conditions of the emergence and implementation of management presuppose the existence of the private property and material responsibility for the decisions taken and measures, which are being implemented.

Thus, environmental management – it is a comprehensive and multi-faceted activities, which can be interpreted in the broad and narrow context. In a broad sense, this type of management basically focuses on the formation and development of ecological production and new environmental quality of life of the people, with the purpose of preservation of the quality of the natural environment, providing the health of the population and the efficient use of resources. A more narrow definition of the understanding of this activity includes the provision of the various aspects of their combination and different levels of the management hierarchy. In accordance with the international standard ISO 14001 environmental management system is part of the overall management system that includes organizational structure, planning activities, responsibilities, practical servant, as well as the procedures, processes and resources for developing, implementing, evaluation of the achieved results of implementation and improvement of environmental policy, on its objectives and tasks (Tendyuk 2010).

Environmental management is based on the fact that all the subsystems, components of the enterprises meet, guaranteed the protection of the environment from the negative effects through the management of their environmental component. He considers and implements variants of the solution of production tasks with minimal negative impact on the environment.

Study of the experience of the leading enterprises in the developed countries shows, that without the implementation and development of the environmental management system it is impossible to operate successfully in the conditions of market economy. If the company carries out strategic planning, in spite of the solution of ecological problems, the guide in advance should understand the possibility of the emergence in the near future a number of problems, connected first of all survival in the conditions of competitive struggle, but also in the fact that the contradiction between the private and common interests.

You have to admit, that in the conditions of modern production and market economy to achieve the necessary rates of economic development, growth of production volumes, profit in conditions of rigid competitive struggle in the external and internal markets, enterprises are forced to carry out modernization of production.

In this connection, environmental management, with his tools of environmental management becomes a prerequisite for effective managerial decisions with the objective of improving the environmental performance of the enterprise, the prevention of threats to the environment in the way of the decision of strategic problems.

4 METHODS OF DEVELOPMENT AND THE MAIN STAGES OF THE IMPLEMENTATION OF THE ECOLOGICAL MANAGEMENT AT THE INDUSTRIAL ENTERPRISES

The first step to creation of system of ecological management at the enterprise there is a clear awareness of the environmental dimension in its activities. Under the environmental aspect understand element of activity of the enterprise of its products or services, which interacts or can interact with the environment (Barkov).

However, as practice shows, for the emergence of the need for an environmental management system is not enough just to identify the environmental aspects of the business. For its implementation at the enterprise should be clearly articulated and consistent environmental policy. It should be documented, well-known and understandable not only to staff and partners, but also to all stakeholders - persons, who have the interest to the environmental aspects of activity of the enterprise, its products and services, as well as persons, who felt the impact of, linked with the ecological aspects of enterprise activity.

To build and to develop the environmental policy and the environmental management system they should use the principle of continuous improvement, that is to be aimed at the achievement of the best results in all environmental aspects of the company in accordance with its environmental policy and the gradual approximation to the goal, the selection of the real goals and definition of real-time achieving them (Galushkina 2010).

One of the main features of the environmental management system is that it is implemented on a voluntary basis, on the initiative of the business entity. While declaring its own environmental policy and implementing it in practice through a system of environmental management, the company liquidates the formalism of administratively regulated environmental activities, and his environmentally focused activities ceases to be forced an "application" to the main activity and should not contradict national standards of environmental activities (Sosnin).

Given the increasing importance of environmental management, its development and implementation, as noted in the international standard of the ISO 14000 series (Sosnin), must take place in the following stages.

Preliminary analysis of the situation that has evolved. Identification of all the requirements of ecological management of the enterprise from the state and environmental management elements, which are already being applied at the enterprise.

Elaboration of the Declaration on the ecological policy of the enterprise, which would be reflected in detail all of the environmental aspects of its activities. Creation of the structure of distribution of duties and responsibilities in the environmental management system.

Assessment of the impact of the plant on the environment. It is necessary to compile a list of the standards, characteristics of the emissions into the atmosphere and water objects, a plan of the placement and utilization of wastes of production and the structure of the environmental impact of enterprises-suppliers. The development of environmental goals and tasks of the enterprise.

Identification of the stages of production, processes and activities that may affect the state of the environment, development of the system of control over these processes.

Development of the program of ecological management, determination of the person responsible for the execution of a person – the management of the enterprise.

A program must be designed so as to take into account not only current, but also the former activities of the enterprise, as well as the likely impact on the environment planned to manufacture products.

Elaboration and publication of a detailed description of the system of ecological management of the enterprise, which would allow the auditor to establish whether the system operates and how all aspects of influence of the enterprise on the environment are taken into account.

The installation of the system of registration of all ecologically important events of environmental activities, cases of violations of the requirements of environmental policy, etc. Installation at the enterprise the system of internal audit according to the standard recommendations. The results of the internal audit may be subject to external audit an independent third party.

Of course, the use of such groups of instruments, which are based on the recognized international legal norms and environmental standards, should contribute to the economic interest of the subjects of management to the implementation of environmental protection, resource-saving and the nature regenerated technologies in existing industrial facilities and in construction of new facilities.

Introduction of ecological management at the enterprise can be considered as economically useful and expedient due to such factors:

– saving of production costs and resources. Thanks to the environmental management can significantly rationalize the consumption of raw materials, water, energy, thereby reducing production costs. In addition, a significant saving of resources and facilities can be achieved due to the manufacturing of products, which is subject to secondary processing. The reduction of emissions of harmful substances helps to avoid fines and penalties;

– improvement of the quality of products. There is a direct relationship between the observance of the principles of ecological policy and environmental management and improvement of quality of products. More and more in the minds of consumers the quality of the products will be associated with its compliance with environmental standards. Not accidentally a large number of subjects of management, which in its time had implemented quality standards ISO 9000, later appealed to the international standards of ecological management ISO 14000;

– the improvement of relations with state authorities. The Declaration of environmental policies and the implementation of environmental management systems usually lead to the easing of administrative pressure on the enterprise by the state control bodies. Moreover, the introduction of environmental management system may provide opportunities for access to certain types of state support;

– expansion of market outlets of products and attracting new customers. Increase of public environmental awareness is directly reflected in the behavior of consumers, which require manufacturers of ecologically safe products and services. For producers access to new markets, especially in developed countries, is impossible without observance of environmental standards;

– output on a new level of technological development and innovation. The search for optimal from an environmental point of view of production solutions leads to the technological modernization of production processes, as well as to the emergence of innovative, that is qualitatively new products.

Research and analytical study of the condition and directions of development of environmental management in enterprises testifies to the fact that its components – environmental management - is a very important tool with the help of which it is possible more effectively and efficiently influence the state of the environment.

But this is possible only in the case if there will be built the system and mechanisms in the management of environmental processes, if the control will be based on the production as such, but to ecological production. And in this sense it is environmental management according to its purpose, is a system of stabilization and harmonization, which is aimed at overcoming the disorder and to achieve coordination of actions of human health and natural-territorial complexes, reduces the risk of contamination of the environment and, accordingly, the expenses for elimination of its consequences.

5 CONCLUSIONS

Thus, environmental management is able to provide the analysis and selection of environmentally acceptable variant of realization of the target function is the cost-effective production, in the process of which it is not the degradation of the environmental performance of the enterprise (decreasing, not increasing the load of the enterprise on the environment). On the other hand, the implementation of the environmental management measures that have an environmental effect, does not lead to a decrease (reduction) of the economic performance of the enterprise.

In itself the environmental management does not mean the replacement of the existing state administration of environmental management, but rather a complement to it, speaking of an independent, initiative activity of the enterprise. Environmental management is actively using the creative potential of entrepreneurship and the opportunities that are provided to a market economy, and is thus the improvement of the existing environmental management at the industrial enterprises.

At the same time that it contributes to the development of production and obtaining more profit is a characteristic and important manifestation of the emergence and development of new, contemporary forms of environmental culture in industrial production, entrepreneurship and market relations. All directions of actions of the enterprise in the process of environmental management at the same time are analysed from the point of view of their environmental and economic efficiency.

REFERENCES

Fostolovich, V.A. 2008. *The environmental management and audit.* Methodical instructions to practical studies on course "Ecological management and audit" for the students of agronomy faculty specialty 6.040.106. «Ecology and environmental protection».

Nikolaenko, V.G. 1981. *Forest and the protection of the environment.* Forestry, 2: 72-77.

Myagchenko, A.P. 1997. *Ecology. The protection of nature. The economy of the rational use of natural resources.* Berdyans'k: Azov Regional Management Institute: 208.

Popova, V.K. & Getmana, A.P. 2001. *Ecological law of the Ukraine.* Kh.: 480.

Moiseev, N.N. 2000. *The fate of civilization. The path of the mind.* Moscow: Languages of Russian culture.

Andreytsev, V.I. 2002. *The right of environmental security: educational and scientific-practical guide.* Kyiv: 332.

Semenov, V.F., Mikhaylyuk, O.L. & Galushkina T.P. 2004. *Environmental management: Textbook. allowance.* The Ministry of education and science of Ukraine. Odessa state economic University. Kyiv: Center of educational literature: 408.

Tendyuk, A.O. 2010. *System of methods and instruments of environmental management. Economic science.* Series "Economy and management": *Collection of scientific works. Lutsk national technical University. Issue 7(26). Part 3. Lutsk [electronic resource]: access Mode:* http://www.nbuv.gov.ua/portal/soc_gum/en_em /2010_7_3/20.pdf. Title from the screen.

Barkov, D.I. *International environmental standards of production quality ISO 14000 and prospects of their implementation in Ukraine [electronic resource]: access Mode:* http://www.erudition.ru/referat/printref /id.49971_1.html. Title from the screen.

Galushkina, T.P. 2000. *Economic tools of environmental management (theory and practice).* Odessa: Institute of market problems and economic-ecological research of the NAS of Ukraine: 280.

Sosnin, O.S. *Standards ISO 14000. [Electronic resource]: access Mode:* http://www.ecolog.spb.ru/article_iso 14000.php. Title from the screen.

Mining of Mineral Deposits – Pivnyak, Bondarenko, Kovalevs'ka & Illiashov (eds)
© *2013 Taylor & Francis Group, London, ISBN: 978-1-138-00108-4*

Fundamentals of highly loaded coal-water slurries

V. Biletskyy & P. Sergeyev
Donetsk National Technical University, Donets'k, Ukraine

O. Krut
Coal Power Technology Institute of the National Academy of Sciences, Kyiv, Ukraine

ABSTRACT: It is demonstrated that the theory of highly loaded coal-water slurries (HLCWS) may be underpinned by an analysis of the energy state of the HLCWS solid phase with the use of basic assumptions of the aggregative stability theory of lyophobic dispersion systems (DLVO theory). Accordingly, an analysis of the energy state of the HLCWS solid phase was performed, which allows to clarify the nature of phenomena taking place with changes to the size and surface potential of coal and mineral particles as well as hydrophilous-hydrophobous balance of their surface and also the HLCWS behavior was examined under dynamic conditions.

1 PROBLEM DEFINITION AND THE PRESENT STATE OF THE ART

In the context of tight fuel resources and changing pricing policy of oil and gas in Ukraine an increase of the coal share in the fuel and energy balance is becoming ever topical. Among promising technologies is the use of highly loaded coal-water slurries (HLCWS) as fuel. On the one hand, it allows producing stable transportable coal-water fuel (CWF) which may be burnt in boiler furnaces with no prior dewatering. On the other hand, this technology features much better environmental safety (Delyagin 1989 & Krut 2002).

Highly loaded coal-water slurry itself is however a complex object characterized by numerous physical and chemical factors which govern its aggregative and sedimentation stability as well as rheological properties.

High stability and fluidity of slurries are the result of their thixotropic properties. Specifically, according to DLVO theory, the inverse thixotropic restorability in turbulent flows is achieved due to coagulation of the disperse solid phase of slurry in the position of a so-called "second potential well" on the curves "combined interaction energy (E_c) – interparticle distance (h)" (Uriev 1988; New chemist's...2006; Uriev 1980).

The theory of thixotropic liquid systems is based on the main provisions of colloid chemistry, formulated in publications (Uriev 1988; New chemist's... 2006; Yefremov 1971; Deryagin 1986). The theory of coal-water slurries is at the moment at the stage of accumulating empiric data and testing working

hypotheses.

The aim of this paper is to assess the basic properties of HLCWS and tendencies in their changes from the viewpoint of the modern stability theory of lyophobic dispersion systems (DLVO theory).

2 RESEARCH RESULTS

Application of DLVO theory as theoretical basis for HLCWS. The main factors governing the behavior of a coal particle in a coagulated structure are: the particle size, a hydrophilous-hydrophobous balance of the particle surface, the total and electrokinetic potential of the latter. The characteristics of a coagulated thixotropic coal-water system are generally governed by the "depth" E_{m2} and coordinate h_{m2} of the second potential well (Figure 1) (New chemist's... 2006).

According to DLVO theory, the combined interaction energy E_c of two spherical particles in liquid has two constituents i.e. ion-electrostatic E_e and molecular-disperse (Van der Waals) E_d:

$$E_c(h) = 2\pi \cdot \varepsilon_o \cdot r \cdot \varphi^2 \cdot ln[1 + exp(-\chi \cdot h)] - \frac{A_r}{12h}, (1)$$

where ε_o – absolute dielectric permeability of water ($\varepsilon_o = 7.26 \cdot 10^{-10}$ F / m); r – radius of spherical coal particles, m; φ – potential of the diffuse double layer (DDL) on the coal particle surface, V; χ – inverse Debye radius (the reciprocal value of the diffusion layer thickness δ), $\chi = 1 / \lambda$, where λ – extension (length) of the diffusion layer DDL (for most cases $\lambda = 1 \cdot 10^{-8} m^{-1}$); h – distance between solid phase particles in slurry; A_r – Hamaker constant, J.

The second potential well of the curve $E_c(h)$ exists because the curve $E_d(h)$ decreases as a power function whereas $E_e(h)$ does so exponentially, i.e. the latter decreases faster than $E_d(h)$ (see Figure 1).

Let us consider the influence of the above factors on the nature of the curve $E_c(h)$. We will take the initial parameters of equation (1) for the purpose (1).

The variation range of the coal particle size (d) is assumed to be 10-100 μm, which corresponds to the main rational size range of the HLCWS solid phase (Delyagin1989 & Krut 2002) and according to (Uriev 1980) corresponds to the size of coarsely dispersed objects of colloid chemistry. It should be noted that such a viewpoint is not flawless but no doubt it exists, at least for some fine grains of the indicated size range.

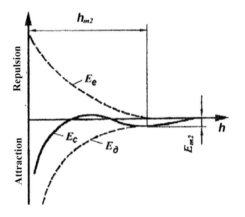

Figure 1. Potential interaction energy curves of two particles as a function of their spacing.

The variation range of the total surface potential of coal particles is assumed to be 20-100 mV according to (Saranchuk 1988 & Zubkova 1973). Due to the different hydrophilous-hydrophobous balance, the Hamaker constant (A_r) for the heterogeneous coal surface varies within $(0.5 - 3.5) \cdot 10^{-19}$ J (New chemist's... 2006).

Figure 2a shows the curves $E_c(h)$ for different coal particle sizes, obtained with the software MathCAD at the potential $\varphi = 100$ mV and the

Hamaker constant $A_r = 3.5 \cdot 10^{-19}$ J.

As seen, the larger the size of coal grains, the higher the energy barrier of repulsion and at the same time the deeper the "second potential well". The first of the above phenomena, the increased energy barrier of repulsion, results in the growing aggregative stability as the barrier prevents the particles from getting into and irreversible coagulation in the "first potential well".

The second phenomenon, the deepening of the "second potential well". contirbutes to a higer stability of the thyxotropic structure. The "deeper" this potential well, the higher the particle interaction energy in the reverse coagulation structures and the higher the HLVWS stability.

Now we will consider the influence of the coal surface heterogenicity factor on the energy state of HLCWS particles. In DLVO theory this heterogenicity (the hydrophilous-hydrophobous balance) is assessed with the help of the Hamaker constant (see equation (1)).

As is known (New chemist's... 2006), the stronger the coal phase interacts with water, the lower the Hamaker constant A_r is. Thie means that interparticle attractive forces get somewhat weaker. In other words, the increase of the constant Ar corresponds to stehgthening of the hydrophobous properties of the coal surface. The analytical curves $E_c(h)$ $A_r = var$, $d_3 = 100$ μm, $\varphi = 100$ mV, which we obtained, confirm this thesis and show (Figure 2b) that the growing hydrophobous properties of coal particles result in some decrease of the repulsion barrier height and hence in the reduced aggregative stability of HLCWS.

Figures 2c,d show the curves $E_c(h)$ for the total potential of the coal particle surface changing within 20-100 mV at the coal grain size $d_3 = 100$ μm and the Hamaker constant $A_r = 3.5 \cdot 10^{-19}$ J.

An analysis of the obtained curves indicates that an increase in the potential of the coal surface gives rise to the energy barrier of repulsion which keeps growing. This barrier appears at $\varphi \approx 50$ mV. At $\varphi < 50$ mV coal-water slurry is aggregatively unstable. Under the influence of dispersive Van der Waals interactions, its grains irreversibly coagulate with each other and slurry laminates.

At $\varphi > 50$ mV there are two typical effects observed. Firstly, the repulsion barrier height significantly grows, which correspondingly increases the HLCWS aggregative stability. Secondly, the coordinate of the second power well h_{m_2} shifts to the right. This results in a larger distance between coal particles which get retained in the second potential well of the thyxotropic structure. As a consequence, the water content of HLCWS for coal with a relatively large surface potential gets increased (and correspondingly its solid phase concentration is reduced).

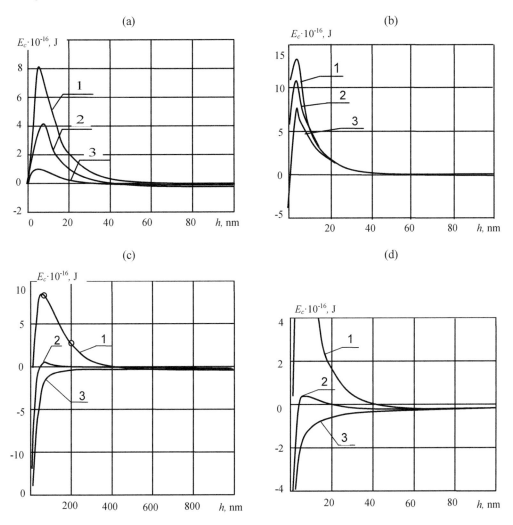

Figure 2. Analytic dependences $E_c(h)$ for coal particles: (a) with the size: $1 - 100 \, \mu m$; $2 - 50 \, \mu m$; $3 - 10 \, \mu m$; (b) at the changing hydrophilous-hydrophobous balance of the coal particle surface, the Hamaker constant: $1 - A_r = 0.5 \cdot 10^{-19} \, J$; $2 - A_r = 1.5 \cdot 10^{-19} \, J$; $3 - A_r = 3.5 \cdot 10^{-19} \, J$; for the coal particle surface potential (c), (d): $1 - 100 \, mV$; $2 - 50 \, mV$; $3 - 20 \, mV$ ((c) general survey scale of the curves; (d) detailed against the axis E_c).

Influence of the mineral constituent on HLCWS thyxotropic properties. Let us analyze the influence of the solid phase mineral constituent on HLCWS thyxotropic properties.

According to data (Yelishevich 1988), the mineral constituent of the Donbass steam coal that can be used for formulation of HLCWS is mainly represented by montmorillonite, kaolinite, hydromica and quartz.

It should be noted here that there can be several types of contact interactions identified in HLCWS as a thyxotropic structure: "coal grain – coal grain", "mineral grain – mineral grain", "coal grain – mineral grain". Respective chains and space structures of these grains may form characteristic local zones in the HLCWS thyxotropic structure. The zones are represented only by contact interactions of the type "coal grain – coal grain", covered above. Let us now consider contact interactions of the type "mineral grain – mineral grain".

We will take the initial parameters of equation (1) for the purpose. The variation range of the mineral particle size (d) is assumed to be 1-10 μm, which corresponds to the actual size range of the HLCWS mineral component (Uriev 1980). According to (Baichenko 1987 & Friedrichsberg 1984), the variation range of the total potential of the mineral particle surface is assumed to be 40-200 mV. According to (Friedrichsberg 1984), the value of the Hamaker constant (A_r) for the hydrophilous mineral surface may be assumed to range $(0.2-2.0) \cdot 10^{-19}$ J.

According to (Friedrichsberg 1984), the inverse Debye radius χ does not depend on the surface charge density and the grain surface potential but is only a function of the charge of the DDL ions and their concentration. According to (Friedrichsberg 1984), the thickness of the diffuse double layer in case of mineral grains in water is within $\delta = 1$-1000 nm. Correspondingly, the inverse Debye radius χ varies within $10^9 - 10^6$ m⁻¹.

Figures 3a,b show the curves $E_c(h)$ for different particle sizes at the potential $\varphi = 100$ mV and the Hamaker constant $A_r = 1 \cdot 10^{-19}$ J.

As seen, the energy barrier of repulsion and simultanepously the depth of the "second potential well" grow with a larger size of mineral grains. The first of the indicated phenomena, the increased energy barrier of repulsion, leads to a higher aggregative stability as the barrier prevents the particles from getting into and irreversible coagulation in the "first potential well". The second phenomenon, the deepening of the "second potential well", contributes to a higher stability of the thyxotropic structure of mineral slurry. The deeper this potential well, the higher the interparticle interaction energy in the reverse coagulation structures and the higher the stability of mineral particle slurry.

The indicated regularities are similar to those we established for coal particles i.e. they are of universal nature for the entire HLCWS soild phase. In practice, however, the repulsion barrier height of mineral particles is much smaller than that of HLCWS coal grains (from several times to 10 times) due to a smaller size of mineral grains and a lower Hamaker constant which reflects physical and chemical properties of the solid surface of substance. This governs their greater tendency to irreversible coagulation at the first energy minimum.

Figures 3c,d show the curves $E_c(h)$ for the total potential of the mineral particle surface changing within 5-200 mV at the grain size $d_3 = 5$ μm and the Hamaker constant $A_r = 1 \cdot 10^{-19}$ J.

An analysis of the curves obtained indicates that an increase in the potential of the mineral surface results in the energy barrier of repulsion which keeps growing. This barrier appears at $\varphi \approx 50$ mV, similarly to coal grains. At $\varphi < 50$ mV mineral slurry is aggregatively unstable. Under the influence of dispersive Van der Waals interactions, its grains irreversibly coagulate with each other and mineral slurry laminates.

At $\varphi > 50$ mV, similarly to the pair "coal grain – coal grain", there are two typical effects observed for the pair "mineral grain – mineral grain" under consideration. Firstly, the repulsion barrier height significantly grows, which correspondingly increases the aggregative stability of mineral slurry. Secondly, the coordinate of the second power well h_{m2} shifts to the right. This results in a larger distance between mineral particles which get retained in the second potential well of the thyxotropic structure. As a consequence, the water content of slurry for the mineral component with a relatively large surface potential gets increased (and correspondingly its solid phase concentration is reduced).

The data obtained demonstrate that with the variation of the total surface potential of mineral and coal particles the behavior patterns of the curves $E_c(h)$ are similar. For the assumed actual conditions of HLCWS the barrier height of repulsion in case of mineral particles is reduced roughly by an order of magnitude versus coal grains, with the surface potential values equal. In other words, it is another confirmation of the fact that mineral slurry tends more to irreversible coagulation at the first energy minimum.

It is also useful to analyze the influence of the nature of the particles comprised by mineral slurry on its energy state. The difference in the material composition of the slurry mineral component is evaluated in DLVO theory with the help of the Hamaker constant.

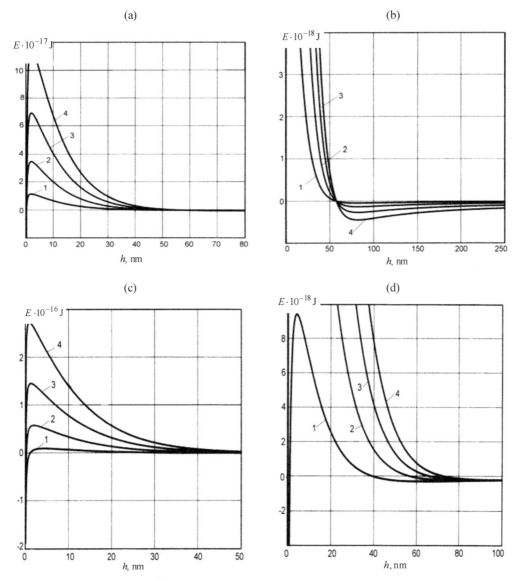

Figure 3. Analytic dependences $E_c(h)$: (a), (b) at the mineral particle size: $1 - 1$ μm; $2 - 3$ μm; $3 - 6$ μm; $4 - 10$ μm; ((a) general survey scale of the curves; (b) detailed against the axis E_c); (c), (d) at the variation of the mineral particle surface potential within 50-200 mV: $1 - 50$ mV; $2 - 100$ mV; $3 - 150$ mV; $4 - 200$ mV ((c) general survey scale of the curves; (d) detailed against the axis E_c).

As known (New chemist's... 2006), the stronger the mineral phase interacts with water, the lower the Hamaker constant A_r is, i.e. the attracting forces between slurry particles get weaker. The increasing value of the constant A_r corresponds to the growing hydrophobial properties of the mineral surface. The analytical curves $E_c(h)$, which we obtained, with

A_r changing over the range of $(0.2 - 2.0) \cdot 10^{-19}$ J, and at $d_3 = 5$ μm, $\varphi = 100$ mV, confirm this thesis and show that the growing hydrophobous properties of mineral particles result in some decrease of the barrier height of repulsion and hence in lower aggregative stability of mineral slurry.

109

As seen in Figure 4, the increase in the Hamaker constant is accompanied by the deepening of the second energy minimum i.e. a higher probability and strength of slurry mineral particle retention at the second potential minimum.

Thus, two opposite tendencies of the influence of the mineral matter nature on slurry characteristics are observed. On the one part, with the increasing A_r the aggregative stability of slurry is reduced to irreversible coagulation. On the other part, thixotropic properties of mineral slurry get stronger (formation of a spatial "grid" of mineral grains retained in the second potential well).

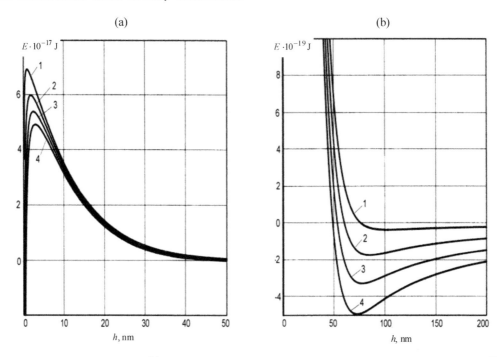

Figure 4. Analytic dependences $E_c(h)$ for mineral particles, with the Hamaker constant changing: $1 - A_r = 0.2 \cdot 10^{-19}$ J; $2 - A_r = 0.8 \cdot 10^{-19}$ J; $3 - A_r = 1.4 \cdot 10^{-19}$ J; $4 - A_r = 2.0 \cdot 10^{-19}$ J; (a) general survey scale of the curves; (b) detailed against the axis E_c.

It should be noted that the second tendency dominates over the first one, that is under similar conditions the second potential well deepens approximately 10 times whereas the height of the repulsion potential barrier is reduced by only 1.5 times.

The essential difference observed in the behavior of mineral grains of different nature in the second potential well is confirmed by data of empiric research (Yelishevich 1988). In other words, the material composition of the HLCWS mineral component is a separate factor influencing thixotropic characteristics of coal-water slurry.

Behavior of HLCWS under dynamic conditions. In our opinion, to allow for dynamic conditions under which HLCWS is hydraulically conveyed, we have to use such a technical approach as comparison of interparticle bonding energies in the HLCWS coagulation structure and the external energy of liquid

flow directed to break this bonding. According to DLVO theory, the combined interaction energy E_c of two spherical particles in liquid is determined by equation (1).

The liquid flow energy may be computed using the following formula (Chugayev 1982):

$$E_f = 0.5\alpha \cdot \rho \cdot \Delta V^3 \cdot S \cdot t, \qquad (2)$$

where α – Coriolis coefficient, ρ is the density of liquid (water), ΔV – differential velocity of flows which run on two adjacent particles in the HLCWS coagulated structure, S – cross-sectional area of flow that runs on an individual particle, t is the duration of flow action.

The HLCWS coagulation structures are destroyed if $E_f > E_c$ i.e.:

$$0.5\alpha \cdot \rho \cdot \Delta V^3 \cdot S \cdot t > 2\pi \cdot \varepsilon_o \cdot \varepsilon \cdot r \cdot \varphi^2 \cdot \delta \cdot l_n \times$$

$$\times \left[1 + exp(-\chi \cdot h)\right] - \frac{A \cdot r}{12h}.$$

Then the critical value ΔV, at which the paired coagulation bonding between adjacent HLCWS particles that may be at the first or second minimum of the potential curves $E_c(h)$ (Figure 1) is observed to be destroyed, is equal to:

$$\Delta V = \sqrt[3]{\frac{E_c}{0.5\alpha \cdot \rho \cdot S \cdot t}}. \tag{3}$$

To make calculations for the first minimum, the value of the Hamaker constant (A) for the surface of HLCWS particles according to (Baichenko 1987) may be assumed to be $1 \cdot 10^{-19}$ J; the interparticle distance in the coagulation structure (irreversible coagulation) $h = 2$ nm, which corresponds to the interval of the interparticle distances at the first potential minimum; the size range of HLCWS coal particles and hence the diameter of liquid flow running on the particles $d = 2r = 1-100$ μm; the variation range of the surface potential of coal and mineral particles is selected within 50-200 mV (Baichenko 1987); Coriolis coefficient $\alpha = 1.15$ (Chugayev 1982); water density $\rho = 1000$ kg / m³.

The curves $\Delta V(d)$ obtained for these conditions are shown in Figure 5. They serve a good illustration of two tendencies typical of disperse systems.

Firstly, the growing surface potential of coal grains results in strengthening of coagulation structures, which in its turn requires a higher liquid flow energy for their destruction. Thus, for example, in case of particles sized $d = 20$ nm the increase of the surface potential from 50 to 200 mV brings about almost a 4-fold increase of the critical value ΔV.

Secondly, a clear dependence of ΔV on the particle size is observed. Furthermore, in the area of ultrafine particles the value ΔV, at which the paired coagulation bonding between adjacent HLCWS particles is observed, drastically grows. This corresponds to classical perceptions of colloid dispersion systems. The smaller the size of the HLCWS solid phase, the stronger coagulation bonding between particles at the first potential minimum.

An important issue in HLCWS preparation and transportation practices is stability of slurry operational properties and determination of conditions under which irreversible coagulation structures may break.

HLCWS is conveyed through long-distance pipe-

lines in the laminar flow mode (Svitlyy & Biletskyy 2009). Figure 6 shows a HLCWS velocity profile in the Belovo-Novosibirs'k pipeline. The long-distance pipeline internal diameter was 0.5 m. The maximal flow velocity is 0.34 m / c. The velocity profile dependence below was obtained with the help of the regression analysis method and the software TableCurve 2D:

$$V = \left(\frac{d}{0.78}\right)^{\frac{1}{0.54}}. \tag{4}$$

Solution of equation (4) shows that the actual value ΔV in a pipeline for the contacting particles sized 100 μm comes to $5.8 \cdot 10^{-5}$ m / s in the area of maximal flow velocity gradients (at a distance of 10-15 cm from the flow axis). At the same time, according to Figure 5, the critical value ΔV for the particles of the same size is $0.5 \cdot 10^{-3}$ m / s, which is higher than the actual value ΔV nearly by a factor of 102.

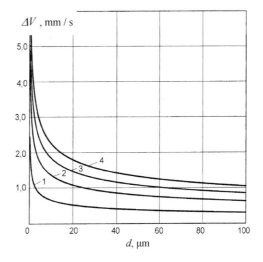

Figure 5. Analytic dependences $\Delta V(d)$ in case of particle retention at the first potential minimum, with the coal particle surface potential changing within 50-200 mV: 1 – 50 mV; 2 – 100 mV; 3 – 150 mV; 4 – 200 mV.

Thus, if irreversible coagulation structures are formed in a pipeline (at the first potential minimum), they will not be destroyed by laminar flow of slurry. This will result in sedimentation of solid phase particles, dehomogenization of the system and deterioration of HLCWS aggregative stability and rheological properties. Recovery of the HLCWS previous (prior to the formation of coagulation structures) characteristics is possible in turbulent flows where the critical value ΔV is exceeded (for example, in pumps).

Another important aspect is study of properties of HLCWS thixotropic structures formed due to retention of coal particles at the second potential minimum (see Figure 1). As known (New chemist's… 2006), the interparticle bonding energy in the second potential "well" is by 1-2 orders of magnitude lower than with retention of particles in the first "well".

To calculate and plot the curve ΔV (d) for the second potential minimum, we assume the same values of the parameters with the exception of the interparticle distance. According to our previous research data, the retention of HLCWS particles at the second potential minimum is observed at the distances $h = 50\text{-}150$ nm. Figure 7 shows the curves ΔV (d) obtained with the help of equation (3) for the second potential minimum.

As seen, the critical value ΔV corresponding to the break of paired coagulation bonding between adjacent particles of HLCWS is reduced, with the interparticle distance increased.

minimum) are on the verge of being destroyed by laminar flow of slurry. The same structures based on the particles sized less than 30 µm are not destroyed by laminar flow of slurry in the area of the maximal flow velocity gradients.

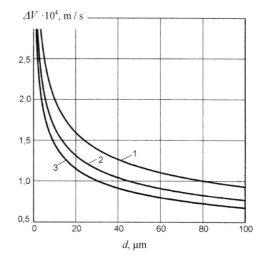

Figure 7. Analytic dependences ΔV(d) in case of particle retention at the second potential minimum, with the interparticle distances of: $1 - 50$ nm; $2 - 100$ nm; $3 - 150$ nm.

As the flow velocity gradients in wall boundary layers and in the central zone of flow are much lower (see Figure 6), the retention probability of the HLCWS thixotropic structure in these areas is significantly higher.

The proposed technical approach may be also applied to chains of grains of the HLCWS thixotropic structure.

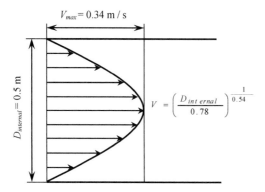

Figure 6. HLCWS velocity profile in the Belovo-Novosibirsk pipeline.

That being said, the critical value ΔV for the particles sized 50-100 µm comes to $(5-9) \cdot 10^{-5}$ m / s, which virtually coincides with the actual values ΔV in a pipeline in the area of the maximal flow velocity gradients (at a distance of 10-15 cm from the flow axis) – $5.8 \cdot 10^{-5}$ m / s. The critical value ΔV for the particles sized less than 30 µm exceeds the actual value ΔV in a pipeline.

Thus, in a long-distance hydrotransport pipeline of HLCWS with particles sized 50-100 µm, reverse coagulation structures (at the second potential

3 CONCLUSIONS

1. Application of DLVO theory for analyzing the energy state of the solid phase of highly-loaded coal-water slurries is a promising instrument for investigating this object and may underpin the HLCWS theory. Specifically, use of DLVO theory allows to explain the nature of phenomena that take place with changes to the size and surface potential of coal and mineral particles as well as their surface hydrophilous-hydrophobous balance.

2. The performed analysis of the energy curves $E_c(h)$ which we obtained shows that at relatively low surface potentials of coal particles ($\varphi < 50$ mV) coal-water slurry is aggregatively unstable.

3. A higher particle potential of coarsely dispersed colloid systems in the area $\varphi > 50$ mV results in a better aggregative stability of coal-water slurries. However, in case of coal with the surface potential $\varphi > 50$ mV the distance between coal particles retained in the "second potential well" of the energy curves $E_c(h)$, is observed to increase, which leads to a larger water content of HLCWS.

4. The growing hydrophobic properties of the coal surface somewhat reduce the HLCWS aggregative stability.

5. The material composition of the HLCWS mineral component is a separate factor influencing thixotropic characteristics of coal-water slurry.

6. A hypothesis of an aggregative stability mechanism of highly loaded coal-water slurries under dynamic conditions is put forward and rationalized. It is demonstrated for the conditions of an actual long-distance hydrotransport of HLCWS that in the laminar flow mode the thixotropic structure, where coal particles sized 50-100 μm are retained at the second potential minimum, is on the verge of being destroyed in the zone of the maximal velocity gradients. The same structure formed by the particles sized less than 30 μm is not destroyed in the indicated zone of velocity gradients. At the flow section, the retention of the HLCWS thixotropic structure is probably observed in the wall boundary and central zones of laminar flow.

In case of particle retention at the first potential minimum the strength of their bonding prevents laminar flow of slurry from destroying these structures throughout the investigated range of the size and surface potential of the HLCWS solid phase. This results in the "aging" effect of HLCWS and the necessity to recover its previous tubular mixing characteristics, for example in pumps.

A promising line of follow-up study is impact assessment of a mechanical and chemical destruction of the solid phase as well as opening of pores on HLCWS properties.

REFERENCES

Delyagin, G.N. 1989. *Coal-water slurries – new type of power fuel.* Stroitelstvo Truboprovodov, 8: 9-12.

Krut, O.A. 2002. *Coal-water fuel.* Kyiv: Naukova Dumka: 172.

Uriev, N.B. 1988. *Physical and chemical fundamentals of dispersion system and material technology.* Moscow: Khimiya: 256.

New chemist's and technologist's handbook. Electrode processes. Chemical kinetics and diffusion. Colloid chemistry. 2006. Saint Petersburg: SIC "Professional": 900.

Uriev, N.B. 1980. *Highly loaded dispersion systems.* Moscow: Khimiya: 320.

Yefremov, I.F. 1971. *Periodic colloid structure.* Leningrad: Khimiya: 192.

Deryagin, B.V. 1986. *Stability theory of colloids and thin films.* Moscow: Nauka: 206.

Saranchuk, V.I., Airuni, A.T. & Kovalyov, K.Y. 1988. *Supermolecular organization, structure and properties of coal.* Kyiv: Naukova dumka: 192.

Zubkova, Y.N., Rodin D.P. & Kucher, R.V. 1973. *Study into electrokinetic properties of fossil coals.* Khimiya tviordogo topliva, 4: 16-19.

Yelishevich, A.T., Korzhenevskaya, N.G., Samoilik, V.G. & Khilko, S.L. 1988. *Investigation of dirt content influence on rheological properties of coal-water slurries.* Khimiya tviordogo topliva, 5: 130-133.

Baichenko, A.A. 1987. Scientific basis and intensive technology of cleaning coal preparation slurries. Dissertation for the degree of Dr. Eng. Kemerovo: 478.

Friedrichsberg, D.A. 1984. *Course of colloid chemistry.* Leningrad: Khimiya: 368.

Chugayev, R.R. 1982. *Hydraulics.* – Leningrad: Energoizdat: 672.

Svitlyy, Y.G. & Biletskyy, V.S. 2009. *Hydraulic transport (study).* Donets'k: Skhidny Vydavnychny Dim, Donets'k Branch of Shevchenko Technical Society, "Mining Encyclopedia Editorial Office": 436.

Genetic classification of gas hydrates deposits types by geologic-structural criteria

V. Bondarenko & E. Maksymova
National Mining University, Dnipropetrovs'k, Ukraine

O. Koval
Deputy of Parliament of Ukraine, Kyiv

ABSTRACT: Classification of gas hydrates natural deposits is described by types depending on different tectonic structures in order to develop methods and technologies of their extraction.

1 INTRODUCTION

Topic of alternative or additional energy sources is becoming more current with each year. It is connected with depletion of such minerals as oil, gaseous condensate, natural gas, coal, reserves of which are not renewable. World oil prices are constantly increasing, hence, the cost of other energy sources is increasing as well. Thus, it is obvious that it is necessary to transfer to alternative, renewable energy sources. Sun radiation, wind power, waves etc. are among them. Nevertheless, intensity of such kinds of energy are proposed to be used by the modern science is not enough. Recently, the science is getting more interested in questions connected with gas hydrates deposits location and their formation conditions (Shnyukov & Ziborov 2004; Bondarenko 2012; Bondarenko & Maksymova 2013; Makogon 2010; Dallimore 1999). In connection with this, underground mining methods and extraction technologies of gas from natural gas hydrates is quite current.

2 MATERIALS UNDER ANALYSIS

To develop methods and technologies of gas recovery from natural gas hydrates from viewpoint of mining production principles and oil and gas industry it is necessary to thoroughly understand mining-geological conditions of the to-be-developed deposit. It is substantiated by the science that oil and gas deposits are related to tectonic structures of the Earth's crust especially to transfer from continents to oceanic troughs that were formed during Mezo-Cenozoic period. Spatial localization of oil and gas is linked with presence of conducting channels in a form of deep cracks along which, as presumed, periodical or continuous replenishment of the deposits is happening (Avilov & Avilova 2009). There three latitudinal belts that include outlines of continents in Atlantic, Indian and the Arctic oceans (Figure 1). Together with two others they form basins in which not less than 70-75% reserves of oil and gas are located (Konyukhov 2009).

Three latitudinal belts are located on the continents that existed for a long time under conditions of calm tectonic regime. The firs tectonic belt – Atlantic shore of the Northern America, Barents Sea, Norway-Greenland and North Sea basins and also basins Kara Sea, Laptev Sea, arctic of Alaska and northern half of Western-Siberian basin. The second and third latitudinal belts of oil and gas deposition are basins of Persian and Mexican gulfs, Caucasian, Turkmen, Brazilian, Western Africa, coast of Indian Ocean and Gulf of Suez.

Except for latitudinal global belts of oil and gas accumulation connected with continents' outlines, there are two meridian belts. One of them is located near active continents' outlines in eastern part of Pacific Ocean and contains oil and gas basins of California borderland, Cook bay and Guayaquil bay. California borderland is the largest area at Pacific Ocean shore in the USA (Figure 2).

There is an entire line of large and small oil and gas basins. Deep-sea origin rocks play the most important within accretionary structures. California borderland – system of elevations and troughs making underwater outline and coast ridges of California, and also fundament within the Great Valley. Rocks with thickness of 6-8 km are accumulated in troughs in form of young Neogene's, more seldom Eocene-Oligocene formations of terrigenous or silica-terrigenous composition. Rapid reduction of oil and gas accumulations age has become the sequence of active tectonic regime and only upper-cretaceous and Cenozoic rocks present the accumulations. Based on literature data of oceanology institute named after P. Shyrshova of Russian Academy of sciences, as of January 1, 2006, oil extraction on California borderland made up 17.8 billion tons and more than 1 billion m^3 of gas.

Figure 1. World map of global tectonic faults (based on materials of "National Geographic Map" National Geographic Society, Washington D.C. Distributed by MapQuest.com, Mountville, Pennsylvania, USA, 1999).

Figure 2. Large barrier reef near California coast in the USA in Pacific Ocean. Picture from space (NASA).

All these global tectonic structures since the 60's of last century are deciphered by geologists on pictures from space. Nowadays, basic principles of aero cosmic researches are a basis of brand new regional geological researches – Earth's surface maps of new type are created – space geological, space tectonic, space geo-dynamic, based on which mineral deposits were discovered (Korchyuganova 1998). As last researches of space pictures show, the higher survey point the deeper and sharper abyssal structure of Earth's crust. Because of a distant survey, small alluvial deposits are not deciphered; general view only covers global structures. This has a tremendous significance as gas hydrate formation is based on gas emission from the Earth's bowels along global tectonic faults.

In 2009 during submerge in Baikal lake on man submersible "Mir" in zone of tectonic fault to examine underwater gas emissions, gas hydrates layer was discovered in marine sediments (Granin & Makarov 2009). It looked as a transparent lake ice without soil inclusions, on top of it there was sand. Notably that methane emission from marine sediments of Baikal Lake was discovered in the 17th century. Then it was confirmed by scientists of Eastern-Siberian department of Russian Imperial Geographical society (Granin & Granina 2002). Gas hydrate in sedimentary strata and mud volcanoes at the lake's bottom were found only in the beginning of XXI century (Kuzmin 1998). In the region there was small-scale survey conducted that has shown presence of tectonic fault. Bottom depth made up 1385 m from northern-western side of the fault and up to 1435 m from its southern-eastern part. Gas flame in this spot was about 950 m high in October

2005. These facts were discovered by scientific expedition that conducts scientific programs far from methane emissions studies but specifically of entire ecosystem of Baikal.

From cosmic pictures, it was known that Caucasus and Middle Asia are dissected by the largest tectonic faults to which oil and gas deposits are coincided including Baku basin. Unfortunately, these faults were discovered later after beginning of oil and gas extraction in these regions. Such a picture is similar to many coal deposits that coincide with gigantic deep faults and large metallic ores deposits are mined today at crossing of such faults.

Permeability of rocks is higher in tectonic faults zones and there are favorable conditions for magma penetration with all following products of metasomatism and hydrothermal activity. And, since gas hydrate is formed under definite temperatures and pressures (Table 1) (Bondarenko 2012; Bondarenko & Maksymova 2013), by penetration of gas molecule into water molecule, then the deposits found on the Earth are, as a rule, connected to these zones. This fact is confirmed by scientific studies of the last 30 years conducted in this direction (Makogon 2010 & Dallimore 1999).

OJSC "Dalmorneftegeophysika" has discovered gas hydrates spread areas equal to 1600 km^2 in Bering Sea at 600-2600 m by regional seismic survey. Basic part of the hydrates coincides with continental incline. The deposits are concentrated in loose unconsolidated sedimentary accumulations. At depth of more than 3500 m the deposit accumulation equal to 26 trillion m^3 was found.

The data of geophysics were confirmed by the results of deep-sea drilling. In 1987 in Bering Sea there was a borehole drilled, collar of which was located at depth of 2110 m. A 15-meter deposit was opened there at depth of 610 m from sea bottom (Petrovskaya & Gretskay 2009).

Russian institute of oil and gas problems researches processes of formation and genesis of oil and gas deposits (Yurkova & Voronin 2009). Under high temperature ($T > 350$ °C) and pressure under the layer of serpentinite fundament of Sea of Okhotsk, methane accumulation and its homologs takes place: ethane, propane, butane, hexane etc. and autoclave situation happens.

By this way, all oil components are formed at the same place. High seismic activity and high porous pressure leads to fundament shield continuity break in central points of earthquakes.

As the fluids are concentrated in compact form, porous pressure in zones of fluids high concentration constantly increases and hydrocarbon extrusions and intrusions migrate along fault cracks, fissures zones and foliation into sedimentary rocks strata in so-called sedimentary traps of pull-apart troughs. This conclusion is confirmed by scientists-geophysics (Dmitrievski & Volodin 2006; Ravdonikas 1990; Obzhyrov 2008). According to these authors' calculation, wave energy pulses constantly rung along the fault that leads to constant gas emission from the Earth's bowels.

When the fluids overcome high temperature zone (Figure 3) then their gas component under certain thermobaric conditions transfers into gas hydrate state (Table 1).

Table 1. Parameters of gas hydrate formation process established under laboratory researches by world leading scientists.

Country, institution	Temperature, °C	Pressure, MPa
Japan	+7…+8	5.8-6.2
Russia, Institute of oil and gas geology and geophysics named after AA. Trophimyuk	-4	4.5
	+3	5.3
Russia, Institute of oil and gas problems, Russian academy of sciences	-5	1.6
Russia, Tyumen state oil and gas university	+1	0.2
	+1	0.5
Russia, Tyumen state university	+2	1.6
USA, State university of Mississippi	-5	0.1
Germany, Scientific-research project "Sugar"	+7	7
China	+3.5	0.1
Ukraine, National Mining University	+20	20
	+19	18.5
	+9	5

Based on the researches of O. Ravdonikasa (Ravdonikas 1990) and at present accumulation of such endogenic fluids in Okhotsk Sea. It is confirmed by the largest deposits of oil and gas hydrates at eastern slope of Sakhalin peninsula and the most powerful earthquakes with magnitude of $M = 7$ balls in this region (Yurkova & Voronin 2009). Multiple measures of gas by volcanologist in erupting volcanoes regions confirms continuous emissions of methane from Earth's bowels. Taking of gas measures in August 2007 that was taken half an hour since an earthquake with magnitude of 6.7 balls by Richter scale in Nevelsk (Russia, Yuzhno-Sakhalinsk) has shown that during the earthquake (August 2 and 9) methane content in atmosphere and reached 28.1%.

Correlating multiple mutually confirming facts of genetic similarities of various gas hydrate deposits and connecting them in one single system multiple world researches of volcanologists, geophysicists, chemists and physicists, the authors propose to rely on geologic-structural features of each concrete deposit during development of technology for gas extraction from gas hydrates.

DEPTH TO SEA BOTTOM 800 м

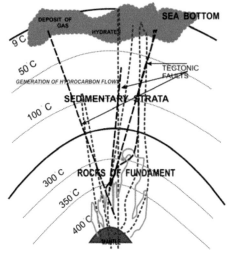

Figure 3. Scheme of gas hydrate deposit generation.

Gas hydrate molecule is described by formula $M \cdot n \cdot H_2O$, where M – gas molecule, n – number showing molecules quantity. These molecules form so-called clathrates – crystal structures in which gas is contained around water molecules. Definition of clathrates was given by Powell in 1948 and derives from Latin "clathratus" that means "put in a cell" (Byakov & Krugliakova 2001). Thus, since gas hydrate forms by way of gas molecules penetration into water molecules then the basis of such deposits development, by the author's opinion, should be the principle of water type in host rocks determination. Hence, the authors have allocated 5 types of gas hydrates deposits development depending on host rocks composition and geological structures.

1 type. Gas hydrate deposits presenting solid layers on sea and oceans bottoms in zones of shelves and troughs. These are formless deposits of gas hydrates in a shape of pure ice.

2 type. Gas hydrate deposits in a shape of solid strata from practically uniform fine structures of gas hydrate massifs bedding in shelves zones and troughs of seas and oceans, predominantly in sands, and on continents of permafrost areas within faults.

3 type. Gas hydrate deposits connected with sabulous and clay-loam depositions, pores of which are saturates with gas hydrate. They can form layered depositions bedding under various dip angles of anticline or syncline folds under bottom of seas and oceans, and also possible in permafrost areas within tectonic faults areas.

4 type. Gas hydrate deposits in fractures rocks of brecciform variations. This kind of a deposits forms under various geodynamic conditions and is characterized by a quite various composition of host rocks. Also forms along planes of rock massifs dislocations under bottom of seas and oceans in permafrost areas.

5 type. Gas hydrate deposits in a shape of vein deposits formed in large massive magmatic rocks along geologic faults. These deposits have mixed structure – from brecciform, fine to formless bedding under the bottom of seas and oceans and also possible in permafrost areas.

3 CONCLUSIONS

1. To consider connection of oil, gas, gas condensate and gas hydrates of sedimentary sheath with tectonic structures of Earth's crust in terms of an additional feature in search for gas hydrates deposits.

2. To classify natural gas hydrate deposits by types depending on their belonging to various tectonic structures, bedding conditions, composition of host rocks;

3. To develop methods and extraction technologies of gas from natural gas hydrate deposits according to corresponding types of such deposits taking into account thermobaric conditions of their stable state and hydrodynamic state of the deposition.

REFERENCES

Shnyukov, E. & Ziborov, A. 2004. *Mineral wealth of the Black Sea.* Scien. Issue NAS of Ukraine.
Bondarenko, V., Ganushevych, K. & Sai, K. 2012. *Substantiation of technological parameters of methane extraction from the black sea gas hydrate.* Krakow: Materiały Konferencyjne "Szkoła Eksploatacji Podziemnej".
Bondarenko, V., Maksymova, E., Ganushevych, K. & Sai K. 2013. *Gas hydrate deposits of the black Sea's trough: currency and features of development.* Krakow: Materiały Konferencyjne "Szkoła Eksploatacji Podziemnej".
Makogon, Y. 2010. *Additional source of energy of Ukraine. Oil and gas industry.*
Dallimore, S. 1999. *Scientific Results from JAPEX / JNOC / GSC Mallik 2L-38 Gas Hydrate research Well.* Canada: Geological survey of Canada, Bulletin.

Avilov, V. & Avilova, S. 2009. *Chemolithoautotrophic source of hydrocarbon*. Geology of seas and oceans: Materials of XVIII International scientific conference (School) on sea geology. Moscow.

Konyukhov, A. 2009. *Outlines of continents – global belts of oil and gas accumulation*. Lithology and mineral deposits. Korchyuganova, I. 1998. *Geologic structures on space pictures*. Scien. Issue. Moscow state geological survaey academy: Earth sciences.

Granin, N., Makarov, M., Kucher, K. & Gnatovski, R. 2009. *Gases emissions on Baikal*. Geology of seas and oceans: Materials of XVIII International scientific conference (School) on sea geology. Moscow.

Granin, N. & Granina, L. 2002. *Gas hydrates and gas emissions in Baikal*. Geology and geophysics.

Kuzmin, M., Kalmychkov, G. & Geletiy, V. 1998. *The first discovery of gas hydrates in sedimentary strata of Baikal Lake*. Report RAS.

Petrovskaya, N. & Gretskay, E. 2009. *New discovery of gas hydrates in Bering region*. Geology of seas and oceans: Materials of XVIII International scientific conference (School) on sea geology. Moscow.

Yurkova, R. & Voronin, B. 2009. *Abiogenic sources of hydrocarbon fluids for deposit formation of oil and gas hydrates in Okhotsk Sea*. Geology of seas and oceans: Materials of XVIII International scientific conference (School) on sea geology. Moscow.

Dmitrievski, A. & Volodin, I. 2006. *Formation and dynamics of energy active zones in geologic medium*. Report RAS.

Ravdonikas, O. 1990. *Fluids geodynamics, oil, and gas content of northern-east outline of Asia*. Report to map. Chabarovsk: DVO USSR.

Obzhyrov, A. 2008. *Migration of hydrocarbon from bowels to the surface and formation of oil and gas deposits and gas hydrates in Okhotsk Sea in the period of seismic tectonic activization*. Earth degassing: geodynamics, geofluids, oil and gas and their paragenesis. Moscow: GEOS.

Byakov, Y. & Krugliakova, R. 2001. *Gas hydrates of sedimentary strata of the Black Sea – hydrocarbon raw of the future*. Survey and protection of resources.

The recourse-saving compositions of backfilling mixtures based on slag waste products

P. Dolzhikov, S. Syemiryagin & P. Furdey
Donbass State Technical University, Alchevsk, Ukraine

ABSTRACT: It had been substantiated the cementless recipes of backfilling mixtures for eliminating the underground voids based on the research of ash and slag suspension properties. It had been provided the technological advices on the formation of backfilling mass, which solve the technical-economic and environmental problems of coal-producing regions.

1 INTRODUCTION

The coal and metallurgical industries are leading in the industry of Ukraine. Currently in the technological process of mining and metallurgy 10-90% of the initial raw materials become wastes. For example, annually the coal industry produces about 1.3 billion tons of rock and about 80 million tons of coal wastes and the output of the ferrous metallurgy slag is about 80 million tons, the ash-slag wastes of HPS are about 70 million tons. Thereby, recycling and disposal of the industrial wastes is a serious environmental problem.

The other hand, during the underground coal mining it had been formed the opening space that causes a deformation of rocks and surface, especially when watering openings. It is known that the coefficient of void residual after the roof collapse of the lava rock is 0.22-0.4 for Donbass mines. For example, for an average mine which is excavated layers with capacity of 0.8-1.2 m the amount of developed space is about 20 million m^3. Therefore it is the very perspective direction of recyclable slag waste is making on the basis of their tamping-backfilling materials. It is caused by that blast-furnace slag, as granular and dumping, includes a wide range of hydraulically active minerals and could be used as an astringent. This question is very important for Ukraine, because there is a deficit of raw materials for the cement production (Kipko, Spichak, Dolzhikov, Pozhidaev & Umanskiy 1998).

2 PURPOSE AND OBJECTIVES OF THE SCIENTIFIC RESEARCH

This article is dedicated to the justification of resource-saving filling mixes based on slag wastes

with high rheological and structural and mechanical properties.

Obtaining a new cementless composition of filling mixtures using only slag wastes unites the solution of problems of coal and metallurgical industries and provides decrease of environmental damage from a harmful production.

The following research objectives are solved in the article:

– the study of the properties of slag wastes;
– the study and the analysis of the rheological and strength properties of slag suspensions;
– the justification of the recipe of ash-slag filling mixes and the technological scheme of their preparation.

3 MATERIALS UNDER ANALYSIS

The research was carried out on the carryover of ash and ash-slag of wet removal on Luhans'k HPS and also on the blast granulated and dumping slag on Alchevs'k Metallurgical Plant. Representative samples were taken in accordance with the rules and techniques of the research are in full compliance with the standards (DSTU B.V 2.7-86-99 1999).

The analysis of the granulometric composition of HPS ash and slag revealed that fineness modulus of dry ash is $M = 0.063$ mm, which corresponds to the grinding fineness of the cement. However, for the ash-slag of wet removal the module is $M = 2.75$ mm at a fraction of up to 10 mm. It had been required screening and selection of the fractions smaller than 1 mm. That had been provided the fineness modulus: $M = 0.134$ mm.

The results of chemical analyzes of ash-slag are illustrated in Table 1.

Table 1. The chemical composition of ash and slag, %.

Sample	SiO_3	SO_3	Fe_2O_3	Al_2O_3	CaO	MgO	K_2O	Na_2O	TiO_2	lb
Dry ash	56.3	0.68	8.48	23.48	4.38	1.25	0.8	0.22	0.34	4.0
Ash of wet removal	57.1	0.51	7.24	22.99	3.34	1.31	0.7	0.25	0.36	6.2

The blast furnace slag of metallurgical production is a lump material that is why the slag was sieved with the selection of fractions less than 2.5 cm. The slag density is 2.6-2.8 g / sm³, the mound density is 1.3-1.5 g / sm³. To be used as an astringent the slag was ground in a ball mill to the specific surface of 4000 sm² / g.

The chemical composition of blast furnace slag AMP is presented in Table 2.

Table 2. The chemical composition of blast furnace slag, %.

Sample	SiO_2	CaO	Al_2O_3	MgO	MnO	FeO	TiO_2	Si	Mn	S	Ti	lb
Blast furnace slag	39.41	45.44	7.17	5.14	0.17	0.34	0.42	0.52	0.17	1.05	0.02	1.15

According to hydraulic properties the slag refers to a solid grade with quality coefficient of 1.2 and the basicity module is 1.25.

Hence the injection of ash and slag suspensions in a watered underground space will not have an eco-logically harmful effect on the rocks and hydro-sphere.

The development of filling mixes on the basis of slag waste was performed by techniques of complex methods of tamping (Kipko, Dolzhikov, Dudlya & etc. 2004). Based on the theoretical and experimen-tal results of the researches of clay-cement slurries it have been developed cementless recipes of backfill-ing suspensions for filling underground cavities, the main parameters of which are the density of the base suspensions, rheological properties and plastic strength (Kipko, Dudlya, Telnikh, Popov & Caplin 2008). An important feature of slag suspensions is their sedimentation stability, provided that the addi-tion of bentonite mud powder. Studies of the rheological characteristics of suspensions proved that the developed backfilling mixtures are vis-coplastic liquids, i.e. are obeyed the Shvedov-Bingham model (Figure 1).

The recipes of developed of ash-slag tamping-filling mixes and their properties are shown in Table 3.

The investigations of the kinetics of filling mixes solidification were performed to increase the plastic strength in time. The dependences of plastic strength on time and composition for slag clay mix-tures are shown in Figure 2. They are subject to general regularity of the solidification of viscoplas-tic liquids (Kipko, Dolzhikov, Dudlya & etc. 2004).

The analysis of the obtained dependencies re-vealed that they are satisfactorily described by the power equations. For example, for the curve 3 in Figure 2, the equation is:

$$P_m = -0.18t^3 + 4.86t^2 + 3.80t + 3.00; \quad R^2 = 0.81.$$

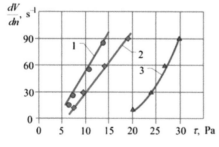

Figure 1. The rheological characteristics of ash-slag sus-pensions: 1 – suspension: slag – 60%, ash – 10%; 2 – sus-pension: slag – 60%, ash – 20%; 3 – slag-clay suspension: slag – 63%, clay – 0.6%.

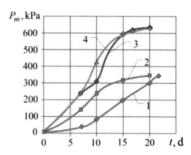

Figure 2. The dependence of the plastic strength of filling mixes on time: 1 – cement 40 kg / m³; 2 – no cement; 3 – clay 10 kg / m³; 4 – clay 20 kg / m³.

Table 3. The compositions and properties of the filling mixes.

#	The parameters of mixtures	The values	
		Ash-slag	Blastfurnace slag
1	The amount of slag, kg / m^3	825	820
2	The water volume, kg / m^3	625	625
3	The amount of ash or ground slag, kg / m^3	120	40
4	The quantity of bentonite, kg / m^3	–	10
5	The density of base suspension, kg / m^3	1450	1450
6	The density of mixture, kg / m^3	1490	1460
7	The spreadability, sm	11	10
8	The dynamic shear stress, Pa	5	22
9	The structural viscosity, Pa·s × 10^{-3}	43	54
10	The plastic strength on the 10th day, kPa	200	420
11	The shrinkage, %	12	2

On the basis of the survey it was determined that ash and slag cementless suspensions start to be structured in the first 4-5 hours and then regardless of the number of clay supplement are gaining the plastic strength over 300 kPa, which fully satisfied the requirements of the filling operations.

Based on the research results it was developed the design technique of parameters suspensions and technological scheme of their filing to the developed-out space (Figure 3).

The main technological operations are: mechanical crushing and sieving of slag wastes; preparation of the slag suspension and the introduction of bentonite; transportation and stacking of backfill material; quality control. The transportation calculations of ash and slag mixtures in gravity flow mode through the pipeline with a diameter of 200 mm at a distance of 2 km demonstrated that the production of backfill complex is 300 m^3 per hour.

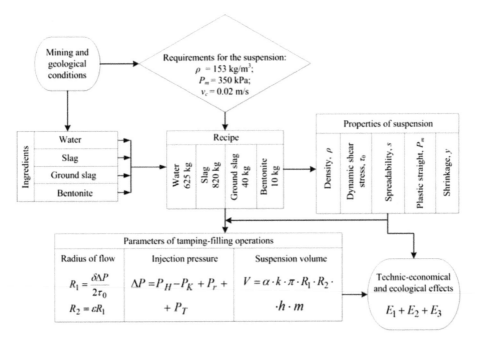

Figure 3. The block-diagram of the design technique of ash-slag backfilling suspensions' recipe.

4 CONCLUSIONS

From the results of the development of filling mixes based on slag wastes it follows: the application of ash and slag suspensions for tamping-filling operations satisfies the technological requirements of filling a developed-out space and solves the disposal problem of metallurgical companies' and power plants' wastes.

REFERENCES

Kipko, E., Spichak, Y., Dolzhikov, P., Pozhidaev S. & Umanskiy, R. 1998. *The use of waste for filling developed-out space of being closed mines.* Journal "Ugol' Ukrainy", 6: 13-14.

DSTU B.V.2.7.-86-99 (GOST 26798.1-96). 1999. *Tamping cements. Test methods.* Derzhbud Ukrainy, 54. Kyiv.

Kipko E., Dolzhikov, P., Dudlya, N., Kipko, A. etc. 2004. *The complex method of tamping in the mines construction*: tutorial, 367. Dnipropetrovs'k.

Kipko, E., Dudlya, N., Telnikh, N., Popov, A. & Caplin, E. 2008. *Design of clay-cement tamping solutions in mining:* monography, 176. Dniprodzerzhyns'k.

Justification of the gasification channel length in underground gas generator

V. Falshtyns'kyy, R. Dychkovs'kyy, V. Lozyns'kyy & P. Saik
National Mining University, Dnipropetrovs'k, Ukraine

ABSTRACT: Justification the length of gasification channel in underground gas generator by oxidation zone calculation and by geomechanics factor are described. The new methods of gasification channel calculation on the base of underground coal combustion process is offered. Dependence the power of gas generator at thin coal seam gasification due to the length and pressure changes is shown. The rocks of roof subsidence are predicted on the base of test researches. The basic technological parameters of underground coal gasification are determined during the results transmission to natural conditions. The change of the rocks subsidence above a combustion face and gasification channel depending on the gasification channel length, combustion face advance and speeds of coal seam gasification are decided. Conclusions are given according to conducted investigations.

1 INTRODUCTION

A basic part of the borehole underground coal gasification system (BUCG) is an underground gas generator. Nascent difficulties at coal seam gasification, related to the management of the thermo-chemical processes, impermeability of gas generator taking into account goaf growth. Decisions of the resulted problems are related to justification and choice the basic technological parameters of underground gas generator, specially the length of active zone of a gasification channel, combustion speed, the rate of gasification products movement, kinetics of exothermic and endothermic reactions in the channel, speed of coal gasification and rocks of roof stability above a gasification channel. These parameters, as it is known by experimental practice of "Pidzemgaz" stations and experimental investigation on the stand models of underground gas generators in a great extend have influence on stability and orientation BUCG process, state the rocks of roof, probability of its destruction and as a result gas generator depressurization. It is set that active zones advance in gasification channel is not evenly, it is related due to the high-quality and quantity indexes and reactions orientation in active zones. Exothermic reactions proceeding in an oxidizing (burning) zone with the active selection of heat and burning gases provide the intensive coal seam burning. In a reducing (smolder) zone endothermic reactions with heat absorption and forming of generator gas with a power or technical orientation.

Based on the results of practical and experimental works, gasification channel advance of oxidizing zone can results to increase the length of channel from 28 to 47%. Such changes of the gasification channel advance affect on disbalance process and origin in the rocks of roof above the gasification channel, the zones excessive rock pressure (Lavrov 1957).

Adjacent strata execute the role of walls limiting motions of a gas and blowing streams, providing impermeability of gasification processes. In this connection at BUCG it is necessary to take into account character and degree of rock-fracture, increase the goaf stowing, with the purpose of diminishing the heat dispersion, losses of blowing and gasification products and also active blowing co-operation with the reactionary surface of combustion face.

2 JUSTIFICATION THE LENGTH OF GASIFICATION CHANNEL IN UNDERGROUND GAS GENERATOR BY OXIDATION ZONE CALCULATION

Rational length of gasification channel in gas generator at borehole underground coal gasification is a parameter which determine stability of thermo-chemical reactions in the oxidation and reduction zone. It also provide the quantitative, high-quality indexes of gasification products and his orientation.

Conventional length of gasification channel at underground coal combustion (UCC) is set due to a combustion zone (oxidizing zone) in a gasification channel $l_{g.c}$ becomes equal to oxidation zone $l_{o.z}$, that $l_{g.c} = l_{o.z}$. If $l_{o.z} < l_{g.c}$ the process of coal combustion can be stopped, in these conditions thermal efficiency of underground gas generator extremely decrease. If $l_{o.z} > l_{g.c}$ a selection of thermal power is maximal. However the oxygen given with air does not have time to be exhausted by reaction with coal seam in combustion face.

Condition $l_{g.c} / l_{o.z}$ is optimum for UCC. In a gasification channel appears the products of complete coal seam decomposition with poor composition of generator gases, at $l_{o.z} = (0.3 \div 0.4) \cdot l_{g.c}$ and by a reducing zone $l_{r.z} = (0.7 \div 0.6) \cdot l_{g.c}$. The temperature of thermo-chemical process determined by recognition of heat selection in an oxidizing zone (exothermic reactions), heat absorption in a reducing zone (endothermic reactions), by the expense of thermal energy on rocks, ash and slags heating and also on water evaporation of external water inflow (Lavrov 1957 & Yanchenko 1989).

For determination the optimal length of gasification channel, it is necessary to take into account intercommunication between the length of oxidization zone and by thermal power of gasification channel from expressions (Yanchenko 1989):

– for a cylindrical gasification channel:

$$l_{c.g} = N \cdot d \cdot \left(\frac{4 \cdot G_{s.c} \cdot V_{m.c} \cdot P_0 \cdot T_C}{v \cdot T_0 \cdot P_{o.z} \cdot \pi \cdot d} \right)^h, \quad (1)$$

– for a rectangular gasification channel:

$$l_{r.g} = W \cdot d^m \left(\frac{2 G_{s.c} \cdot V_{m.c} \cdot P_0 \cdot T_C}{(a+b) \cdot T_0 \cdot P_{o.z} \cdot v} \right)^c, \quad (2)$$

where N, W, h, c – numerical coefficients, $G_{s.c}$ – mass speed of coal combustion, kg / s; $V_{m.c}$ – volume of moist products of combustion from 1 kg of coal in the air blowing, determined from expression:

$$V_{m.c} = V_{d.p} + V_{s.w},$$

where $V_{d.p}$ – volume of dry products of combustion from 1 kg of coal, m³; $V_{s.w}$ – volume of steam from a water inflow, m³.

The specific gases volume in gasification channel is determined as average index $V_{c.c}$ and $V_{d.p}$:

$$V = 0.5 \cdot (V_{c.c} + V_{d.p}),$$

where $V_{c.c}$ – an air volume, necessary for complete combustion of 1kg of coal, m³ / kg,

$$V_{d.p} = \frac{1.1 \cdot G_I^2}{4186},$$

$$V_{c.c} = \frac{1.1 \cdot G_I^2}{4186} + 1.243 \cdot q_{w.i},$$

where $q_{w.i}$ – specific external water inflow, kg.water / kg.coal; G_I – lower calorific value from 1kg of coal, MJ / kg; p_0 – pressure in gasification channel, Pa; T_c – temperature of combustion products in outlet from an oxidizing zone, T_c:

$$T_C = \frac{G_I^2}{V_{d.p} \cdot C_{h.c}},$$

where $C_{h.c}$ – isobar by volume heat capacity of combustion products, MJ / m³.

$$C_{h.c} = 1.98 \cdot 0.0001784 \cdot T_c,$$

where v – coefficient of kinematics viscosity in outlet from the combustion zone m³ / s; T_0 – air temperature, given in a gasification channel; °C; $P_{o.z}$ – pressure in the combustion channel (oxidizing zone), Pa; a, b – average high and width of gasification channel, m; d – equivalent diameter of gasification channel, $d = \frac{4S}{P}$; m – experimental coefficient.

$$m = \frac{2.86 \cdot a}{2a + b}.$$

Analyzing expressions (1, 2) it is possible to establish that dependence between the sizes of the channel $l_{c.g.c}$ and $l_{r.g.c}$ is directly-proportional. The increase of coal speed combustion in 2 times results in growth $l_{o.z}$ on 10-15 %.

Taking into account conditions: $G_{s.c.c} = \frac{A_\kappa}{G_I^2}$ the length of gasification channel is limited by the longitudinal sizes of gasification block l_b and thermal power of gasification channel is determined (Rumshinskyy 1971):

– for a cylindrical gasification channel

$$A_{c.g.c} = \frac{v \cdot P_K \cdot T_0 \cdot \pi \cdot d \cdot G_I^2}{4 \cdot V_{m.c} \cdot P_0 \cdot T_C} \cdot \sqrt[h]{l_B / N \cdot d},$$

– for a rectangular gasification channel

$$Ar.g.c = \frac{(a+\text{\emph{в}})\cdot v \cdot P_K \cdot T_0 \cdot G_I^2}{2\cdot V_{BC}\cdot P_0 \cdot T_C}\cdot\sqrt[c]{l_B / W \cdot d^m}\ .$$

Parameters of change the power of gasification channel due to the pressure and length changes are shown on Figure1.

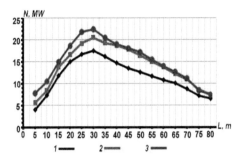

Figure 1. Dependence the power of gas generator at thin coal seam gasification ($m = 0.94$ m) due to the length and pressure changes: 1 – 0.14-0.37 MPa; 2 – 0.2-0.4 MPa; 3 – 0.44-0.62 MPa.

The power diminishing is related with blowing and gas losses which are increased due to the pressure growth in the gasification channel. At pressure 0.45 MPa and the length of combustion channel more 60 m become critical power efficiency (25-28%).

On a Figure 2 the diagram characterizing the changes between the length of gasification channel and temperature of gasification process from a water inflow is shown.

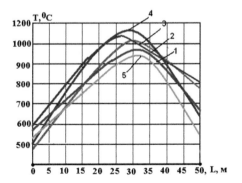

Figure 2. Temperature changes on the length of gasification channel at thin coal seam gasification depending on water inflow (g_w) and speed of gasification product movement in a channel (V_g): 1 – $g_w = 1.67$ kg water / kg coal; V_g=1.82 m / s; 2 – g_w= 1.04 kg water / kg coal; V_g=1.21 m / s; 3 – g_w= 1.18 kg water / kg coal; V_g = 1.3 m / s; 4 – g_w = 0.98 kg water kg coal; V_g = 0.85 m / s; 5 – g_w = 1.85 kg water / kg coal; V_g = 2.2 m / s.

The speed of gasification products movement on the length of combustion channel is determined from expression (Yanchenko 1989):

$$V = \frac{1.27\cdot V_{m.c}\cdot G_{s.c.c}\cdot T_c \cdot P_0}{d^2 \cdot P_{o.z}}\ .$$

Determination the Reynolds (R_e) criterion in outlet from oxidization zone:

$$R_e = \frac{V\cdot d}{v}\ .$$

– for middle composition of UCG gases:

$$\frac{R_e}{R_e} = \frac{v\cdot V}{v\cdot V} = \frac{T_c \cdot V}{T_c \cdot V}\ .$$

Difference in calculations on determination the length of combustion (oxidizations) zone $l_{o.z}$ taking into account the geometrical form of gasification channel during coal seam gasification, for a cylindrical reactionary channel 4-6%, rectangular form 8-10%.

At the practical calculations the length of oxidization zone in gasification channel is accepted taking into account the change of hydro-geological conditions and aerodynamic parameters of the gas generator system and gasification process.

3 JUSTIFICATION THE LENGTH OF GASIFICATION CHANNEL IN UNDERGROUND GAS GENERATOR BY GEOMECHANICS FACTOR

Deformations and rocks subsidence in an underground gas generator take a place under the action of two factors. A primary factor is mining pressure. The second factor is a high temperature of gasification process at gas generator exploitation. These two factors result in rocks of roof destruction. Resistance of immediate roof after the first collapse is equal to the zero. This fact means that under act of mining pressure before burning rocks of roof lose discontinuity. Distribution of the heat in deep into broken rock mass mainly takes a place not due to a conductive heat exchange, but by means of high temperature products convection on cracks and cavities in strata (Kazak 1960).

In fact, character of rocks deformation at BUCG and conventional mining are approximately identical. Distinctions appear if rock mass under act of high temperatures increase in a volume. In this case a degree and size of deformation zone changes insignificantly, but character of rocks moving above goaf remains the same as at conventional mining.

Experience of underground gas generator exploitation and experiments on stand options show that change the state of mining rocks, in a great deal depends on mechanical properties of rocks, their structure, humidity, fracturing, excessive rock pressure and goaf (Falshtyns'kyy 2011).

Immediate rocks of roof subsidence in a time at coal seam gasification experimental research, are presented on a Figure 3.

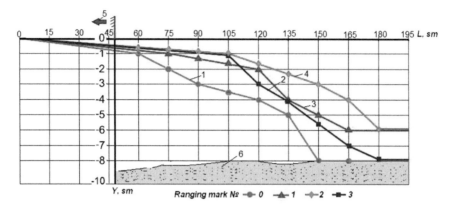

Figure 3. Rocks of roof subsidence at coal seam gasification.

Rocks of roof subsidence in a maximum made 8 sm, that is related with low rocks subsidence on persistent ash (17-22% power of coal seam) and rocks swelling above gasification channel with the coefficient of swelling K_s = 1.4-1.9.

Application of calculation method of the tensely-deformed state of rocks, developed in NMU by Professor A.V. Savostianov, will allow to define geomechanical parameters, related to the conduct of rock mass during gasification of coal seam (Falshtyns'kyy 2010).

The advantage of this method of calculation is taking the features of rock mass stratification with formation of cavities stratification, unevenness of loadings distribution in rock seams, dynamics of rocks behaviour. The methods of calculating of parameters supporting pressure ahead of wallface mean the determination the width of supporting area and degree of tensions in it at the mining seam level.

Description of seam area and determining maximum tensions requires mine calculations.

In the process of coal seam gasification modeling the speed of combustion face advance was changed. This situation is connected with some hazards which can be detected and simulated (Kabiesz 2002). As a result the parameters of combustion face in oxidation and reducing zone, the rocks of immediate roof subsidence above gasification channel and also the size equivalent tension SPR and maximal tangent tensions T_{max} taking into account the temperature coefficient K_t and coefficient of rocks swelling K_s at the strain state calculation was also changed. Durability of rocks property on a simple compression $R_{s.c}$ and on splitting R_s made 26 and 11 MPa (Falshtyns'kyy 2011).

In the examined conditions, it is set that the geometrical parameters of supporting zone practically do not depend on length of gasification channel, but dependence of these parameters is set on combustion face advance.

Table 1. Parameters of supporting pressure depending on combustion face advance.

Advance, m / day	Geometrical parameters of supporting pressures, m				Physical parameters of supporting pressure, MPa		
	a_{max}	a	$d_{0\,max}$	d_0	S_1	$S_{1\,max}$	GHP
0.5	21/16.5	6.4	19.2/14	4.2	17.4/20.9	24.5	16.7
1	18.8/13.6	5.9	13.6/7.4	3.5	18.1/23.8	25.9	19.4
1.5	16.4/12.4	4.0	11.2/5.8	2.6	19.5/25.2	27.4	20.6
2	14/10.9	3.3	9.7/4.4	2.3	20.4/28.2	30.2	22.4

Numerator: parameters of supporting pressure created by rocks layer with a power 26 m and bedding from a layer in the distance 39 m. Denominator: parameters of supporting pressure created by rocks layer with a power 17 m, bedding from a layer in the distance 10-15 m.

Analysing the information from Table 1 and parameters presented on a Figure 4, it is possible to establish that with the combustion face advance increase the zone of supporting pressure narrow and normal loadings increase. The changes of geometrical and power parameters in a supporting zone depend on gasification speed.

For estimation the rocks state the calculation of subsidence and rocks tensions are executed depending on length of gasification channel, speeds of combustion face advance taking into account temperature influences.

The results of calculation are shown in Table 2 and on Figure 5.

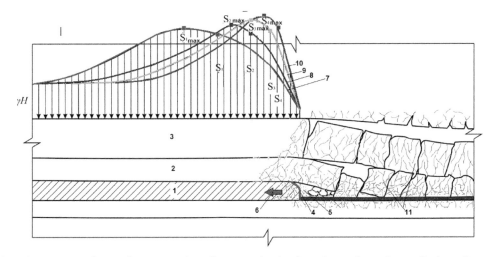

Figure 4. Parameters of supporting pressure depending on combustion face advance: 1 – coal seam; 2 – immediate roof; 3 – main roof; 4 – combustion face; 5 – combustion channel; 6 – direction of combustion face advance; 7 – 0.5 m / day; 8 – 1 m / day; 9 – 1.5 m / day; 10 – 2 m / day; 11 – zone of the rocks failure.

Table 2. The state of rocks depending on length of gasification channel and combustion face advance.

Length of combustion face, m	Combustion face advance, m / сут	Roof subsidence, mm		Tensions, MPa		Coefficient margin of safety	
		Above a combustion face	At a border of gasification channel	SPR	T_{max}	$R_{s.c}$ /SPR	$R_{s.c}$ / T_{max}
30	0.5	211	649	4.1	6.4	3.8	1.5
	1	205	536	5.6	7.0	3.2	1.1
	1.5	197	440	8.2	9.8	2.6	0.9
	2	189	368	12.4	11.9	2.2	0.7
40	0.5	202	624	4.5	6.5	3.4	1.2
	1	195	518	6.1	7.6	2.8	1.0
	1.5	188	425	8.4	10.1	1.2	0.7
	2	180	346	12.7	12.2	0.9	0.6
50	0.5	197	623	4.7	6.8	2.8	1.1
	1	191	514	6.4	7.9	2.5	1.0
	1.5	184	422	8.7	10.5	1.0	0.7
	2	175	345	12.8	13.6	0.8	0.5
60	0.5	188	620	5.0	7.3	2.6	1.0
	1	180	515	6.8	8.4	1.5	0.9
	1.5	173	421	9.1	10.9	0.9	0.6
	2	164	343	13.2	14.0	0.6	0.4

From the Table 2 and Figure 5 it is obvious that with the combustion face advance increase the rocks subsidence decrease (Figure 6). Thus the most subsidence are observed at combustion face advance near 2 m / day. State of rocks of roof at coal seam gasification it is possible to estimate equivalent tensions and maximal tangent tensions, comparing them with durability of rock on a monaxonic compression and splitting. On the basis of this comparison in Table 2 the coefficients of margin safety are resulted. It ensues from the analysis of these coefficients, that the most high probability of destruction rocks above the combustion face in the examined terms is related to maximal tensions and tension from temperatures in gasification channel.

Increase the violations in the rocks of roof from tangent tensions, will be observed at combustion face advance from 1 to 2 m / day. These violations as cracks, will be disposed in perpendicular direction under a corner to the plane of combustion face.

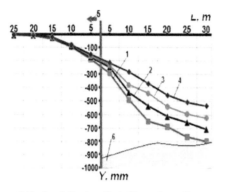

Figure 5. Rocks of direct roof subsidence depending on combustion face advance at the length 30 m: 1 – 0.5 m / day; 2 – 1 m / day; 3 – 1.5 m / day; 4 – 2 m / day.

Figure 6. Changes of equivalent tensions SPR depending on combustion face advance and the length of combustion face: 1 – 30 m; 2 – 40 m; 3 – 50 m; 4 – 60 m.

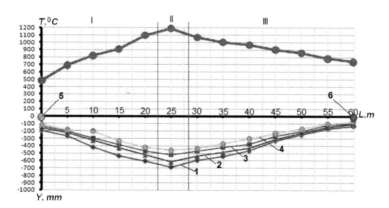

Figure 7. Rocks of roof subsidence on the length of gasification channel depending on the temperatures parameters in gas generator: I – oxidizing zone; II – transition zone; III – reducing zone; 1 – at combustion face advance 0.5 m / day; 2 – 1 m / day; 3 – 1.5 m / day; 4 – 2 m / day; 5 – injection well; 6 – production well.

More intensive fracturing with the slow rocks of roof subsidence will be observed on the injection well coupling with an oxidizing zone, that is related to influence of mining excessive pressure and growth of thermal stress. Above the middle of gasification channel there will be slow rocks subsidence in the zone of transition and reducing (Figure 7).

Rocks of immediate roof destructions above a combustion channel from equivalent tensions in the examined terms will be not observed, because the gasification channel is too small (Falshtyns'kyy 2010). Increase of rocks violations above gasification channel can take a place in gas generator with a length 50-60 m at combustion face advance speed 2 m / day and more.

In Table 5 the parameters of soil rocks in supporting zone of gas generator depending on combustion face advance speed.

The soil of gasification coal seam (slate) can swelling, that is related to the presence in their composition of chemical components: $SiO_2 - 68.2\%$, $Al_2O_3 / SiO_2 - 0.32\%$, $C - 4\%$, here the swelling coefficient makes $K_s = 1.1\text{-}1.18$ (Figure 8).

Analyzing Table 5 and Figures 8, 9 follows, that with the increase of combustion face advance speed the depth of rocks destruction decrease. Raising of soil also decrease and tensions go down from a flexion moment above the middle of gasification channel.

Table 5 The state of soil rocks at coal seam gasification.

Combustion face advance, m / day	Raising of soil rocks, mm	Depth of rocks destructions, m	Tensions, MPa		Stocked to durability	
			σ_c	σ_p	R_c / σ_c	R_p / σ_p
0.5	302	1.84	118.2	30.4	0.12	0.23
1	296	1.55	71.6	18.5	0.19	0.36
1.5	274	1.32	42.5	9.2	0.42	0.85
2	249	1.16	20.7	5.1	0.56	1.05

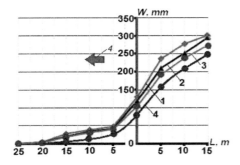

Figure 8. Change of excessive rock pressure depending on combustion face advance speed due to soil rocks swelling: 1 – 0.5 m / day; 2 – 1 m / day; 3 – 1.5 m / day; 4 – 2 m / day.

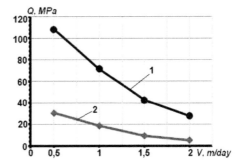

Figure 9. Change of stress depending on combustion face advance speed due to soil rocks swelling tensions: 1 – compression; 2 – strain.

4 CONCLUSIONS

The parameters of oxidizing zone in gasification channel depend on coal seam and blowing composition, coal seam thickness in the stable and balanced gasification mode does not exceed general length of channel 30-32%.

Steady length of oxidizing zone in gasification channel on condition of the balanced blowing composition, provided by intensity of exothermic reactions with complete oxygen combustion on length of oxidizing zone with passing to the reducing zone.

Combustion face advance speed of oxidizing zone in 1.18-1.47 time passes the speed of reducing zone. The combustion face of oxidizing zone distorts and increased on length that's why it is necessary to hold reverse blowing regime.

Presence of heavy rocks layer, bedding near-by gasificated coal seam provide two maximums in the supporting zone formed by strata.

The rocks of roof deformations in a gas generator are exposed to mining pressure further to high temperatures of gasification process.

Rocks swelling become due to presence of chemical components in rocks of roof.

The increase of a supporting pressure is observed with the gasification channel length increase.

Increase the gasification channel length of underground gas generator at speed advance more than 2 m / day result in intensive rocks of roof violation.

The parameters of gasification channel in underground gas generator in a greater extend determine the stable exothermic and endothermic reactions and thermal balance exchange.

REFERENCES

Lavrov, N.V. 1957. *Physical and chemical bases of fuel combustion and gasification.* Moscow: 40.
Yanchenko, G.A. 1989. *Material balance of gasification process.* Moscow: MMI: 57.

Rumshinskyy, L.Z. 1971. *Mathematical treatment of experiment results*. "Certificate manual". Moscow: 192.

Kazak, V.N., Orlov, G.V. & Popov, V.I. 1960. *Deformation of mining rocks above goaf*. Moscow: Nedra: 162-165.

Kabiesz, J. 2002. *Principles of modeling associated* hazards. Archives of Mining Sciences, 47: 1-15.

Falchtys'kyy, V.S., Dychkovs'kyy, R.O., Lozyns'kyy, V.G. & Saik, P.B. 2012. *Research an Adaptation Processes of the System "Rock and Coal Massif –*

Underground Gasgenerator" on Stand Setting.

Falchtys'kyy, V.S., Dychkovs'kyy, R.O., Svetkina, O.Yu. & Lozyns'kyy, V.G. 2010. *Mathematical model of roc mass behaviour at underground coal gasification*. Szkoła Eksploatacji podziemnej: 604-607.

Cheberiachko, S., Yavors'ka, O. & Morozova, T. 2012. *Analyses of test methods determining antidust respirator quality*. Technical and Geoinformational Systems in Mining. Taylor & Francis Group. London: 123-127.

Mining of Mineral Deposits – -Pivnyak, Bondarenko, Kovalevs'ka & Illiashov (eds)
© 2013 Taylor & Francis Group, London, ISBN: 978-1-138-00108-4

Assessment of auger mining application in Polish hard coal deep mines

Z. Lubosik
Glowny Instytut Gornictwa, Katowice, Poland

ABSTRACT: Although the review of the world current best practices associated with auger mining, highlighted that the augering is a potential method for extracting hard coal reserves uneconomic or technically not suitable for longwalling (or room and pillar) in deep mines the auger mining is not used in Polish hard coal mining industry. The paper provides a description of the initial assessment of auger mining application possibility in conditions of one Polish hard coal deep mine. At first the world current best practices and hazards associated with auger mining are described. Next the results of undertaken detailed examination of the reserves at the colliery with the intention of identifying areas of coal that could be recovered by auger are shown. Finally a conception of a feasibility study of auger application in selected coal seam panel containing an assessment of technical constrains and expected economic results of the project is briefly presented.

1 INTRODUCTION

Significant part of hard coal reserves in the Polish mining is deposited in conditions, which do not allow the application of longwalling method of coal extraction, due to technical, economic or work safety reasons (Cebula at al. 2012; Turek & Lubosik 2008a). In particular, this pertains to the deposits in small shapeless panels, tectonically disturbed panels (eg. faults, pinching), in thin seams or steeply inclined ones. These reserves usually remain not recovered, what negatively influences the mining conditions in the concerned seam and in the neighbouring ones, as well as the financial results and life of the mine. Thus, it seems rational to recover these deposits using mining systems other than longwalling. There is a number of such systems (Turek & Lubosik 2008b; Lubosik 2009), one of them being the auger mining method.

Auger mining consists in drilling of large-diameter holes, over 100 m long, from the headings, by means of a coal augering machine. In the process of drilling, auger flights are placed behind the cutterhead to increase the length of drilling duct and to haul the mined coal to the mouth of the hole. The length, diameter and spacing between the holes depends on mining-geological conditions. Augering facilitates extraction of coal from hard coal panels of any shape and any seam thickness.

2 EXAMPLES OF COAL AUGERING SYSTEM APPLICATION IN UNDERGROUND MINES

The first attempts of hard coal seams' augering in underground mines were made in the forties of the past century in the United States (Hibbert 2001). The first augering machine was of a simple construction, equipped in a drill of 0.76 m diameter and 0.91 m length, further the machine had been equipped in walking skids, a conveyor used for loading of mined coal into mine cars, later on the drill diameter got increased, as well as its length (Figure 1).

Coal seams 0.8 m thick have been augered in Myra mine in USA using Cardox-Hardsocg machine (Figure 2), by means of holes 64 cm diameter and 35 m length (Berry 1951), also the Elkhorn seam in Pike Country in USA (Newell & Storey 1952). In the UK, augering was attempted in Hucknall mine (Young 1954), where-owing to methane hazard, the construction of the machine was modified in view of facilitating the ventilation of the hole.

In 1954 a self-propelled augering machine, type UCA-201 was manufactured, equipped with skids and anchor jacks, which consisted of the supply unit and the drilling unit (Forlington at al. 2001a). This machine has been applied in Wind Rock Coal & Coke Co. mine in the USA for augering of coal seam by means of holes 86 cm diameter and ca 30 m length (Anonymous author 1959). It took a two-men crew ca 53 minutes to drill a 30.5 m long hole in coal, as well as to set the machine.

At the turn of the sixties/seventies of the 20th century, the augering machines have been introduced in the Ruhr Basin in Germany, where drills used to get continuously jammed in the holes, in spite of increasing power of particular models of the machines (UCA-400 – 112 kW, UCA-1000 – 350 kW) – probably due to the depth reaching even 600 m.

The machine constructed in 1972 by the Badger Manufacturing Company (USA), in which a sliding frame designed for shifting of the drilling module to a subsequent hole and for storing the drill elements, turned out to be a constructional breakthrough. Further, the machine had been equipped with hydraulic props used for its spragging in the heading, in order to ensure the desired direction of drilling. The machine was driven by motors of 112 kW power and facilitated drilling of holes 0.91 m diameter and up to 30 m length.

Figure 1. Coal seam auger manufactured in ca.1950 by Compton Augers from USA (source: Follington at al. 2001a).

In 1974, the Badger Manufacturing Corp. from USA designed the coal seams' augering machine, called the CoalBadger (Mason 1974). This machine facilitated drilling of holes up to 25 m length and 106 cm or 137 cm diameter, and its capacity amounted to 250 t / shift. A few machines of this type had been manufactured, they were in use in underground mines with varying success (Chironis 1979).

Figure 2. Augering machine Cardox-Hardsocg during exploitation of Elkhorn seam at Pike County (source: Newell and Storey 1952).

In 1979, a modified version of the CoalBadger machine was applied for augering of the Pittsburgh No 8 seam in Ireland mine in Moundsville, USA (Chironis 1979). The seam up to 1.83 m thick, was deposited at the depth of ca 300 m and had been augered by means of holes 106 cm diameter and length up to 30 m. Coal production at the level of 182 t per shift was accomplished.

Figure 3. Coal augering machine CoalBadger (source: Mason 1974).

In 1988 the augering of Dundas coal seam in Hlobane mine was carried out (Collins 1988) by means of holes 0.6 m diameter and depth up to 25 m (the machine also provided for drilling of holes of 0.5 and 0.7 m diameter and length up to 70 m). Power of the machine amounted to 147 kW and length of single auger flight – 2.0 m. The trial operations conducted underground consisted in recovery of coal from pillars 16 × 25 m, left behind during bord and pillar extraction. Depth of seam deposition was ca 45 m and its thickness 1.67 m; the 0.7 m thick layer next to the floor had been augered. The augering operation had been carried out from the heading of 5.0 m width and height equal to the seam thickness. Width of coal pillars left behind between particular holes and between the roof and floor of the seam amounted to 0.2 and 0.5 m respectively, what facilitated the recovery of ca 40% of the panel reserves. In the course of drilling only a few holes did not reach the predesigned depth, as some disturbances occurred with maintaining the direction of drilling. The costs of extraction were considerably lower compared to other mining methods, yet the production level did not prove satisfactory, it reached only 15 t per shift, what resulted in 300 t / month.

In 2000 a prototype machine for augering of coal seams 1-1.8 m thick by means of holes up to 80 m in length – BryDet BUA 600 (Figure 4), has been applied in Matla colliery near Witbank in RSA (Follington at al. 2001). The minimum dimensions of the heading from which the augering operation was to be carried out should have amounted to 6.0 × 1.8 m.

Trial operation of the BryDet BUA 600 seam augering machine has been undertaken in 2000 in the Matla opencast mine, yet despite positive results of these tests (30 holes of depth up to 80 m were drilled), no tests had been undertaken in the underground mine.

In the Ukrainian mining industry in some cases augering is the basic mining method, due to large majority of coal reserves deposited in seams with thickness less than 1.5 m (ca 90% of total reserves) (Korski 2006). For example, the large diameter drilling machine type BGA-2M had been applied, facilitating the drilling of holes with 1070 mm diameter (Juniewicz 1991) or the BUG-3 augering machine (Ławrow at al. 1990) which was used for coal augering with simultaneous deposition of waste rock in the drilled hole. The augering machine type BZM-1M1, produced by the Morozow company from Charkov (Figure 5), has a similar application. It is designed for augering of coal in seams of 0.8-0.9 m thickness by means of holes up to 60 m long and up to ca 2 m wide and for deposition of waste rock in the drilled holes. This machine is operated by 4 persons and it can reach the capacity of 100 t / day.

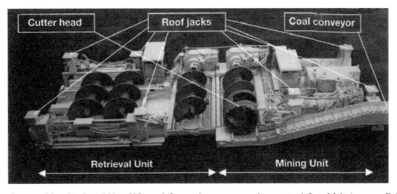

Figure 4. Augering machine BryDet BUA 600 used for coal seams augering up to 1.8m thick (source: Follington at al. 2001a; Follington at al. 2001b).

Another equipment used in the Ukrainian mining is the BShK-2DM machine(Figure 6-7), applied for drilling of holes 1.9-2.1 m wide and up to 85 m long by means of drills of 625, 725, 825 mm diameter. These holes can be drilled in seams 0.6-0.9 m thick, with ±15° inclination. This machine is operated by 4-5 persons.

An assessment of the possibility of application of BShK-2DM machine in the Karvina-Ostrava Basin conditions, also in gassy seams, prone to bumps, gas and rock outbursts, has been done by (Stonis & Hudecek 2009). The modified BShK-2DM machine, called VS-SEAL-625 P1 (Figure 8), has been equipped in 3 drilling bits facilitating the drilling of holes over 100 m long, 1905-2105 mm wide and ±20° inclination, where CH_4 concentration can reach up to1.5%. The machine is operated by 4-5 persons (Kosnovsky 2008).

Figure 5. Coal augering machine type BZM-1M1 produced by Morozow company from Charkov, Ukraine (source: www.morozov.com.ua).

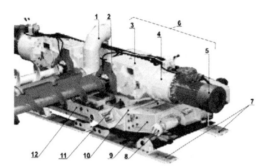

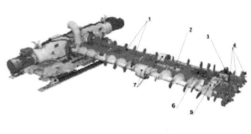

Figure 6. Driving unit of BShK-2DM machine (source: promotional materials of Malyshev company). Key: 1 – ventilation feeder branch pipe; 2-lock wrench; 3-reduction gear, 4 – hydraulic coupling; 5 – electric motor; 6 – driver unit; 7 – slides; 8 – thrust hydraulic cylinders; 9 – orientation unit; 10 – hydraulic feeder unit; 11 – drive frame; 12 – centering unit.

Figure 7. Auger drill of BShK-2DM machine (source: promotional materials of Malyshev company). Key: 1 – auger; 2 – ventilation duct; 3 – support; 4 – drill bits; 5 – reduction gear; 6 – control section; 7 – support section.

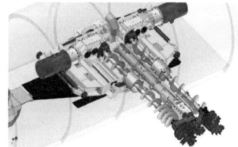

Figure 8. Coal seams augering machine VS-SEAL-625 P1 (source: http://www.se-mi.cz/download/vrtaci-stroj-en.pdf).

Underground trial operations of the VS-SEAL-625 P1 augering machine were conducted in 2008 in the Paskov mine, belonging to OKD Ostrava (Kosnovsky 2008). This equipment was used for recovery of coal remaining in safety pillars of transportation galleries (Figure 9). Length of the drilled holes was 25, 48 and 80 m, slope up to 8° and their dimensions – 0.625×2.5 m. Augering was conducted from the heading with cross-section 12 m², and the distance between consecutive holes was 1.0 m.

Coal seam augering has also been undertaken in the Australian mining. First attempts took place in the fifties of the past century in the Ayrfield Colliery, where holes with 0.75 m diameter were drilled

(Buddery and Hill 2004). In 1993 the Austalian company Horizontal Boring Pty Ltd and Coal Augering Pty Ltd constructed a machine for augering of seams, which enabled the drilling of holes with 1.8 m diameter and 50 m length (Mc Kinnon 1997 & Coffey 1998). In 2002 the Coal Recovery Australia Pty Ltd carried out augering in the German Creek coal seam, 2.7 m thick in the Southern Colliery, deposited at the depth of 70-145 m. Panels with dimensions 52×105 m have been augered by means of holes with 1.6 m diameter and up to 100 m length, from the heading 5 m wide. Pillars 0.9 m wide have been left between the holes, whereas pillars 8m wide – between the panels.

Augering was done with the application of the UA 560 machine, which consists of the drill unit (Figure 10), rom haulage unit (Figure 11) and the flight retrieval unit (Figure 12).

The UA-560 coal seam augering machine facilitates the drilling of holes of 1.6 m diameter (Figure 13) and length even over 100 m. It is supplied with 1000 V current, the drilling unit motors' power is 560 kW and mass 53.39 t.

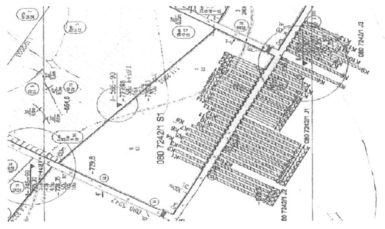

Figure 9. Section of seam map with marked holes drilled by the auger machine VS-SEAL-625 P1 (source: Kosnovsky 2008).

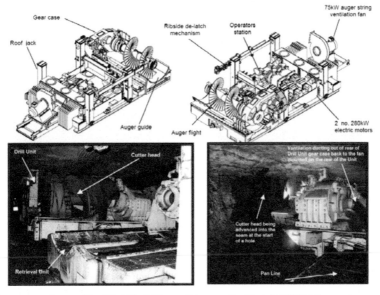

Figure 10. Coal seam augering machine UA 560: drilling unit (source: Buddery & Hill 2004).

137

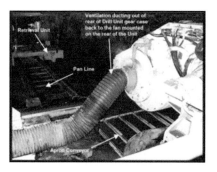

Figure 11. Coal seam augering machine UA 560: rom haulage unit (left) and drilling unit, drill and transfer arm (right) (source: Buddery & Hill 2004).

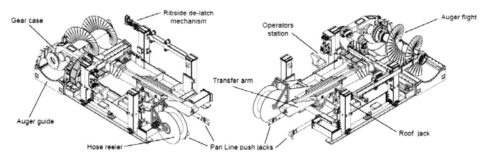

Figure 12. Coal seam augering machine UA 560: flight retrieval unit (source: Buddery & Hill 2004).

In the course of augering, due to practical reasons connected with the time of flights' retrieval from the holes and their storage in the heading, it has been decided to shorten the length of the holes to 60 m and to increase the width of pillars between the holes to 1-1.8 m. The convergence of holes did not exceed 3 mm.

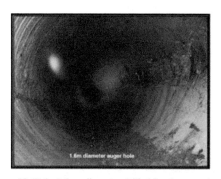

Figure 13. Hole 1.6 m diameter drilled by the augering machine UA 560 (left) and the view of heading side-wall after seam augering and holes sealing (right) (source: Buddery & Hill 2004).

2.1 *Identification of reserves suitable for augering on the example of a Polish hard coal mine*

Coal augering can be carried out practically in any coal panel, where the only technical limitation is the coal seam inclination (drilled holes), which should be in the range ±20°. Fulfilling the remaining requirements, i.e. minimum dimensions of the heading from which augering is carried out, ventilation of the working by circulating air current as well as outfitting of the working (power supply, suspended monorail, haulage equipment, lighting, air ducts) is technically viable and depends entirely on the expected economic effect.

Obviously, the most suitable are the coal panels in the area of which headings with the desired dimensions are located, ventilated by circulating air current, with existing haulage facilities. In case of the mine, which has been analyzed, 16 such panels have been identified, the overall reserves totaling ca 700 thousand tons. The major group – as much as 11 panels, i.e. 68.75% of the total, comprised small panels with reserves up to 50 thousand tons, further – panels with 50-100 thousand tons (18.75%) and last-the biggest panels with reserves over 100 thousand tons (12.5%).

2.2 Feasibility study of auger application in selected coal seam panel (assessment of technical constraints and expected economic results)

The possibility of augering a selected coal panel (Figure 14 – green colour), located in the vicinity of Incline B-32 has been assessed. This incline is an existing heading supported by ŁP10/V32 supports (steel arch yielding support) placed every 0.8 m, ventilated by a circulating air current, equipped in haulage facilities and suspended monorail.

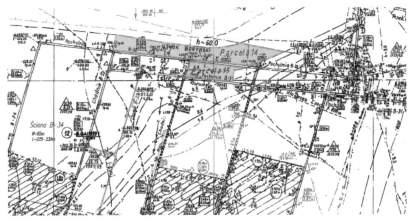

Figure 14. Section of seam map with outlined panel meant for augering (green colour).

Augering will be carried out with the use of an augering machine VS-SEAL-625 (Figure 8), by means of holes with dimensions $0.8 \times 2.5 \times 40$ m, leaving a safety pillar 1.0m wide between the holes

(Figure 15). Diagram of augering elaborated with the application of numerical modelling using FLAC 3D programme.

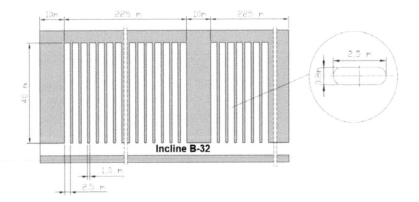

Figure 15. Diagram of augering of the analyzed panels.

139

In the model works the following were determined: vertical stress values, vertical displacement values, range of destruction area along the heading, from which augering will be carried out (before and after augering), along the plane vertical to the axis of heading in the middle of the created model, i.e. at the distance of 25.5 m from the boundary of the model, along consecutive holes, along the plane vertical to the drilled holes at the 13th m of their length. Assumptions for the model are presented on Figure 16, whereas a sample result on Figure 17.

The highest values of vertical displacements and vertical stresses for all the analyzed situations (cross section planes) are:

– vertical displacements: roof subsidence up to 13.0 cm and floor uplift up to 5.70 cm;
– vertical stresses: compressive stresses up to 54.70 MPa and tensile stresses up to 1.36 MPa.

Daily manpower for one augering machine, in case of 4 – shift work system, with division of labour into particular shifts is given in Table 1.

For the analyzed case it has been assumed that during one day there will be one hole of 40 m length drilled, what-considering the dimensions of the hole 0.80×2.50 m, would result in coal output of ca 106.40 t.

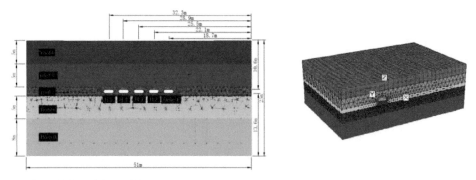

Figure 16. General view of the model.

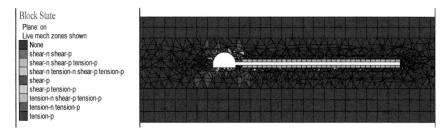

Figure 17. Destruction zone along one of the drilled holes – after drilling the first hole.

Table 1. Daily manpower for 1 augering machine.

| Position | No of workers | | | |
| | shift 1 | shift 2 | shift 3 | shift 4 |
	side-wall exposure, augering	augering, hole sealing	machine shifting, maintenance works	stabilization of support, removal of arches
overman (augering unit operator)	1	1	1	1
flight retrieval unit operator, transport of materials	2	2	–	–
haulage crew	2	2	–	–
carpenter	–	–	2	–
miner	–	–	–	1
Sum	**5**	**5**	**3**	**2**
Total	**20 mandays**			

Analysis of financial effectiveness of the planned venture, carried out with the application of dynamic discount methods, with consideration of change of money value in time, has indicated the effectiveness of this investment project. It has been assumed for the purpose of calculation, that the investment period will be 4 years (i.e. after recovery of coal from a selected panel, augering will be continued at other locations), discount rate is 7%, and the depreciation rate of augering machine is 25%. The average cost of manday is 540 zł / day and coal selling price is 482 zł / t. Capital expenditure for the project in question is equal to the purchase price of the augering machine, which has been estimated at 4.5 mln zł. Remaining costs, i.e. costs of preparation of the mining face and costs of production were determined on the basis of the elaborated time schedule of works, as well as basing on experience in the calculation of particular component costs. The results of economic analysis of effectiveness of augering the analyzed coal panel are presented in Table 2.

The results of economic analysis indicate that already in the first year of the undertaking, the cumulated value of discounted cashflow is positive. This is due to the fact, that practically the only expenditure required for realization of the project is purchase of the augering machine, and no costs are borne in connection with development works. In the fourth – last year of the project execution the NPV of the project reaches the value of 14.6 mln zł, return on investment is 3.25, and the payback period calculated from the moment of starting the project is less than a year (Table 2).

Table 2. Economic parameters of coal panels augering project execution.

Specification	Value
Capital expenditure, zł	4 500 000
Costs of development works, zł	1 157 160
Total production costs, zł	9 558 747
Total coal production, tones	74 052
Total revenue on coal sale, zł	35 693 064
Net present value NPV, zł	14 616 189
Capital expenditure rate of return	3.25
Payback period calculated from the moment of coal production start, –	< 1 year

The sensitivity of the project – coal reserves recovery from coal panels using the augering method in relation to the change of coal selling price level and costs of production, has also been analyzed (Figure 18).

The analysis of sensitivity proved, that the project – coal panel augering, in the assumed range of volatility (±50%), is characterized by positive values of NPV, thus the conclusion is that it is not sensitive to rise of coal selling price and increase of coal production costs.

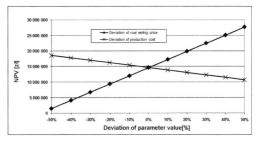

Figure 18. Analysis of sensitivity for the panel extraction project by augering – NPV values considering deviation of coal selling price and production costs.

3 CONCLUSIONS

The augering system facilitates the exploitation of hard coal seam panels of any shape and any seam thickness. The potential places for application of the augering system are these coal seam panels, which cannot be extracted by other mining methods. The most suitable ones are the coal panels in the area of which headings with the desired dimensions are located, ventilated by circulating air current, with existing haulage facilities.

The advantages of augering system are: possibility of exploitation of low seams deposited in difficult mining and geological conditions, where the crew, as well as the equipment operate in well protected headings, possibility of coal recovery from safety pillars; possibility of extraction of these sections of coal seam, which cannot be mined by longwalling; low capital expenditure connected with purchase of equipment (as compared to longwall system); low manpower required for running the augering operations.

The disadvantages of augering system cover: difficulties in maintaining the horizontal and vertical direction of the holes; lack of possibility of clean extraction of coal from a particular panel; requirement of running the augering operations from a heading with adequate dimensions; inclination of augering holes limited to ca. ±20°.

For the purpose of proper running of augering and maintaining the stability of holes, pillars and heading, from which the augering is being operated, it is necessary to design proper dimensions of the boundary pillars of the panel, as well as of the safety pillars between the panels. The presented example of model works and the obtained preliminary results have proved the suitability of FLAC 3D programme for this type of works. Further, the accuracy of model net should be increased and the range of interaction of particular holes should be analyzed in order to optimize the width of pillar.

The results of analyses, as well as the elaborated technical design and economic assessment confirm, that safe and effective augering of seams in the Polish hard coal mining is possible. Augering should supplement the longwalling system commonly applied in the Polish mining, what should improve the economic results of mines (eg. through increase of coal production from panels opened out for longwalling, with utilization of the existing headings and infrastructure), increase the safety of work by elimination of zones of increased stresses in the rock masses as a result of clean extraction of seam and increase the life of the mine (increase the extraction ratio).

REFERENCES

Anonymous author. 1959. *Underground Augering Today.* Coal Age. November.

Berry, J. K. 1951. *Auger Mining Underground.* Coal Age. September.

Buddery, P. & Hill, D. 2004. *Extraction Panel Guidelines for High Production Underground Auger Mining in Australian Conditions.* Australian Coal Association Research Program (ACARP) No. C11041.

Cebula, C., Słota, M. & Wrana, A. 2012. *Inwentaryzacja resztek pokładów węgla oraz analiza mechanizmów ich powstania na przykładzie kopalni "Piekary".* Katowice: Wiadomości Górnicze, 2.

Chironis, N.P. 1979. *Underground Auger Mines Pillars.* Coal Age. April.

Coffey, S. 1998. *Coal Augering Project.* Australian Coal Association Research Program. Report, 4016.

Collins, R.E. 1988. *Low-Seam Auger Mining at Hlobane Colliery http://www.thevirtualminemuseum.org.za.*

Follington, I.L., Deeter, R., Share, D. & Moolman, C. 2001a. *A new underground auger mining system.* The Journal of The South African Institute of Mining and Metallurgy. January/February.

Follington, I., Deeter, R. & Share, D. 2001b. *Economic Extraction.* World Coal. February.

Hibbert, D.E. 2001. *Scavening Remnant Coal Reserves by Underground Auger Mining. Coal International.* March/April.

Juniewicz, S. 1991. *Nowe technologie eksploatacji pokładów silnie nachylonych i stromych.* Katowice: Prace GIG Seria dodatkowa.

Korski, J. 2006. *Eksploatacyjne zwiercanie pokładów węgla w kopalniach podziemnych.* Przegląd Górniczy, 6.

Kosnovsky, V. 2008. *Technologie dobywani zbytkowych zasob sirokoprumerovym vrtanim pomoci vrtne soupravy VS SEAL-625 P1.*

Lubosik, Z. 2009. *Geoinżynieryjne i ekonomiczne kryteria eksploatacji węgla kamiennego z resztkowych parcel pokładów.* Katowice: Prace Naukowe GIG Górnictwo i Środowisko, 3.

Ławrow, W.W., Elczaninow, E.A. & Szor, A.I. 1990. *Perspiektiwa primenenija na szachtach buroszniekowogo sposoba wyjemki uglja.* UGOL, 9.

Malyshev Plant: promotional materials of Malyshev company http://www.malyshevplant.com/content/bshk2dm.

Mason, R.H. 1974. *Underground coal auger claims practical success.* Coal Mining & Porcessing. March.

Mc Kinnon, B. 1997. *Underground Augering.* The Australian Coal Review.

Newell, J.P. & Storey, R.W. 1952. *Experiments with an underground auger.* Mining Engineering. July.

Stonis, M. & Hudecek, V. 2009. *Mining of Coal Pillars Using the Drilling Method.* Acta Montanica Slovaca: 241-249.

Turek, M. & Lubosik, Z. 2008a. *Identyfikacja resztkowych parcel pokładów węgla kamiennego.* Katowice: Wiadomości Górnicze, 3.

Turek, M. & Lubosik, Z. 2008b. *Sposoby wybierania resztkowych parcel pokładów węgla.* Katowice: Wiadomości Górnicze, 5.

Young, M.H. 1954. *Ventilated Auger Mining at Hucknall Colliery.* The Colliery Guardian. Sept. 9.

Mining of Mineral Deposits – Pivnyak, Bondarenko, Kovalevs'ka & Illiashov (eds)
© 2013 Taylor & Francis Group, London, ISBN: 978-1-138-00108-4

The enhancement of hydrocarbon recovery from depleted gas and gas-condensate fields

O. Kondrat
Ivano-Frankivs'k National Technical University of Oil and Gas, Ukraine

ABSTRACT: Most gas and gas-condensate fields in Ukraine are at the final stage of their development. Still they contain significant remaining deposits of gas and condensate, the production of which under the conditions of formation pressure depletion is a very difficult problem. The methods of remaining hydrocarbons recovery from depleted fields are characterized. Concerning the depleted gas drive fields it is suggested the analytical dependences of final formation pressure and gas recovery factor on features of formation bottomhole zone and technological parameters of well operations and gas gathering system which enable to ground the most effective methods of gas recovery enhancement for certain fields. The relations are tested for the gas deposit of horizon ND-4 of the Oparsk gas field.

1 INTRODUCTION

Most gas and gas-condensate fields in Ukraine from which the bulk volume of gas and condensate has been produced recently or is still being produced are at the final stage of development and are characterized by a drop-down production of hydrocarbon raw materials. Depleted fields contain significant remaining deposits of hydrocarbons in the form of gas in gas-saturated beds including low permeable and poorly drained areas, as well as trapped gas in watered beds and condensate hydrocarbons in gas-condensate fields which are developed under formation depletion drive. Under the circumstances of importing a major part of gas consumed in Ukraine the increase of own gas production is of burning importance. One of the ways of its solution is increasing the degree of hydrocarbon recovery from depleted fields. According to the industrial data after the completion of the field development the average gas recovery factor of the gas drive fields is 80-85%, under the conditions of water drive – 70-85%, and condensate recovery factor of the gas-condensate depleted fields – 13-40%. Thus, after completion of the field development, beds still contain a significant quantity of gas and condensate that are the important source of hydrocarbon raw materials.

2 ORIGIN OF PROBLEM

The main problem of fuel and energy industry of Ukraine is increasing own hydrocarbon raw materials production. Along with discoveries and acceleration of bringing into the development of new fields including in the shelf area of the Black Sea and the Sea of Azov, and unconventional natural gas fields development (shale gas, tight gas, methane of coal beds, gas of gas hydrate deposits on the seabed of the Black Sea etc.) one of the ways of its solution is the increase of full gas and gas-condensate recovery from depleted fields which are on operation. Because of slow development of new (conventional and unconventional) natural gas fields depleted fields will define the gas production rate in Ukraine in the nearest decades.

The possible ways of increasing the final gas recovery factor of gas drive fields include (Zakirov 1998; Kondrat 1992 & Zakirov 2000): provision of uniform drainage encompassment of the whole gas-saturated porous volume and equal (close) values of final formation pressure in different bed areas; minimization of the final formation pressure values; driving (displacement) of the remaining gas part with the non-hydrocarbon gases, in particular, nitrogen, liquid-and-gas mixtures based on non-hydrocarbon gases and different liquids.

The trapped gas recovery from depleted fields can be reached with their formation pressure decrease by different methods of trapped hydrocarbon gas displacement with non-hydrocarbon gases.

For condensated hydrocarbons recovery from depleted gas-condensate fields it is suggested to use dry hydrocarbons and non-hydrocarbon gases, liquid-and-gas mixtures based on non-hydrocarbon gases, carbon dioxide fringe, hydrocarbon solvents, surfactant solutions, surface active polymer-containing systems with further injection of water-and-gas mixtures or water.

The paper examines the issue of well operation parameters' influence and gas gathering system and bottomhole formation zone characteristics on the final gas recovery factor of gas drive fields.

3 THE EQUATIONS AND DEPENDENCES USED WHILE RESEARCHING

To examine the geological-and-physical and lithological parameters influence on the decrease of final gas recovery factor of the depleted gas fields, the following formulae and dependences are used:
Binominal formula of gas inflow

$$P_f^{\,2} - P_b^{\,2} = A \cdot q + B \cdot q^2 ; \qquad (1)$$

G.A. Adamov's formula establishes the contact between the bottomhole and wellhead pressures and gas flow in the operating well.

$$P_b = \sqrt{P_w^{\,2} \cdot e^{2S} + \theta \cdot q^2} ; \qquad (2)$$

Formula for flow efficiency of the linear horizontal gas-pipeline (exhausting line)

$$q_g = 0.32 \cdot E \cdot \sqrt{\frac{\left(P_w^{\,2} - P_{le}^{\,2}\right) \cdot d_l^{\,5}}{\lambda_l \cdot \bar{\rho}_g \cdot Z_{av.l} \cdot T_{av.l} \cdot L_l}} ; \qquad (3)$$

$$A = \frac{\mu_f \cdot Z_f \cdot P_{atm} \cdot T_f}{\pi \cdot k \cdot h \cdot T_{st}} \cdot \left(ln\left(\frac{R_c}{r_d}\right) + C_1 + C_2 \right); \qquad (4)$$

$$B = \frac{\rho_{st} \cdot Z_f \cdot P_{atm} \cdot T_f}{2 \cdot \pi^2 \cdot h^2 \cdot l \cdot T_{st}} \cdot \left(\frac{1}{r_d} - \frac{1}{R_c} + C_3 + C_4 \right); \qquad (5)$$

$$S = 0.03415 \cdot \frac{\bar{\rho}_g \cdot L_w}{Z_{av.w} \cdot T_{av.w}} ; \qquad (6)$$

$$\theta = 0.0133 \cdot \lambda_w \cdot \frac{Z_{av.w}^{\,2} \cdot T_{av.w}^{\,2}}{d_t^{\,5}} \left(e^{2S} - 1\right); \qquad (7)$$

$$T_{av.w} = \frac{T_f - T_w}{ln\dfrac{T_f}{T_f}} ; \qquad (8)$$

$$P_{av.w} = \frac{2}{3}\left(P_b + \frac{P_w^{\,2}}{P_b + P_w} \right); \qquad (9)$$

$$\mu_{av.f} = \frac{\mu_f + \mu_b}{2} ; \quad Z_{av.f} = \frac{Z_f + Z_b}{2} ; \qquad (10)$$

$$E = 1 - \frac{0.15 \eta_c^{\,0.25}}{W_{av}^{\,0.5}} ; \qquad (11)$$

under $0 < \eta_c \le 180 \ cm^3/m^3$ and $2 < W_{av} \le 11 \ m/s$

$$E = 1 - \frac{0.1 \eta_c^{\,0.25}}{W_{av}} ; \qquad (12)$$

under $0 < \eta_c \le 1500 \ cm^3/m^3$ and $1 < W_{av} \le 6 \ m/s$

$$\eta_c = \frac{\left(q_w + q_c\right) \cdot 10^{-6}}{q_g} ; \qquad (13)$$

$$\lambda = f\left(Re, \varepsilon\right), \qquad (14)$$

where P_f – the formation pressure; P_b – bottomhole pressure, MPa; P_w – wellhead pressure, MPa; P_{le} – end of exhausting line pressure, MPa; $P_{av.w}$ – average pressures in the borehole , MPa; $P_{av.l}$ – exhausting line average pressure, MPa; T_f – formation temperature, K; T_w – wellhead temperature, K; $T_{av.b}$ – average temperature in the borehole, K; $T_{av.l}$ – average temperature exhausting line, K; q – gas flow under standard conditions, thousand m^3/d; A, B – filtration resistance factors of bottom-hole zone; S and θ – complex parameters; L_w – length of tubing (distance from the wellhead to the mid-perforation point), m; L_l – length of the exhausting line, m; d_t – inside diameter regarding tubing; d_l – inside diameter regarding exhausting line, cm; k – formation permeability factor, m^2; h – formation thickness, m; R_c – supply contour radius (well drainage zone radius), m; r_d – drillbit well radius, m; l – formation macroroughness parameter, m; R_e – Reynolds number; ε – relative tube roughness; $\bar{\rho}_g$ – relative gas density; ρ_{st} – gas density under standard conditions, kg/m^3; Z_f, Z_b – gas compressibility under formation temperature and under pressures P_f and P_b relatively; μ_f, μ_b – dynamic gas viscosity factor under formation temperature and P_f and P_b relatively, MPa·sec; $Z_{av.w}$, $Z_{av.l}$ – gas compressibility under $T_{av.w}$ and $P_{av.w}$ and under $T_{av.l}$ and $P_{av.l}$ relatively; λ_w, λ_l – hydraulic resistance factor relative to the tubing and well exhausting line;

C_1, C_3 – well imperfection factors according to the level of formation uncover; C_2, C_4 – well imperfection factors according to the character of formation uncover; E – correcting factor that considers the liquid influence on gas-pipeline flow rate decrease; η_c – gas-condensate ratio, cm^3/m^3; q_w, q_c – water and stable condensate yields relatively, m^3/d; w_a – average speed of gas movement along the gas-pipeline, m / sec.

$P_{atm} = 0.1013 \cdot 10^6$ Pa; $T_{st} = 293$ K.

From the joint equations solution (1)-(3) the following formulae for the formation pressure calculating are got:

$$P_f = \sqrt{P_w^2 \cdot e^{2S} + A \cdot q_g + (B + \theta) \cdot q_g^2} \; ; \qquad (15)$$

$$P_f = \sqrt{\begin{array}{l} P_w^2 \cdot e^{2S} + 1.157 \cdot 10^{-14} \cdot \dfrac{\mu_{av.f} \cdot Z_{av.f} \cdot P_{atm} \cdot T_f}{\pi \cdot k \cdot h \cdot T_{st}} \cdot \left(ln\left(\dfrac{R_c}{r_d}\right) + C_1 + C_2 \right) \cdot q_g + \\[3mm] + \left(\begin{array}{l} 1.339 \cdot 10^{-16} \cdot \dfrac{\rho_{st} \cdot Z_{av.f} \cdot P_{atm} \cdot T_f}{2 \cdot \pi^2 \cdot h^2 \cdot l \cdot T_{st}} \cdot \left(\dfrac{1}{r_d} - \dfrac{1}{R_c} + C_3 + C_4 \right) + \\[3mm] + 0.0133 \cdot \lambda_w \cdot \dfrac{Z_{av.w}^2 \cdot T_{av.w}^2}{d_t^5} \left(e^{2S} - 1\right) \end{array} \right) \cdot q_g^2 \end{array}} \qquad (16)$$

– without considering pressure losses in exhausting line of a well

$$P_f = \sqrt{\begin{array}{l} P_{le}^2 \cdot e^{2S} + 1.157 \cdot 10^{-14} \cdot \dfrac{\mu_{av.f} \cdot Z_{av.f} \cdot P_{atm} \cdot T_f}{\pi \cdot k \cdot h \cdot T_{st}} \cdot \left(ln\left(\dfrac{R_c}{r_d}\right) + C_1 + C_2 \right) \cdot q_g + \\[3mm] + \left(\begin{array}{l} 1.339 \cdot 10^{-16} \cdot \dfrac{\rho_{st} \cdot Z_{av.f} \cdot P_{atm} \cdot T_f}{2 \cdot \pi^2 \cdot h^2 \cdot l \cdot T_{st}} \cdot \left(\dfrac{1}{r_d} - \dfrac{1}{R_c} + C_3 + C_4 \right) + \\[3mm] + 0.0133 \cdot \lambda_w \cdot \dfrac{Z_{av.w}^2 \cdot T_{av.w}^2}{d_t^5} \left(e^{2S} - 1\right) + \\[3mm] + \dfrac{\lambda_l \cdot \overline{\rho}_g \cdot Z_{av.l} \cdot T_{av.l} \cdot L_l \cdot e^{2S}}{0.32^2 \cdot E^2 \cdot d_l^5} \end{array} \right) \cdot q_g^2 \end{array}} \qquad (17)$$

According to the formation pressure value, the gas recovery factor of the field is defined by the formula:

$$\beta = 1 - \dfrac{P_f \cdot Z_{in}}{P_{in} \cdot Z_f}, \qquad (18)$$

where P_{in}, P_f – initial and current formation pressures relatively, MPa; Z_{in}, Z_f – gas compressibility factor under the formation temperature and the initial and current formation pressures relatively.

The obtained dependences (15)-(17) allow us to assess the impact on the final formation pressure and, correspondingly, on the gas recovery factor (according to the dependence 18), bottomhole formation zone characteristics and technological parameters of wells operation and their exhausting lines (pipelines) to the gas treatment plant.

The analysis of (15)-(17) equations shows that for the minimization of the formation pressure value P_f, it is necessary to decrease the wellhead pressure value P_w, filtration resistance factors A and B of bottomhole formation zone and the complex parameter θ that characterizes the pressure losses in the process of the gas movement in tubing.

Thus, the main ways of final formation pressure decrease comprise:

– decrease of the operating pressure at the wellhead, P_w;

– decrease of hydraulic pressure losses in tubing (decrease of the hydraulic resistance factor λ_w);

– decrease of pressure loss in the bottomhole formation zone by increasing the permeability factor k and bed macroroughness l;

– increase of the hydrodynamic well enhancement

145

by the stage and character of formation completion (decrease of the wells enhancement factors C_1, C_2, C_3, C_4).

The minimum values of the wellhead pressure can be provided with the commissioning the boosting compressor station, pressure loss decrease in well exhausting lines, near wellhead ejectors and submerged vacuum compressors use, gas supply to local consumers and gas processing at the place of production To decrease the pressure loss in the exhausting lines the hydrating inhibitor is injected into the gas-and-liquid flow at the beginning of the exhausting line in a case of hydrates formation but in a case of fluid collection the foam surfactants, different types of treatment facilities (pistons) and expansion chambers (water collection headers) must be used near the wellhead or in the low pipeline sections. Pressure losses in the wellbore can be reduced with the prevention of clay-and-sand and fluid plugs formations at the wellhead, hydrates and salting in tubing and making the homogeneous highly dispersed gas-and-liquid flow in tubing if fluid is available in the formation products. Pressure losses in the bottom-hole zone can be reduced by conducting appropriate special chemical, physical, mechanical and complex formation treatment in the bottom-hole zones in order to clean it from pollution, to increase the permeability compared to the natural value and to increase the well improvement due to the level and character of formation completion. The well improvement due to the character of formation completion can be increased by making additional perforation channels to join the formation with the well and involving in the process of operation non-producing gas bearing layers. The well improvement due to the level of formation completion can be increased by deepening the well borehole if there are necessary conditions or by drilling additional branch holes.

4 RESULTS OF RESEARCH

Using the suggested analytical dependences (15)-(17) we've conducted the research concerning the conditions for the deposits of horizon ND-4 of the Oparsk gas field.

Horizon ND-4 occurs within the range of 520-560 m depth and is composed of alternating layers of sandstones and siltstones. Gas accumulation in horizon ND-4 forms an arched gas deposit of the bedded type. The deposit was put into operation in 1940. Since 01.01.2012 the deposit has been developed by 9 wells. Gas flow varies from 0.3 to 8.5 m³ / d (average gas flow rate is 4.22 thousand m³ / d), the op-

erating pressure at the wellhead varies from 0.5 to 0.6 MPa (on average it is 0.57 MPa). Since the beginning of development the deposit pressure has decreased from 5.39 to 0.6 MPa. The average values of filtration resistances factor of bottomhole formation zone are equal to $A = 0.099$ MPa² d / th. m³, $B = 3.75 \cdot 10^{-4}$ (MPa d / th. m³)².

By the field data the refined initial drainage gas deposits are 4036 million m³. Current gas recovery factor is 85.8%. Assessed by the field data the forecast final gas recovery factor is 86.73% in accordance with the existing system of deposit development. This justifies the necessity of conducting the geological and technical measures to increase it.

For the conditions of deposit of horizon ND-4 in the Oparsk gas field we've estimated the impact on the final gas recovery factor at the wellhead, gas flow rate and filtration resistance factor at the bottom-hole formation zone A and B. Calculations are performed for different values of pressure at the well head P_w – 0.503; 0.5; 0.4; 0.3; 0.2; 0.1 MPa, gas flow rate q_g – 4.622; 4,0; 3,0; 2,0; 1,0; 0,5; 0,1 thousand m³ / d and the degree of reduction of filtration resistance factors A and B in 2, 4, 6, 8, 10 times compared with the actual values.

In the calculations the following actual field data are used: initial formation pressure $P_i = 5.39$ MPa, the current formation pressure $P_f = 0.6$ MPa, the relative gas density $\rho_{st} = 0.559$; formation temperature $T_f = 306$ K, the gas temperature at the wellhead $T_w = 291$ K, the current flow rate of a typical well $q_g = 4.22$ thousand m³ / d, the current pressure at the wellhead of a typical well $P_w = 0.503$ MPa, the length of penetration of tubing $L_w = 506$ m, internal diameter of tubing $d_t = 0.062$ m; the coefficient of tubing hydraulic resistance $\lambda_w = 0.024$.

In studies taking into account the set values of gas flow rate q_g and wellhead pressure P_w, the bottomhole pressure P_b was determined by the formula (2), and then the formation pressure for given values of filtration resistance factors at the bottomhole formation zone by the formula:

$$P_f = \sqrt{P_b^{\,2} + A \cdot q_g + B \cdot q_g^{\,2}} \; . \tag{19}$$

According to the values of formation pressure the gas recovery factor was determined by the formula (18).

In accordance with the results of the conducted

calculations we've plotted the relation of the final gas recovery factor to the degree of reduction of filtration resistance factor of the bottom-hole formation zone for different gas flow values and wellhead pressure. Figure 1 shows an example of dependence for the gas flow rate 4.22 thousand m^3/d. Other dependences are the same.

The research result analysis shows that the reduction of the values of all the significant factors leads to the increase of final gas recovery factor. We've examined the impact of each of them.

For the conditions of the deposit of horizon ND-4 in the Oparsk field the final gas recovery factor increases substantially with decreasing the values of filtration resistances factors at the bottomhole for-

mation zone A and B in 4 times. Further reduction of values of factors A and B affects less significantly the gas recovery factor. The greater the degree of increase of the final gas recovery factor along with the decrease of the values of the factors A and B is, the greater the flow gas rate is. Therefore, with the purpose of the filtration resistance factors A and B reduction the well head formation treatment should be done at the early stages of deposit development under the conditions of high gas flow rates. At low gas flow rates (in this case 0.1-0.3 m^3/d) the filtration resistance factors A and B reduction hardly affects the values of final gas recovery factor.

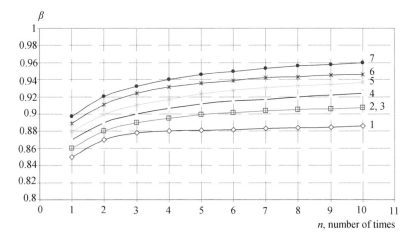

Figure 1. The dependence of the final gas recovery factor on the degree of the reduction of filtration resistance factors A and B for the gas flow rate 4.22 thousand m^3/d and different values of the wellhead pressure: $1 - 0.6$; $2 - 0.503$; $3 - 0.5$; $4 - 0.4$; $5 - 0.3$; $6 - 0.2$; $7 - 0.1$ MPa.

The impact of pressure reduction at the wellhead and the decrease of the minimum efficient gas flow rate on the increase of the final gas recovery factor grows up with decreasing the filtration resistance factors A and B values. The greater the impact of well head pressure reduction on the increase of gas recovery factor is, the lower the minimum efficient gas flow rate is, and the influence of minimum-efficient gas flow rate reduction becomes greater with the well head pressure decrease.

The analysis of the calculated data shows that with the well head pressure reduction the rate of the increase in the final gas recovery factor gradually decreases. The well head pressure decrease significantly affects the values of final gas recovery factor at the completion stages of deposit development under the conditions of low gas flow rates. Therefore, it is very important for obtaining high values of final

gas recovery factor to reduce the maximum pressure at the wellhead, for example by commissioning the boosting compressor station, and operating wells with the lowest possible efficient gas flow rate.

According to research data at the final stage of gas fields development the well head pressure reduction significantly affects the increase of final gas recovery factor compared to the reduction of filtration resistance factors A and B values at the bottomhole formation zone. This stresses one more time the feasibility of measures application to ensure minimum values of the well head pressure in depleted natural gas fields.

5 CONCLUSIONS

The suggested analytical dependences for determining the formation pressure in depleted gas drive fields takes into account the current state of field development, the wellhead characteristics and technological parameters of wells operation and gas gathering system. For certain field conditions they allow evaluating the influence of separate determining factors and their combination on the final gas recovery factor and proving the most effective ways to increase it. Dependences have been tested for the conditions of gas deposits of horizon ND-4 in the Oparsk gas field. We've obtained a series of significantly important results concerning the application of different methods to increase the gas recovery factor in the depleted gas fields.

REFERENCES

Zakirov, S.N. 1998. *Development of gas, gas condensate and oil and gas condensate fields*. Moscow: Struna: 648.
Kondrat, R.M. 1992. *Gas condensate recovery factor of formations*. Moscow: Niedra: 255.
Zakirov, S.N. 2000. *Improvement of technologies of oil and gas field development* (Edited by S.N. Zakirova). Moscow: Hraal.

Study of rock geomechanical processes while mining two-level interchamber pillars

V. Russkikh & A. Yavors'kyy
National Mining University, Dnipropetrovs'k, Ukraine

S. Zubko
CJSC "Zaporizhzhya iron ore integrated works", Dniprorudne, Ukraine

Ye. Chistyakov
Kryvorizhskyi National University, Kryvyi Rig, Ukraine

ABSTRACT: The research of deformation processes of the rock mass adjacent to the worked-out areas during working two-level interchamber pillars between stowed chambers at 640-840 m is carried out on models of equivalent materials.

1 INTRODUCTION

Currently Yuzhno-Belozyorsky deposit is mined by two fronts within 301-303 m and 640-840 m using chamber system with consolidating stowage.

Depths 340-840 m have been mined since 1970. Division of influence zones of upper and lower mining zones as well as drainage of ore rock mass within 301-330 m was in progress within this period. 301-330 m mining takes place under protective pillar of high-pressure water-bearing level at low levels of stress condition of the mass while the one of 640-840 m is mined at levels close to ultimate ones as for stability conditions in the southern limb and low ones in the northern limb.

Work objective is to assure stability of structural elements of the system "chamber-pillar" while mining deposits with high chambers.

Large-scale mining of iron-ore deposits by the systems with consolidating stowage implemented at Zaporizhzhya iron-ore integrated works is a pilot one. That is why searching for means of securing production profitability at the expense of cost reduction and ore dilution related to chamber stability is a burning problem.

2 DEVELOPMENT OF SIMULATION SCHEMES AND PARAMETERS ON EQUIVALENT MATERIALS

Research using models of equivalent materials which meet natural conditions (Kuznetsov & Budko 1968, Kozina & Rutkovskaya 1973, Kabiesz & Makowka 2009) allows obtaining the data considering stability of elements of mining systems at various forms and dimensions of mined-out space simulated on the models.

The essence of simulation by means of equivalent materials is in the following: to obtain mechanical similarity the model is made of the materials which properties are similar to physical and mechanical properties of natural materials. Such correlations are determined on the basis of general law of force similarity considering simultaneous influence of both gravity and internal stress forces.

Geometrical similarity between the model and nature can be found only if all the dimensions of the model space occupied by the studied system as well as its separate elements will be changed by definite quantities of times occupied by the similar natural system. Characteristics of the geometrical similarity will be indicated as the correlations of such linear dimensions as:

$$\alpha_e = \frac{L}{l}, \tag{1}$$

where α_e – geometrical simulating scale; L – linear dimension of the element under simulation in nature; l – linear dimension of the corresponding simulated element.

To ensure the similarity of the mechanical processes in the model to the natural ones mechanical characteristics of the equivalent material should meet the following requirements:

$$N_m = \frac{l}{L} \frac{\gamma_m}{\gamma_n} N_n, \tag{2}$$

where N_m and N_n – mechanical characteristics of

the equivalent material of the model and natural material correspondingly; γ_m and γ_n – volume weight of the material of the model and nature correspondingly.

Considering the characteristics of geometrical similarity and similarity of mechanical processes and possessing the data about mechanical characteristics of mine rocks under simulation expressed by numerical values of N_n of the corresponding characteristics of the mechanical properties of model materials necessary to ensure the similarity are calculated for the specified modeling scale $\dfrac{\gamma_m}{\gamma_n}$.

Simulation using equivalent materials is performed on a flat stand with the transparent front wall. Model parameters are as follows: length is 140 cm, height is 125 cm, and width is 24 cm.

Figure 1 shows the scheme of the model of equivalent materials at 1 in 500 scale. The scale is selected so that the parameters of mined-out space ensure the required accuracy for determining deformation values in rock massif taking into account boundary conditions as well as mechanical and geometrical similarity.

Thus while simulating the whole rock massif at a scale of 1 in 500 stand height should be 170 cm. Part of rock mass being missing in the model should be compensated by surcharging.

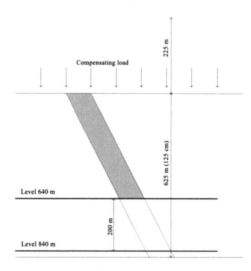

Figure 1. Scheme of the model of two-level interchamber pillar mining between stowed chamber at 640-840 m depth (1 in 500 scale).

In the model superincumbent rock (of 225 m high in nature or 45 cm high on the model) is compensated by the load of 18 kg being applied to each plate. Load values compensating lack of superincumbent rock are calculated taking into consideration use of the lever of the second kind according to the formula:

$$P_1 L_1 = P_2 L_2 + \ldots + P_n L_n, \qquad (3)$$

where $P_1 \ldots P_n$ – load being applied to the plate on the arm of the lever, kg; $L_1 \ldots L_n$ – length of the arm, cm.

Value of the load being applied to each plate is determined according to the formula:

$$P_n = \left(H_n M - H_m\right)\gamma_m S_M, \qquad (4)$$

where P_n – load being applied to the n plate, kG; H_n – thickness of rock mass under study, cm; M – scale of the model; H_m – height of the stand, cm; γ_m – specific weight of the equivalent material, g/cm^3; S_M – the plate area, cm^2.

Figure 2. Scheme of the lever of the second kind for load compensation.

Calculation of compensating loads is given in Table 1.

Table 1. Compensating loads.

Parameter	Scale 1 : 500
Natural / simulated height to be compensated, m / cm	225 / 45
Missing weight per plate, kg	18
Compensating torque for the arm $L_1 = 65$ mm, kg·cm	117
Load for arm $L_2 = 650$ mm, kg	1.8
Load for arm $L_3 = 715$ mm, kg	1.6
Load for arm $L_4 = 780$ mm, kg	1.5
Load for arm $L_5 = 845$ mm, kg	1.4
Load for arm $L_6 = 910$ mm, kg	1.3
Load for arm $L_7 = 975$ mm, kg	1.2

To select the material depending on the linear scale of simulation, characteristics of equivalent materials are determined according to the following formulas:

$$\sigma_{utsm} = \frac{l}{L}\frac{\gamma_m}{\gamma_n}\sigma_{utsn} ; \qquad (5)$$

$$\sigma_{ucsm} = \frac{l}{L}\frac{\gamma_m}{\gamma_n}\sigma_{ucsn} ; \qquad (6)$$

$$C_m = \frac{l}{L}\frac{\gamma_m}{\gamma_n}C_n ; \qquad (7)$$

$$tg\varphi_m = tg\varphi_n ; \qquad (8)$$

$$\mu_m = \mu_n ; \qquad (9)$$

$$E_m = \frac{l}{L}\frac{\gamma_m}{\gamma_n}E_n , \qquad (10)$$

where L, l – natural linear dimensions and on the model correspondingly; γ_n, γ_m – rock volume weight both natural and simulated; σ_{utsn}, σ_{utsm} – ultimate tensile strength of the rocks both natural and simulated; σ_{ucsn}, σ_{ucsm} – ultimate compressive strength of the rocks both natural and simulated; C_n, C_m – adherence of the rocks both natural and simulated; φ_n, φ_m – angle of internal friction both natural and simulated; μ_n, μ_m – Poisson ratio both natural and simulated; E_n, E_m – Young modulus both natural and simulated.

Sand, solid oil, paraffine, talc, alabaster, cement were equivalent materials. That has allowed ensuring the equality of internal friction angles of both the model and nature.

Materials are tested for strength by means of crushing cubes with the side of 4 by 7 cm. Cube tests are the basis for selecting the correlation of the components at which equivalent material would have the strength calculated according to the formulas (5-10).

3 MANUFACTURING, DEVELOPMENT AND TESTING THE EQUIVALENT MATERIAL MODELS

Simulation is performed on a flat stand with the front transparent wall. Model dimensions are as follows: length is 140 cm, height is 125 cm, and width is 24 cm.

To simulate deposit mining, average parameters of the deposit and enclosing rock are adopted (representative geological section for the conditions for occurrence mode of ore body in resistant rocks):
– deposit thickness is 50 m;

– deposit dip is 75°;
– inclination angle of chamber bottom of hanging wall is 55°;
– chamber height is 200 m;
– hanging wall is represented by resistant rocks (quartzites);
– lying wall is represented by shale rocks.

To study the resistance of structural elements of the development systems geometrical scale of modeling such as 1 in 500 has been adopted for the model being appropriate to meet the limiting conditions. To meet the third theorem of Kirpichyov, geometrical similarities and sequence of stoping and stowing operations in viscoelastic-plastic media are observed. Third similarity theorem establishes the following rules of physical modeling: original object and its model should geometrically equal; processes both in the model and original object should belong to one and the same class and be described using the same differential equations; initial and limiting conditions for the model and original object should be similar; determining dimensionless parameters should be similar both for the model and original object (Kuznetsov & Budko 1968, Kabiesz & Makowka 2009). Mixtures with different correlation of sand, solid oil, paraffin, talc, alabaster, cement, and micanite (Kuznetsov & Budko 1968, Kozina & Rutkovskaya 1973, Kabiesz & Makowka 2009) were used as the equivalent materials of the model. It has allowed ensuring equality of internal friction angles of the materials as well as similarity of the processes of deformation and destruction of both the model and nature.

Since the conditions of mining deposit up to the depth of 840 m were simulated than according to the adopted scale (1:500) model height should be 168 m. Missing part of gravitational load is compensated by external surcharging according to the calculations represented in Table 1. After applying the load the model should be kept up to the displacement stabilization.

To control deformations of geometrical parameters of the chamber indicating rulers were used (Figure 3).

Model for mining two-level interchamber pillar at the depth of 640-840 m was tested as the following:

1. Model formation. Ore deposit is mined by the systems of consolidating stowage according to the schemes from "chamber-six pillars" to "chamber-pillar". Ore deposit is mined from the upper sublevel to the contact of ore body with the stowage.

2. Upper sublevel of mined-out space is stowed.

3. Middle-sublevel ore deposit is mined to the contact of ore body with the stowage.

4. Middle sublevel of mined-out space is stowed.

5. Ore deposit of the lower sublevel is mined to the contact of ore body with the stowage.

6. Mined-out space of the upper sublevel is stowed.

7. Ore deposit between two slowed chambers within 640-840 m throughout the whole chamber height (two levels) to the contact of ore body with the stowage is mined.

Indicating rulers cannot record deformations and displacements in the rock mass adjacent to the worked-out space both while two-stage mining (from the upper sublevel to the lower one) and while single-stage mining (with two-level chamber height) (Figure 4).

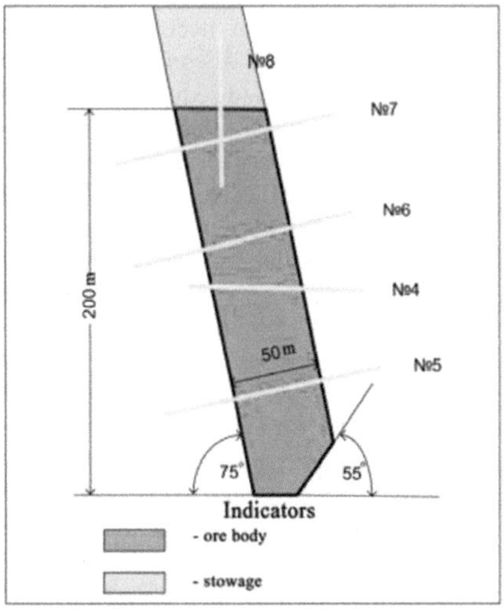

Figure 3. Scheme of positioning indicating rulers #4, 5, 6, 7, 8, on the model simulating mining the chamber of 200 m depth, deposit thickness is 50 m.

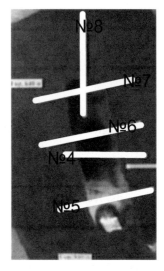

Figure 4. Model after mining throughout the whole height of inter-chamber two-level pillar between stowed chambers at the depth of 640-840 m to the contact of ore body with the stowage.

4 CONCLUSIONS

Simulation by means of equivalent material obser-ving geometrical, force, and mechanical similarity has shown:

– artificial three-month year-old mass stowed with consolidating mixtures is not subject to stresses exceeding standard power of resistance and keeps its competence. Possible cases of resistance loss are due to the dimensions of undercutting of soft rocks of hanging wall;

– zonation according to the resistance of hanging wall is the reason for differential approach as for the establishing both the sequence of mining operations and geometrical parameters of chambers. In nor-thern wing where hanging wall does not cave at the applied parameters and shapes the procedure of chamber mining is not limited. The alternative of deposit mining by chambers of 200 m high is pos-sible It is recommended for southern wing to start mining in the middle part, then at the lying wall, and to finish it at the hanging one. Mining near the hanging wall is recommended to carry out by the chambers of reduced width (10 or 15 m depending on hanging wall resistance). The alternative of mi-ning with leaving ore triangle in the hanging wall is possible.

REFERENCES

Kuznetsov, G.N., Budko, M.N., Vasiliev, Yu.N., Shklyar-sky, M.F. & Yurevich, G.G. 1968. *Modeling of rock pressure manifestation.* Leningrad: Nedra: 280.

Kozina, A.M. & Rutkovskaya, Ye.P. 1973. *Guidance in selecting and testing equivalent materials for modeling.* Moscow: IGD named after Skochinsky: 23.

Kabiesz, J & Makowka, J. 2009. *Selected elements of the rock bursts assessment with some case studies from the Silesian coal mines.* Mining Science and Technology. Formerly Journal of China University of Mining & Technology. Vol. 19, #5:660-667.

Guidance in Studying Rock Pressure Manifestation on the Models of Equivalent Materials. 1976. Leningrad: VNIMI: 37.

Mining of Mineral Deposits – Pivnyak, Bondarenko, Kovalevs'ka & Illiashov (eds)
© 2013 Taylor & Francis Group, London, ISBN: 978-1-138-00108-4

Heat pumps for mine water waste heat recovery

V. Samusya, Y. Oksen & M. Radiuk
National Mining University, Dnipropetrovs'k, Ukraine

ABSTRACT: A pilot heat pump power plant that utilizes the mine water waste heat for mine's hot water supply system on "Blagodatna" mine that belongs to PC "DTEK Pavlogradugol" has been designed and built. The results of the experimental research of the heat pump plant have been shown that the measured values of the heat transfer agent volume flow rate in the circuits, temperatures and plant's heat rate well agreed with the calculated values. The real value of the performance coefficient of the heat pump modules calculated experimentally is 3.5.

1 INTRODUCTION

Because of fuel resources depletion, the problems of rational fossil fuel consumption, renewable sources utilization and power plants' waste heat recovery are becoming urgency. Recently the heap pump technology that utilizes low-grade waste heat as well as the heat of the natural sources and makes it appropriate for being used in heating and hot water supply systems has been becoming more and more popular. The advantages of the technology are its energy efficiency and environmental-friendliness. The main disadvantage is its high price not only the heat pump equipment itself but also the systems of the waste heat collection. Due to that, the heat pump technology implementation for waste heat recovery from mine water seems to be very promising since the waste heat collection system costs are low. Especially, the heat pumps can be useful for mines where there are no any high-grade waste heat sources. For those mines the need for hot water is satisfied by means of coal boiler plants that have very low efficiency and strong influence to the environment (Andrew Hall 2011).

2 IMMITATION MODELING

The main objective of this study is to design a mine water waste heat recovery system for mine's hot water supply on "Blagodatna" mine that belongs to PC "DTEK Pavlogradugol".

The heat pump is similar to a refrigerator machine. Those machines remove the heat from the sources with low temperature, increase its temperature potential and transfer the heat to the source with higher temperature potential. To make

this process possible, the energy in the form of a work is needed. This work is transformed into the heat that also passed to the high temperature potential source. The difference between a heat pump and a refrigerator is determined by the purpose of their usage. The refrigerator machine is used to cool down the low temperature source, but increasing the high temperature source that is considered to be a side effect of the process. The heat pump is used to heat up a high temperature source, but decreasing the low temperature source that is considered to be a side effect in this case.

A schematic representation of a vapor compression heat pump system and temperature-entropy diagram of its thermodynamic cycle are shown in Figure 1.

The heat pump thermodynamic cycle is shown on the diagram against the background of the saturated liquid line $x = 0$ and dry-saturated vapor line $x = 1$ converging in the critical point K, as well as isobars $p_1 = const$ and $p_2 = const$ that correspond to the inlet and outlet pressure of the compressor (Kyrychenko Y, Samusya & Kyrychenko V 2012). Line 1-2 determines the process of working fluid compression in the compressor, 2-3 – the process of working fluid cooling and condensation in the condenser, 3-4 – throttling and 4-1 – boiling in the evaporator. The shaded region corresponds to the heat that was removed from the low temperature source, (specific evaporator heat load q_{ev}) and specific compressor work l_c (Samusya, Oksen & Radiuk 2010). The sum of these areas determines the specific condenser heat load q_{cn} that is the amount of heat that is transferred to the high-temperature source.

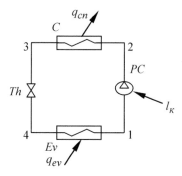
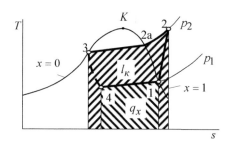

Figure 1. Schematic diagram of the heat pump and its thermodynamic cycle: C – condenser; PC – heat pump's compressor; Ev – evaporator; Th – throttle valve.

Energy analysis of the heat pump system:

Dry vapor density ρ_1, enthalpy i_1 and entropy s_1 of the working fluid at point 1:

$$\rho_1 = f(p_1, t_1);$$

$$i_1 = f(\rho_1, t_1);$$

$$s_1 = f(\rho_1, t_1).$$

At point 3 the working fluid is in saturated state. For this state the main thermodynamic parameters calculated as:

$$p_3 = f(t_3);$$

$$i_3 = i_{2a} - r_0(t_3);$$

$$s_3 = s_{2a} - \frac{1}{273.15 + t_3} r_0(t_3),$$

where $r_0(t_3)$ – heat of vaporization for the working fluid at temperature t_3, kJ / kg.

At point 2 the working fluid is superheated. Since the process 1-2 is isobaric $p_2 = p_3$.

$$i_2 = i_1 + \frac{\Delta i_s}{\eta_s}.$$

Specific vapor, condenser heat load and compressor work calculated as:

$$q_{ev} = i_1 - i_4;$$

$$q_{cn} = q_{ev} + l_c;$$

$$l_c = i_2 - i_1.$$

Their total values:

$$Q_{ev} = m_{wf} q_{ev};$$

$$Q_{cn} = m_{wf} q_{cn};$$

$$N_c = m_{wf} l_c,$$

where Q_{cn}, Q_{ev}, N_c – total condenser heat load, evaporator heat load and heat pump's compressor work correspondingly, kW; m_{wf} – working fluid mass flow rate, kg / s.

The heat pump efficiency is determined by the coefficient of performance as:

$$k = \frac{q_{cn}}{l_c} = \frac{Q_{cn}}{N_c}.$$

Since the source of the waste heat, mine water, polluted with suspended particles, and the water for the hot water supply system contains hardness salts, to protect the heat transfer surfaces of the heat pump's evaporator and the condenser from pollution the scheme with the intermediate circuits has been designed. In these intermediate loops the clean water circuits. The scheme of the heat pump system with the intermediate circuits is shown in Figure 2.

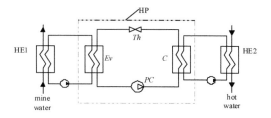

Figure 2. Schematic representation of the heat pump system with the intermediate circuits: HE1 – mine water heat exchanger; HE2 – hot water heat exchanger; HP – heat pump.

154

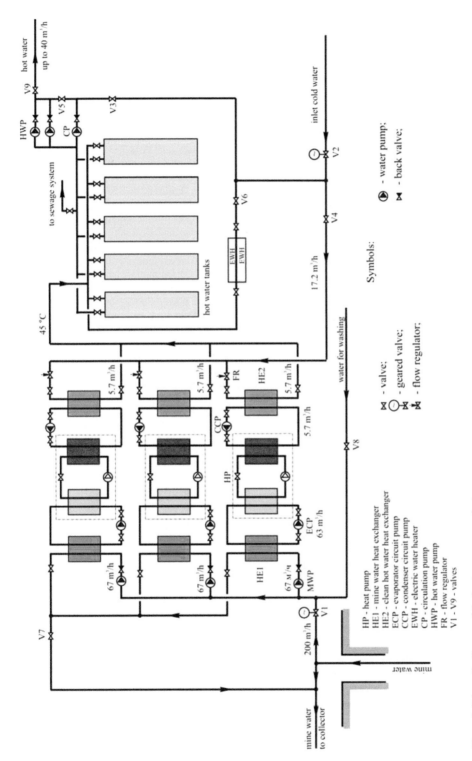

Figure 3. Schematic diagram of the heat pump power plant.

155

In this case only HE1 and HE2 are exposed to pollution. The cost of these heat exchangers are considerably lower that the heat pump's condenser and evaporator.

3 HEAT PUMP POWER PLANT

In "Blagodatna" mine the mine water temperature does not drop below 16-17 °C, mine water volume flow rate is 200 m³/h. To decrease the electricity costs the mine water is pumped at night when the electricity cost is the lowest. Accordingly, the heat pump system operates only at that time. The hot water supply system temperature is 42 °C; hot water volume flow rate is 120 m³/day. Supply pipeline of the household water use has been considered to be the source for the hot water supply system.

To provide the independent work of the system of water preparation and water consumption, it has been suggested to accumulate the hot water in five tanks of 100 m³ of capacity.

To increase the flexibility of the regulation, reliability and diagnostics of the heat pump power plan several point have been stated.

Firstly, the required heat rate is provided by three heat pump modules that are connected parallel and operate according to the scheme with the intermediate circuits.

Secondly, to provide the possibility for the heat pumps to heat up the water collected in the reservoirs to the required temperature in case of its cooling-down.

Finally, to provide the possibility of the electrical heating of the water in the reservoirs.

As the result of the research conducted, it has been carried out that to heat the water with the volume flow rate of 120 m³/day from 5 °C till 45 °C for 7 hours the required heat rate is 798 kW. To provide that the heat rate of each heat pump must be 266 kW.

In the heat pumps HKT-270 produced by "Refma" plant the screw compressors "Bitzer", plate-type evaporators "Alfa Laval" have been used. As working fluid refrigerant R407C has been chosen. The compact size of the compressors and heat pump equipment ensured the compactness of the heat pump construction which made it possible to install all equipment in one building of the size 10.5×12 view in plan and working area of 5.5 m in height. The heat pump plant room is located in the subtable building where mine water pumped to. That made to provide the minimum distance from the mine water collector to the heat exchangers. The scheme of the heat pump power plant is shown in Figure 3.

Mine water heat exchangers are made shell-and-tube. The mine water flows inside tubing. The water of the intermediate circuits flows in tube space. The heat exchangers are tree-pass. The tube bundle consists of 300 caper tubes with the diameter of 20 mm each and the thickness of 2 mm. The surface area is 34 m².

Clean water heat exchangers are made plate-type and eight-pass. The surface area of the heat exchanger is 16 m². The general view of the heat pump modules is show in Figure 4.

Figure 4. The heat pump modules (general view).

The calculated volume flow rate of the heat-transfer agents in each module are as follows: in the circuit of the mine water is 66.7 m³/h; in the circuit of the hot water for the hot water supply system is 5.7 m³/h; in the intermediate circuit of the evaporator is 63.2 m³/h; in the intermediate circuit of the condenser is 5.7 m³/h. The heat rate of the heat pumps HKT-270 at the boiling temperature of the working fluid in the evaporator that equals +6.5 °C and at the condensation temperature that equals +55 °C amounts to 256 kW. The plant operates in the automatic mode and does not require the maintenance staff.

4 CONSLUSIONS

The results of the experimental research of the heat pump power plant have been shown that the measured values of the heat transfer agent volume flow rate in the circuits, temperatures and heat rate of the plant well agreed with the calculated values. The real value of the performance coefficient of the heat pumps calculated experimentally is 4.2. For the heat pump modules performance coefficient calculation the electrical power consumed by the water pumps has been taken into account. Therefore, the performance coefficient for the modules is 3.5. These data state that for each 1 kW·h of the electrical energy consumed, 3.5 kW·h of heat energy produced, and that 2.5 kW·h of that heat energy consists of the heat that removed from mine water low-grade waste heat. That fact proves the high energy efficiency of the heat pump plant. The heat pump plant enables to exclude the need of the coal boiler plants and brings considerable environmental benefit. The expected payback time of the capital costs is 3.2 years.

The heat pump power plant build in "Blagodatna" mine, that belongs to PC "DTEK Pavlogradugol", is the first and the biggest mine water waste heat recovery plant in the countries of CIS.

REFERENCES

Andrew Hall, John Ashley Scott & Helen Shang, 2011. *Geothermal energy recovery from underground mines.* Renewable and Sustainable Energy Reviews. Volume 15. Issue 2: 916-924.

Kyrychenko, Y., Samusya, V. & Kyrychenko, V. 2012. *Experimental investigation of aeroelastic and hydroelestic instability parameters of a marine pipeline.* The Netherlands: CRC Press/Balkema. Geomechanical Processes During Underground Mining: 163-167.

Samusya, V., Oksen, Y. & Radiuk, M. 2010. *Efficiency estimation of heat pump waste heat recovery.* State Higher Educational Institution National Mining University. Dnipropetrovs'k: Naukovyi visnyk NGU, 6: 78-82.

Mining of Mineral Deposits – Pivnyak, Bondarenko, Kovalevs'ka & Illiashov (eds)
© 2013 Taylor & Francis Group, London, ISBN: 978-1-138-00108-4

Research of stress-strain state of cracked coal-containing massif near-the-working area using finite elements technique

I. Kovalevs'ka, G. Symanovych & V. Fomychov
National Mining University, Dnipropetrovs'k, Ukraine

ABSTRACT: Results of stress-strain state (SSS) of cracked layered massif researches in roof, walls and lying wall were shown. Recommendations concerning stability increase of rock uncovering by means of its hardening with help of bolt and frame-bolt supports, which operate in united bearing system were given.

1 INTRODUCTION

There is a question of stress and strain interaction calculation using non-linear law, which is strain diagram appearing in the process of SSS calculation in view of plastic components of rock massif strain. This diagram must be plotted at real stress and strain, in "stress intensity – stain intensity" system of reference. General strain diagram is considered fair for any point stress. However, destruction "cuts" this diagram in different points, depending on a kind of stress state. Usually, the bigger average stress against intensity is, the less plastic strain intensity at the moment of destruction is. Intensity of stresses and strains is described through their main components

$$\sigma_i = \frac{1}{\sqrt{2}}\sqrt{(\sigma_1 - \sigma_2)^2 + (\sigma_2 - \sigma_3)^2 + (\sigma_3 - \sigma_1)^2} \ ;$$

$$\varepsilon_i = \frac{\sqrt{2}}{3}\sqrt{(\varepsilon_1 - \varepsilon_2)^2 + (\varepsilon_2 - \varepsilon_3)^2 + (\varepsilon_3 - \varepsilon_1)^2} \ .$$

For most of non-brittle materials, strain diagram, received experimentally using uniaxial tension matches general diagram. Equation of general strain diagram in elastic zone looks like:

$$\sigma_i 3 = G\varepsilon_i \ ,$$

where $G = \dfrac{E}{2(1+v)}$ – elasticity of shear module,

and for uniaxial stress state the equation takes such form:

$$\sigma_x = E\varepsilon_x \ .$$

However, for materials "massif-support" system,

the most often met in the process of geomechanical issues solution, it is necessary to put real, experimentally received strain diagram into calculation. There are several methods with different degree of approximation. Quite often approximate bilinear diagramm is used. It is described by formulas:

$$\sigma = E\varepsilon \ \text{ where } \ \varepsilon \le \sigma_{fl} \ ;$$

$$\sigma = \sigma_{fl} + E(\varepsilon - \varepsilon_{fl}) \ \text{ where } \ \varepsilon > \varepsilon_{fl} \ ,$$

where $\sigma_{fl} = E\varepsilon_{fl}$ – determines strain ε_{fl} at yield stress σ_{fl}; E_{fl} – hardening module in elastoplastic area $0 \le E_{fl} \le E$.

The other method – polygonal approximation of strain diagram over coordinates of diagram separate points. For that purpose let us put relative coordinates

$$\bar{\sigma} = \frac{\sigma}{\sigma_{fl}} \ ; \ \bar{\varepsilon} = \frac{\varepsilon}{\varepsilon_{fl}} \ .$$

In this case elastic part of diagrams is described by equation $\bar{\sigma} = \bar{\varepsilon}$. In every separate strain interval $\bar{\varepsilon}_n \le \bar{\varepsilon} \le \bar{\varepsilon}_{n+1}$ equation of strain curve can be noted in following way:

$$\bar{\sigma} = \bar{\sigma}_n + \frac{\bar{\sigma}_{n+1} - \bar{\sigma}_n}{\bar{\varepsilon}_{n+1} - \bar{\varepsilon}_n}(\bar{\varepsilon} - \bar{\varepsilon}_n) \ .$$

In such case, strain curve is described by inscribed brake and at every interval is displayed as direct. At the initial area of the diagram, strain intervals are usually considered as not identical, but as equal to 1; 1,25; 1,5; 2; 3; 4; 5.

Good agreement with experiments results makes up strain diagram gradual approximation

$\bar{\sigma} = \bar{\varepsilon}$ where $\sigma \leq \sigma_{fl}$;

$\bar{\sigma} = \bar{\varepsilon}^m$ where $\sigma > \sigma_{fl}$,

where $\varepsilon_{fl} = E\sigma_{fl}$; σ_{fl} – estimated proportionality limit.

Based on the above, formula of hardening index expression is received

$$m = \frac{0.75 lg\left[\dfrac{\sigma_B}{\sigma_{0.2}}(1 + 1.4\psi_K)\right]}{lg\left[\dfrac{1}{0.002 + \sigma_{0.2}/E}ln\dfrac{1}{1 - \psi_K}\right]}.$$

Given real strain diagram approximation techniques allow to describe elastoplastic characteristics of materials on location, which are used in limit and out-of-limit state zones calculations in geomechanics issues very accurately. The process of cracks formation in working massif is done on basis of stress intensity factor. Stress and strain in proximity to the crack tip are proportional $K/\sqrt{r}$, where K – stress intensity factor; r – distance of selected point to the tip.

During the process of stress intensity factors calculation using finite elements technique (FET), accuracy of the results will depend on the quality of modeling of stress and deformation field in area of crack tip. As usual finite elements being build in accordance with displacement field notion with help of polynominal with integral power, such modeling is considerably complicated. Such trouble can be removed by implementation of one or several elements into the net, which simulate stress singularity. Displacement functions within special element have member, proportional $\sqrt{r}$. That is why intermediated node of isoparametric quadric element is placed not in the middle, but moved by one forth of the side lengths in the direction of the crack tip. As the result displacement distribution along given side is described by equation

$$u_i = C_1 + C_2\sqrt{r} + C_3 r. \tag{1}$$

Assume $u_i(0)$, $u_i(L/4)$, $u_i(L)$ – components of units displacement in the crack tip, of intermediated and corner ones respectively. Here L – the lengths of the side, located along half line which begins in the crack tip and coordinate r, which is counted out of this tip. Then, constants for displacements distribution (1) will be described as

$C_1 = u_i(0)$, $C_2 = \dfrac{1}{\sqrt{L}}\left[4u_i(L/4) - u_i(L) - 3u_i(0)\right]$;

$C_3 = \dfrac{1}{L}\left[2u_i(L) - 2u_i(0) - 4u_i(L/4)\right].$

Value C_1 describes crack tip displacement, C_2 – the part of displacement, which corresponds to asymptotic behavior of displacement, C_3 – constant strain and body rotation as the entire. After doing some transformations and taking up $u_i(0) = 0$, we receive a formula used to find out stress intensity factor through direct method

$$K_I = \frac{2G\sqrt{2\pi}}{F_i(\theta)\sqrt{L}}\left[4u_i(L/4) - u_i(L)\right],$$

which gives greater freedom to implementation of higher-ranking real diagram approximation of subject of inquiry in computing experiment.

2 COMMON DATA OF MODELING

Observation analysis of jointing and rocks bedding elements, which were received using acoustic logging method allows to underline three systems of cracks with 60-75% angles of slope on the measured surface.

I system of cracks is north-west oriented and parallel to faults or close to their direction. Cracks, not typical for sand-flags and filled mostly with quartzite, cracks widths varies within the range of 0.1-5.0 mm. The distance between the cracks is between 0.10 and 0.8 m. In case of sandstones, distance between cracks is severe and puts together 0.05-0.2 m. For shale rocks it can be variated between 0.1-0.5 m. Cracks surface is rough.

II system of cracks is parallel to cracks of the I system, but it has alternate angle of slope (southward). Cracks characteristic of this system is similar to those of the I one, the difference is in differently directed angles of slope.

III system. Angles of slope are eastward oriented. Cracks are enclosed. In case of sandstones, cracks are seen everywhere and they occur again every 0.1-0.6 m.

According to the data of geological examination and chart of extraction working, geomechanical model of system "massif-support" interaction was built. It reflects its basic features.

Firstly, real structure of rock layered massif was modeled:

– the first rock layer of roof with heights $m_1^r = 12.8$ m is represented by rock bind (in the bottom part of the layer) with heights 3.5 m and crushing strength $\sigma^r_{comp1} = 50.6...54.2$ MPa and by sand slate (in the top part of the layer) and $\sigma^r_{comp1} = 78...97$ MPa;

– the second rock layer of roof with heights $m^r_2 \geq 16.5$ is represented by sandstone with $\sigma^r_{comp2} = 94...127.2$ MPa.

Coal layer is modeled with height $m^{coal} = 0.95$ m and $\sigma^{coal}_{comp} = 10...20$ MPa.

In the immediate lying wall underlay:

– argillaceous slate with height $m^L_1 = 4.35$ m $\sigma^L_{comp1} = 15.6...20.8$ MPa;

– rock bind with height $m^L_2 = 0.6$ m and $\sigma^L_{comp2} = 50.6...54.2$ MPa;

– coal seam with height $m^L_3 = 0.6$ m and $\sigma^L_{comp3} = 10...20$ MPa.

In the main lying wall underlay:

– sand slate with height $m^L_4 = 6.75$ m and $\sigma^L_{comp4} = 82.3...91$ MPa;

– sandstone with height $m^L_5 \geq 10$ m and $\sigma^L_{comp5} = 108...112$ MPa.

Secondly, real sizes of working, support and parameters of its location in relation to coal seam are modeled:

– angle of layer slope is accepted $\alpha = 4°$;

– extraction working is driven using combined ripping – of roof by 0.5 m and lying wall by 1.65 m;

– typical working cross section with neat area till 11.2 m^2 settlement with support KMP-A3-11.2;

– real geometry of SVP-22 special profile was built;

– frame support density – 1.25 frames / m;

– material of special profile – steel St. 5 with yield stress $\sigma_{fl} = 270$ MPa, elastic modulus $E^P = 21 \cdot 10^4$ MPa, Poisson's ratio $\mu^P = 0.3$;

– barrier between frames – reinforced concrete lagging of standard size $800 \times 200 \times 50$ mm with mechanical characteristics $\sigma^{lag}_{cr} = 20$ MPa, $E^{lag} = 2 \cdot 10^4$ MPa, $\mu^{lag} = 0.2$;

– fixed space 150 mm wide, backfilled with broken rocks with mechanical characteristics:

$E^b = 50$ MPa, $\mu^b = 0.25$.

– working laying depth – 800 m.

3 RESULTS

Vertical stresses σ_y de-stressed zone appears in roof with the appearance of tension stresses σ_y up to 2.5 MPa on the working contour. De-stressed zone form is dome-shaped, symmetric about the axis and extracted along the coordinate Y : its sizes in the context of double de-stressed value ($\sigma_y = 0.5\gamma H$) make up 2.8-3 m height and 3.5-4 m across the width of the working. De-stressed zone is fully localized in the first rock layer of the roof, and stress activity level σ_y does not cause stress damage even in rock bind, which underlays in the bottom part of the first rock layer of the roof. However, tension stresses σ_y appear in near-the contour-rocks of the roof. Rocks resist them very weakly, especially, if we take into account their disturbance by natural jointing and man-made one, formed during working construction. That is why unstable rock area, disposed to falls occurs in the roof. Sizes of this area are evaluated according to the law $\sigma_y = 0$: heights of unstable rock area – 1.0-1.3 m, width – 2.1-2.4 m.

Horizontal stresses diagram σ_x clearly demonstrates roof rock layers deflection in cavity of the working. However, caused by fairly big height of the rock layer of the roof ($m^r_1 = 12.8$ m), flexural stress does not so substantially affect its stability. Nevertheless, local area of softened rocks caused by tension stresses σ_x appears somewhere in the region of the key arch: height – 0.5-0.6 m; width – 0.7-0.9 m. Shown sizes of the area of possible fall are smaller, than the same ones caused by activity of vertical tension stresses σ_y. However, cumulative action of tension stresses σ_x and σ_y allowing for natural and man-made near-the-contour rocks of the roof disturbance allows to predict the area of possible fall in roof with width, corresponding to the width of the working in drivage, and heights up to 1.5-2.0 m.

Vertical stresses diagram σ_y unambiguously points at abutment zone formation in working walls. It is concentrated mainly in the first lying wall rock layer of the coal seam m^L_1 from any height of the ripping with concentration $\sigma_y \geq 2\gamma H$ in the areas of footings and the top part of their rectilinear part; the width of the areas with $\sigma_y \geq 2\gamma H$ is no more than 0.4-0.6 m in working walls. However, underlying in the lying wall of coal seam argillaceous slate has low resistance to compression (up to 20.8 MPa) and there is a possibility of its softening in the working walls and more significant distances. For example, $\sigma_y = (1.25...1.5)\gamma H = 27.5...30$ MPa are entirely capable to cause softening of argillaceous slate with 3 m width spreading in every side of the working. Similar specifics can also have place in a coal seam, compression resistance of which is also no more than 20 MPa.

Horizontal stresses diagram σ_x points at intensive deflection of the first rock layer of the coal seam lying wall, in which working walls lay. Tension intensities σ_x up to 5 MPa take place on the place of contact with coal seam. They destroy the top part of the argillaceous slate layer for 1.5-1.8 m width. Concentration of compression stresses σ_x 1.2-1.6 m width in the area of frame support footings can also cause argillaceous slate softening.

In the abutment zone located on each side of the wall compression stresses σ_y and σ_x are mainly taking place. Cumulative action of them in geomechanical researches is traditionally evaluated according to the volume of indicated stresses σ and compared with compression resistance of one or another lying wall layer or coal seam in accordance with More's strength theory. In this context diagram of indicated stresses σ is demonstrative, analysis of which points out extensive areas of limit state in working walls of the first rock layer of coal seam lying wall (width up to 3.1-3.4 m) and in coal seam (width 2.8-3.0 m). Such significant areas of massif limit state provoke not only intensive side displacement of frame footing, but also rock heaving directly in the working.

Vertical stresses diagram σ_y points at more extensive de-stressed zone formation in the working lying wall, which spreads across the entire width of the working, and in the stress depth $\sigma_y = 0.5\gamma H$ can reach the rock layer of the main bottom. That means, that intensive de-stress acts at depth of no less than 4 m from the contour of the working lying wall. Area of tension stresses activity σ_y is also bigger than in the roof. It corresponds to the width of the working and spreads 1.3-1.4 m depth. This borders delimit area of softened rock in the lying wall of the working from activity of vertical tension stresses.

On horizontal stresses diagram σ_x deflection in cavity of the first rock layer of the lying wall is clearly observed: tension stresses σ_x up to or more 5 MPa take place in near-the-contour rocks of the lying wall. Compressing stresses $\sigma_x \geq 10$ MPa appear at the border of rock bind.

By analyzing obtained data, two conclusions can be done. On the one hand, the depth of the softened rock caused by tension stresses σ_y и σ_x is relatively smallish and it do not provoke significant rock heaving of the working development. On the other hand, a vast zone of softened rocks is forming in the walls of the working they are "squeezed out" by harder and more holistic layers of the roof into the working cavity. They also move in the direction of horizontal displacements of the rocks under frame support footings bearers. This displacement transform into sloping and vertical ones when coming closer to the middle of the working lying wall.

The second factor can have decisive action, especially in influence zone of stoping, when area of working is increasing and bearing pressure in working walls is growing. Besides, soft argillaceous slate in immediate bottom is sensitive to wetting and its compression resistance is significantly reduced. In such case, lying wall rock will be softening not only because of tension stresses, but also compressing σ_y and σ_x, and the size of this area will be largely increased and lead to intensification of working lying wall rocks heaving development.

4 CONCLUSIONS

1. Near-the-contour rocks around driven extraction working are softening at different depth all around its perimeter.

2. Unstable rocks area, disposed to fall 1.0-1.3 m depth and 2.1-2.4 m width appears in the roof of the working. Fall height can reach 1.5-2.0 m caused by weakening factors of natural and man-made disturbance of the massif.

3. The fall can be prevented using one of two recommended ways:

– hardening of near-the-contour roof rocks with bolts, which develop high bearing ability, even in softened rocks;

– installing KVT-2-11.7 support instead of KMP-A3-11.2. It has reduced width and reduced height of arching, form of which increases roof rocks stability.

4. Vast area of massif limit state appears in working walls: in coal seam up to 2.8-3.0 m width, in the first rock layer of the lying wall up to 3.1-3.4 m width. It determines development of significant side displacements of near-the-contour rocks with in-

creased side pressure on frame support footings.

5. Improving stability of working walls is recommended to realize by forming frame-bolt support during bolts installation in walls and their connection into one bearing system with frame footings using space-facile communication units.

6. The depth of softened rocks can reach 1.3-1.4 m and fill all the width of the working in its lying wall. Argillaceous slate compression resistance reduces 2-3 times in case of it wetting and softened rocks area can be broadened all over first lying wall rock layer height. This factor together with vast zone of limit state of rocks in working walls intensifies working lying wall heaving.

Mining of Mineral Deposits – Pivnyak, Bondarenko, Kovalevs'ka & Illiashov (eds)
© *2013 Taylor & Francis Group, London, ISBN: 978-1-138-00108-4*

The results of instrumental observations on rock pressure in order to substantiate complete excavation of coal reserves

Yu. Khalymendyk, A. Bruy & Yu. Zabolotnaya
National Mining University, Dnipropetrovs'k, Ukraine

ABSTRACT: The protection of main roadways at Ukrainian coal mines in accordance with normative documents is carried out by pillars. It is possible to obtain the maximal plenitude of coal reserves excavation when set-up entry is conducted skin-to-skin to the roadway. The normal functioning of roadway is provided by support reinforcement in the period of longwall face retreat on distance equal to the width of spacing of roof caving or width of abutment pressure for these conditions. The overworking of main mine roadways without pillars is practicable solution. It is necessary to reinforce them with additional support before the initiation of abutment from longwall face. The stop of longwall face is carried out in the distance, greater than widths of abutment pressure, then the additional support is dismounted and the roadway is exploited in the discharged area.

1 INTRODUCTION

Repair-free exploitation of mine roadways is an actual problem for coal mines of Ukraine. Protection of main mine roadways in most cases, in accordance with regulative documents (SOU 10.1.00185790. 011:2007; KD 12.01.01.201-98 1998; KD 12.07.301-96 1996), is carried out by leaving pillars of coal. Extraction panels are recovered from protective pillars left behind on the side of main mine roadways and up to them.

2 FORMULATING THE PROBLEM

Dimensions of protective pillars are accepted equal or larger than the size of the abutment zone. Increase of mining depth leads to the growth of rock mass stress condition and widening of the abutment pressure zone in the seam selvedge from 30-35 m at the depths of 250-300 m and to 100-120 m at the depths of 600-1000 m. This leads to irrational utilization of coal reserves in the subsoil. While at the small depths the coal losses in pillars do not exceed 10-12%, with mining deepening to 1000 m they reach 25-30% (Hudin et al. 1983).

Coal pillars left for mining protection create high stress zones in the rock mass which creates danger of unexpected bursts of coal, gas and bounces and worsens conditions of maintenance of roadways when mining superimposed seams.

Condition of main mine roadways with the depth growing remains unsatisfactory (Babenko 2008). As a rule a large scope of works on preservation of roadways in the chart condition is performed, there are some difficulties in provision of aeration and solution of transportation problems.

Developing of roadways in the regional zones of discharge (Zborschik & Nazimko 1991) makes it possible to solve the problem before mining of neighboring mine sections takes place.

Main mine roadways, when being driven, are in the zone of low stresses. With development of mine operations the worked-out areas of longwalls join and full or partial rock pressure recovery takes place inside them. It is significant that the high rock pressure zones (HRP zones) are formed not only in coal pillars but also in the zone discharged earlier as well as in the middle part of a worked-out area (Babenko 2008).

After full extraction of a panel the stresses in the regional zone of discharge increase. Concentration of stresses with respect to pressure inside the virgin rock reaches 1.6 (Babenko 2008) which adversely affects stability of main mine roadways initially located in the regional zone of discharge. Thus the problem of actual unsatisfactory condition of main haulage roadways with development of mine operations remains unsolved.

Modern notions (Hudin et al. 1983; Khalymendyk & Begichev 2002) about geomechanical processes in the massif and based on this knowledge approach to maintenance of main roadways makes it possible to significantly increase the fullness of coal reserves extraction and simultaneously provide normal functioning of mine roadways.

The simplest and the most effective solution is development of gateroads after completion of longwall mining in skin-to-skin way or in the discharged area. To carry out pilarless extraction of coal reserves at the existing roadways some technological solutions with taking into account directions of the longwall face movement have been developed.

The most reasonable from the positions of geomechanics is to make a retreat of the longwall face from the protected roadway. The following variants of development and protection of roadways are possible (Khalymendyk & Begichev 2002):

• skin-to-skin development heading after the longwall face retreat;

• roadway drivage in the discharged area after the longwall face retreat;

• short-term overworking of the entry;

• retreat of the longwall face from the roadway or directly out of the roadway with its maintenance;

• leaving behind triangular pillars of small dimensions.

When the longwall faces advance on the roadways the following cases are considered:

• the longwall face advance with the reduced size of the pillar;

• the longwall face advance without leaving the front pillar;

• the longwall face overworking of the cross-strata roadway;

• cross-strata roadway development after completion of the panel extraction.

It is an established fact that with the longwall face retreat from the set-up entry pressure in the near-face part of the seam increases (Jacobi 1981). To the same extent pressure increases on the remaining wall of the set-up entry (even if it is extinguished). Dimensions of the front abutment pressure zone ahead longwall face depending first of all on the depth of mine works, extracted thickness of a bed and hardness of rocks are the subject of many researches. As for the parameters of the abutment pressure zone in the pillar at the longwall face retreat from the set-up entry, they have been studied very little so far.

The objective of this paper is the analysis of abutment pressure manifestation during the longwall face retreat from the set-up entry and ahead of longwall face.

3 IN-SITU OBSERVATIONS DURING LONGWALL RETREAT

The observations were carried out under the conditions of Zapadno-Donbasskaya Mine of OJSC "Pavlogradugol" during extraction of the coal reserves of the 830th panel (Figure 1). The observing station was installed in the Southern main haulage roadway #3 (SMHR #3) located in the bottom of seam C_8 to the beginning of its overworking. Every 10 m measurements of the roadway cross-section were carried out with affixement to the seam plane.

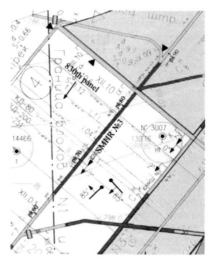

Figure 1. Print-out of a plan of mining operations on seam C_8 of Zapadno-Donbasskaya mine of OJSC "Pavlogradugol".

Development of abutment pressure with the longwall face retreat from the set-up entry by results of processing of measurements at the station is presented at Figure 3.

Analysis of the results of observations of the state of SMHR #3 denotes formation of the following four zones with the longwall face retreat:

I – an abutment pressure zone to the side opposite to the direction of movement; dimensions of the zone to the maximum value of the abutment pressure in this case is estimated to approximately 15 m after which stresses in a pillar reduce;

II – a zone free from bearing pressure (a rock pressure arch in the roofing of longwall); extension under these conditions is estimated to 30 m;

III – a zone characterized by constant increase of convergence of the overworked roadway (a zone of abutment pressure formation). Extension of the 3rd zone is about 100 m.

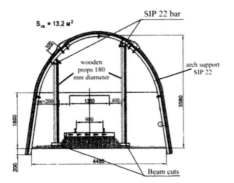

Figure 2. SMHR #3 maintenance chart.

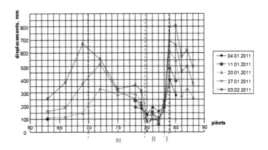

Figure 3. Total vertical convergence development of SMHR #3 during the longwall face retreat from the set-up entry.

4 FULL-SCALE EXPERIMENT OF ROADWAY OVERWORKING

Proceeding from the results of observations, to provide full extraction with retreat from the main roadways, that is, without leaving pillars behind, and to maintain the main roadways in the passport condition, their drivage in the seam bottom with subsequent short-time overworking is expedient. To minimize influence of the bearing pressure it is necessary that the roadways of the main direction be located within zone II. With such location of roadways relative to the set-up entry, increase of pressure on a support due to the powered support resistance should be expected at the moment of transfer. That is why, before occurrence of surcharge from moving longwall face, it is necessary to additionally reinforce the protected roadways using a reinforcement support.

Condition of the overworked roadway was examined during overworking the Eastern intermediate roadway of level 400 m (EIR 400 m) by the 157th panel of Stepnaya Mine (Figure 4).

The roadway was driven at the depth of 400 m; the roadway width was 4.3 m, height – 3.4 m, the estimated hardness of rock $\sigma_{cr} = 24$ MPa.

The estimated height of the arch of laminated rock amounts to 6.5 m, which is approximately equal to the distance between the floor of the mined layer and the roof of the roadway, i.e., lamination of rocks above the roadway may reach the floor of the mining face. In this case the zone of inelastic deformations of rock at the panel and minable roadway could unite. When studying the arch of laminated rock development an effect of the shield support was taken into consideration (its pressure on the floor of the mining face).

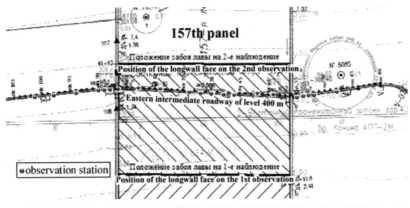

Figure 4. General layout of the observation station in the Eastern intermediate roadway of level 400 m.

Given these premises, calculation of reinforcement of a roadway support was carried out and measures on protection of the minable EIR 400m

were offered (Figure 5).

Support reinforcement of the minable drift by elements (Figure 5):

- installation of beam cuts;
- installation of wood props no less than 22 cm in diameter along the axis of the roadway up to the wooden cut of the beam, 2 pieces between each frame of the arch support;
- installation of wood props no less than 22 cm in diameter under a hold-down beam from SIP 22 (specially interchangeable profile 22 kg / m).

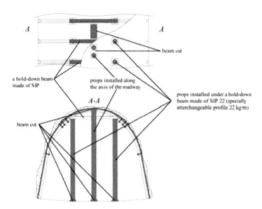

Figure 5. Diagram of support reinforcement of minable roadway of lev. 400.

Control of geometric parameters of the EIR 400 m was carried out at the observation stations.

By results of surveying observations at the EIR 400 m it has been established that the total vertical convergence in the eastern intermediate roadway of lev. 400 m did not exceed 350 mm. The experiment of overworking of the roadway using reinforcement of the support confirms possibility of its further operation.

5 CONCLUSIONS

Planning of mining operations with taking into account location of main roadways in discharged zones and short-time influence of second working with reinforcement of a support makes it possible to preserve roadways in good working order and substantially reduce losses of coal. Effective realization of this trend is possible under the condition of utilization of mine working supports and means of their reinforcement with high bearing capacity.

REFERENCES

SOU 10.1.00185790.011:2007. *Gateroads on flat seams. Support, methods and facilities.* Kyiv: Ministry of Coal Industry. Ukraine: 113.
KD 12.01.01.201-98. 1998. *Attitude, protection and maintenance of mine workings at working off of coal seams on mines.* Kyiv: Ministry of Coal Industry: 150.
KD 12.07.301-96. 1997. *Scheming of discharged areas and high rock pressure zones for conditions of deep mining.* Donetsk: UkrNDMI NAN Ukraine: 46.
Hudin, Yu. L., Ustinov, M.I. & Braitsev, A.V. et al. 1983. *Pilarless extraction of seams.* Moscow: Nedra: 280.
Babenko, E.V. 2008. *Features of evolution of zones of the high rock pressure at development of a second working.* Transactions of UkrNDMI NAN Ukraine, 2: 211-226.
Zborschik, M.P. & Nazimko, V.V. 1991. *Protection of mine workings of deep mines in discharged areas.* Kyiv: Tehnika.
Khalymendyk, Yu.M. & Begichev, S.V. 2002. *About full pilarless extraction of coal resources.* International Mining Forum. Krakow: 333-340.
Jacobi, O. 1981. *Praxis der Gebirgsbeherrschung. Verlag Gluckauf.*

The study of ecological state of waste disposal areas of energy and mining companies

A. Gorova, A. Pavlychenko & O. Borysovs'ka
National Mining University, Dnipropetrovs'k, Ukraine

ABSTRACT: Features of the impact on the state of the environment of fuel and energy sector waste as well as mining industry waste are considered. The necessity of developing environmental monitoring system of industrial waste disposal areas is proved.

Power objects of the cities are the main source of supply of heat and electricity for population. At the same time, they cause significant damage to the environment and their negative impact takes place during the extraction and use of fuels as well as processes of transformation and transmission of energy. Thermal power plants (TPPs) for solid fuels are especially dangerous. In the process of their activity varied products of combustion in solid, liquid and gaseous states are emitted into the atmosphere. In addition, the sources of contamination of the environment are fugitive emissions from the storage of coal, fly ash and slag dumps etc. (Belyavskiy 2002).

Development of coal industry has particular importance for Ukraine, because in the XXI century coal as a fuel resource becomes the principal source of energy, and in the future – the main source of energy. Development of energy sector, therefore will improve the country's energy supply (Shildovskiy 2001).

At the same time, long-term development of powerful mining and fuel and energy complex (FEC) has led to high levels of anthropogenic impact in industrialized regions of Ukraine. As a result of these economic sectors activity significant amounts of waste were formed, dangerous geological processes were activated, levels of pollution of ecosystems components and also morbidity of local population were increased (Lubchik 2002 & Gorova 2009).

Production and use of fuel and energy resources leads to the formation of large amounts of waste that negatively affects the condition of ecosystems components and has potentially toxic and mutagenic properties (Gorova 2008). It should be noted that the negative consequences of their influence lie not only in the transformation of the components of ecological systems, but also in general toxic effects on living systems, which leads to inhibition of viability

and premature death of individuals. In addition, the disposal of waste in the environment leads to the alienation of large areas and to the creation of risk of technogenic emergency situations (Koskov 1999 & Gorova 2012).

Currently existing system of assessing the quality of the environment at the local level differs little effectiveness as it is based on sanitary regulations, which do not include many important criteria, namely:

1. The actual toxicity and hygienic hazard of combinations of chemical elements both among themselves and other inorganic and organic compounds.

2. The difference between forms of the harmful elements in the environmental objects and their forms set out under the definition of the norm.

3. The effect of deposition and long-term effect of low concentrations.

Experience of monitoring studies of waste disposal areas indicates the need for additional use of statistical parameters of the distribution of chemical elements, which is calculated on the results of regional studies. Otherwise, the assessment of the ecological state of area that is under intense man-made pressure is one-sided and does not consider the effect of aggregate contiguous objects.

Analysis of the monitoring studies in the areas of stockpiles of energy sector companies waste as well as mining industry waste indicates that existing there network of observation posts is planted without taking into account the possibility of regional natural and mining systems formation, and regardless the breadth of pollutants range. That is why necessary reliable information cannot be obtained. Thus, the magnitude and specificity of the negative impact of waste on all components of the environment requires special studies.

Application of modern physical, chemical and

biological methods, methods of cartographic modeling and scientific forecasting and statistical data using computer software allows determining the ecological status of environmental objects. The use of biological methods of environmental quality evaluation can identify and assess the response of biological systems to complex effects of man-made factors and can determine the toxic and mutagenic activity of environmental objects. The high sensitivity to minimum concentrations of toxic substances and very quick and specific response to their presence in the components of ecosystems is the basis for introduction of cytogenetic methods in environmental monitoring system. Setting the sustainability of different biological systems to complex action of toxicants contained in the waste, will allow to develop effective conservation and rehabilitation measures aimed at reducing toxic and mutagenic activity of the environment. Creation of scientific basis of monitoring research involves identifying regularities of anthropogenic pollution of the environment in order to develop principles of environmental monitoring system and to supplement it with highly sensitive bioindication methods (Gorova 2009, 2012). This approach is innovative in the field of theoretical foundations of environmental monitoring in the impact zone of energy sector waste and mining industry waste.

The aim of the work is to study the ecological state of the waste disposal areas of energy and mining industry enterprises to develop recommendations to reduce their negative effects on the environment and public health.

To achieve this goal following tasks have been met:
– the features of waste formation, accumulation and disposal depending on the technology of mining were investigated;
– volumes of slags from coal-fired thermal power plants were defined;
– existing systems for environmental monitoring of enterprises of fuel and energy complex and mining industries were analyzed;
– the features of the waste impact on the state of environment components and public health were analyzed;
– ecological status of environmental objects in areas of waste disposal, using physical, chemical and bioindication methods was assessed;
– health statistics of the population living in areas adjacent to hazardous environmental enterprises and industrial waste disposal areas were analyzed;
– the levels of environmental and ecological and genetic risks for environmental objects on the waste disposal territories were defined.

As a result of many years of research ecological state of the environment in areas adjacent to the placement of energy sector and mining industry waste have been identified. Studies were conducted on the territories next to the waste dumps of coal mines as well as fly ash and slag dumps of thermal power plants.

For example, in 2012, study of the ecological state of the environment in the zone of the Prydniprovska thermal power plant in the Lyubimovka village of Dnepropetrovsk region was conducted. Ecological conditions of air and soil were estimated at 34 monitoring points that were fixed at different distances from the Prydniprovs'ka TPP. Environmental state of air quality was evaluated by "Pollen sterility of plants-indicators" test. Toxicity and mutagenicity of soil was determined by tests "mitotic index" and "frequency of aberrant chromosomes" in meristematic cells of phyto-indicators. The obtained data about the level of toxicity and mutagenicity of the soil and the air allowed to build a series of environmental maps.

Solving the problem of the waste impact on the environment objects is possible through the introduction of comprehensive system of environmental monitoring of areas of waste disposal. This system will allow efficient and timely assessing the real environmental threat of man-made objects and components of waste, to predict changes in the environment and social and ecological systems and to suggest ways of ecological and safe operation of industrial enterprises and disposal of hazardous industrial waste.

As a result of conducted studies the impact regularities of power plants waste and mining industry at the level of damage of biological systems and changes in population health status were established. The fixed regularities should be used for the development of scientific bases of environmental monitoring of technogenic-hazardous areas of waste disposal and for implementation of biological assessment methods to monitoring system.

Taking into consideration that long-term development of the mining and energy industries leads to high levels of anthropogenic impact on the environment, the accumulation of a large amount of industrial waste (including radioactive), there is a need for the development and introduction of national target programs of industries greening.

Greening program must include:
– the system of comprehensive evaluation of environmental quality and determination of levels of environmental hazards to humans and biota;
– recommendations for improving the technology of mining to reduce the number of industrial wastes;
– recommendations for processing and extracting useful components from waste that can be used in

different sectors of the economy;

– complex of environmental measures designed to improve the state of the environment in areas affected by waste of energy sector companies and mining industry.

Thus, the priority of environmental safety requirements, mandatory observance of environmental standards and regulations to protect the environment and natural resources should be the strategic directions of energy resources production and use.

REFERENCES

Belyavskiy, G.A., Varlamov, G.B. & Getman, V.V. 2002. *Assessment of the impact of energy facilities on the environment.* Kharkov: KNUME: 359.

Shildovskiy, A.K. & Kovalko, M.L. 2001. *Fuel and Energy Complex of Ukraine on the eve of the third millennium.* Kyiv: Ukrainian encyclopedic knowledge: 400.

Lubchik, G.L. & Varlamov, G.B. 2002. Resource and environmental issues of global and regional energy consumption. Energy and Electrification. 9: 35-47.

Gorova, A. & Pavlychenko, A. 2009. *Integral assessment of the socio-ecological state of mining regions of Ukraine.* Mining Journal. 5: 49-52.

Gorova, A., Kotelevets, O. & Pavlychenko, A. 2008. *Cytogenetic evaluation of environmental hazard of coal mining industry waste.* Scientific Bulletin of Chernivtsi University: Collected Essays. Biology. Chernivtsi: Ruta: 417: 194-199.

Koskov, I.G., Dokukin, O.S. & Kononenko, N.A. 1999. *Conceptual Foundations of Ecological Security in the Regions of Mines Closure.* The Coal of Ukraine, 2: 15-18.

Gorova, A., Pavlychenko, A. & Kulyna, S. 2012. *Ecological problems of post-industrial mining areas.* Geomechanical processes during underground mining. Leiden, The Netherlands : CRC Press / Balkema: 35-40.

Gorova, A.I., Ryzhenko, S.A. & Skvortsova, T.V. 2007. *Guidelines 2.2.12-141-2007. Survey and Zoning by the Degree of Influence of Anthropogenic Factors on the State of Environmental Objects with the Application of Integrated Cytogenetic Assessment Methods.* Kyiv: Polimed: 35.

The main technical solutions in rational excavation of minerals in open-pit mining

M. Chetverik, E. Bubnova & E. Babiy
M.S. Polyakov Institute of Geotechnical Mechanics, Dnipropetrovs'k, Ukraine

ABSTRACT: The main alternative technologies, technical and technological solutions in rational excavation of minerals were analyzed. On the example of Petrovsky quarry contact areas by type of host rocks were grounded, ferruginous quartzite losses according to the planned mode of mining operations for the period 2013-2025 years were calculated, the amount of rock to be lumpy magnetic mechanized sorting was defined. The impact of surface mining on rational land use and the alteration of the hydrogeological environment and ecological condition of the mining region was analyzed.

1 INTRODUCTION

Ukraine has a high resource potential, but in spite of this there is a question of rational use of mineral resources and the maintenance of natural components in recent times, including land cover, mineral resources, surface and ground water. Nature management issues repeatedly raises by scientists of different disciplines, international and national programs are devising, but their practical result is extremely small since as a rule all developments are aimed at solving specific problems in certain enterprises. Therefore, this problem remains acute, and its solution is relevant.

The object of this study is the mining industry that realize surface mining of mineral resources, as it is surface mining that has irrational usage of the resources in our country. In view of the fact that Ukraine can not refuse surface mining, the efforts of all scientists and government programs should be aimed at the development and implementation of effective resource-saving technologies.

2 RESEARCH AND PUBLICATIONS ANALYSIS

In the mining industry one of the areas of resource-saving management is the land and subsoil use, which is reduced to the delicate cycle of mining operations, decrease of agricultural land confiscation for slag-heap and tailings dump, and also to minimize losses and dilution of mineral resources. There are different ways of rational extraction of minerals and improving the quality of the lode rock:

1. Systems development parameters changing: reducing the height of the step (bench), the selection under-benches.

2. Mining planning: developing of new and improving of existing technologies of extraction and processing of minerals, blasting to the conservation of the geological environment, selective mining, mining regimes planning, career characteristic, developing of optimization of the formation of quarry traffic criteria, ore averaging before enrichment and metallurgical processing during production, on warehouses, in bunkers or continuous flows, define the boundaries of quarries and evaluate economically substantiated cut-off grade of useful component, use pre-enrichment at the processing plant or in a quarry.

3. Change of mining equipment parameters: perform a selection of excavating equipment, use of power shovels with a small bucket capacity (up to 4-5 m^3), use of rotating buckets.

The analysis of the existing methods to improve the quality of ore in breakage face showed that, despite of modern scientific and technical development, the most common ways in manufacture are selective mining and averaging of minerals. However, the selective extraction of ore was effectively implemented in 1980-1990 years, when excavators with bucket capacity 3-5 m^3 sometimes 8 m^3 (with dump truck capacity 45-75 tone) were used. In the 2000s, in the faces excavators with bucket capacity 8-12 m^3 (in accordance with the increase of dump truck capacity up to 110-150 tone) are used and it is planned to use excavators with a capacity of 20 m^3. This trend of increasing excavating bucket equipment capacity leads to heightened ore mass impoverishment or large losses (Babii 2011).

Thus, all above-listed methods of rational extraction of mineral resources do not provide the desired results, as each of them operates on single tasks of technical or technological, and for production it is necessary to dispose the problem of rational nature management fully through the development of efficient technologies and losses minimization.

As it was mentioned, one of the areas of resource-saving is the protection of the environment. Numerous studies, that are impossible to list, take place in order to reduce the anthropogenic impact on the environment. The main purpose of this work is the rational use and protection of natural objects (land resources, free air, water bodies). But the effectiveness of development is poor, as it marked the annual environmental degradation of anthropogenic laden regions.

Problem definition. The purpose of this paper is to analyze the causes of irrational use of resources in surface mining and justification of resource-saving areas.

3 PRESENTATION OF THE MATERIAL AND RESULTS

The problem of rational mining advised to dispose through the use of technology of ore preconcentration in the quarry to reduce losses and rock impoverishment. According to this technology in the quarry or close to it running lumpy sorting of individual cargo traffic (Babiy 2011). With the help of experts of IGTM NAS of Ukraine previous studies was found that the choice of technological ore preconcentration, including staging of crushing rock, it is necessary to consider the type and value of the ore rock. Conditioning ore with a high content of useful component it is rational to subject to all three stages of crushing and only after this preliminary enrich through dry magnetic separation. In consideration of the fine crushing process dustiness and the difficulty of transporting the fine fraction, it is better to process the ore in the processing plant. Ore off-balance stock subquality reserves or ore impoverishment submagnetic inclusions of contact zones to be ore-picking after secondary crushing in a open-cast or in close proximity of it. After primary crushing expose to ore-picking rationally ore impoverishment draw rock and overburden containing magnetite, therethrough increasing the productivity of ore open-cast, reduce the amount of overburden, reduce the loss of minerals.

Therefore, it is efficiently to direct traffic flows on the complex lump ore-picking in open-cast after a mechanical primary crushing:

1. Overburden of the contact zones "ore – stripping" to reduce mineral resources loss.

2. Ore mass with strongly magnetic properties at contact zones "ore – overburden" development with barren rocks to reduce the impoverishment.

3. Ore mass at mining: (a) barren layers up to 10 m, which according to (The branch instruction... 1975) have to take in minerals, (b) complex structure of mining faces, where the alternation of ore layers and barren with more than two layers and (c) during nip of ore-shoot into the enclosing rock for reduce impoverishment;

4. Overburden at mining the ferruginous horizons that are not included in the productive strata, for the rational management of mineral resources, minimize minerals loss and waste technologies creation. For instance, in the iron ore stata of Pervomaisky deposite allocate seven ferruginous and seven shale horizons. While productive series are only silicate-magnetite quartzites of the fifth and sixth iron horizons. And the rest of ferruginous horizons, which have non-permanent thickness or body condition of useful components not included in the balance reserves.

Preliminary enrichment of stripping soils to extract the magnetic component allow first reduce the loss of balance of mineral reserves. So for example, analyzed the Petrovsky ore-picking contact zones "ore – enclosing rock", ferruginous quartzite losses are calculated according to the planned mode of mining for the period 2013-2025 years, defined the amount of rock subject to be dry magnetic separation.

Losses were determined according to the classification of contact zones (Babiy 2011) by a factor – the type of rocks adjacent to the ore body. The adjoining rocks are divided into: oxidized quartzites, subquality reserves ore, barren rock and overburden containing magnetite. During the analyzed period of Petrovsky open-cast, contact zones of magnetite quartzite represented by pegmatite, gneisses and magnetite-silicate quartzites. Magnetite-silicate iron quartzites with iron cut-off grade are not good ore. Whereas pegmatites and gneisses are barren rocks, which have crystals of iron that difficult to remove in chemical compounds (total iron content of 1-3%). To justify the cost-effectiveness of a package of lumpy ore picking on Petrovsky open-cast were held bench testing of the samples of rock in "Research and Technology Center of magnetic separation "Magnis Ltd". The obtained test results showed that from the overburden ore / rock (the content of magnetic iron 7.5-14%) magnetic product output is 62.7-72.5% magnetic iron content of 25-35% and 27-37% of dry tailings with an iron content of less than 2% of the magnetic iron. Consequently, if on the lumpy mechanized ore picking complex process volumes of the contact zone, it is possible to minimize the loss of minerals. In addition, when a good performance indicator of the complex work, it is rational to process the ore mass volumes from the surface area for dilution (impoverishment) reduce.

Calculating the number of ferruginous quartzite losses was made in accordance with the "Branch instructions... " (The branch instruction... 1975). We took into account the direction of mining contacts ore deposits, which are determined by the direction of the front advancing mining for each of the calendar year of mining plan according to project "Developing and opening deep quarry horizons #3 (Third turn of quarry deepening #3)"

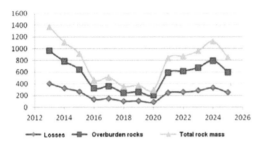

Figure 1. Distribution of the rock mass in kt, that are subject to lumpy ore-picking by years.

Volumes variability explained to the selected scheme development of working horizons, mode of ore blocks mining and intermittency of the ore bed with adjacent strata.

Precious resource of our country is fertile land, much of which is withdrawn from the crop rotation for objects of the mining industry. Used technologies of mining and mineral processing are aimed at a more complete extraction of the useful component with the lowest material cost and the preservation of natural resources such as land and water, is taken into account the last thing (last of all).

Consequently to open-pit mining an open excavation with large area and depth formed, the fertile layer is removed and stored for decades what reduces its value. In addition, as a result of dehydration of the massif is a violation of the geological strata, including land resources and surface subsidense (Chetverik & Bubnova 2004). Such violations are not taken into account until now. Along with direct violation of the earths surface land withdrawal for location of anthropogenic objects (waste rock dumps, tailing and slime storages, gathering ponds, clamps with fertile layer) takes place.

The influence of anthropogenic environment on soil conditions by the following:
– exclusion from economic turnover of large areas of land under waste products;
– destruction or degradation of land through drifts

of dust from the dumps and tailings surfaces;
– environmental pollution (soil, surface and ground water, air) by the heavy metals and salts in concentrations often exceed the permissible limits.

The issue of reducing the impact of mining on land cover can be divided into two areas: the exclusion of further exposure of the operations and the issue of already disturbed land.

Because the causes of disturbed land are equal for all enterprises producing minerals by surface mining, then the directions of improving of rational land use are the same.

As an effective solution to further exclude of the impact of mining on land resources is the following: (a) storage of tailings and overburden in the area of coffins, and (b) the development of waste dumps and tailings with the extraction of useful components, which will provide further capacity for waste disposal without additional set-aside and increase resource intensiveness by involving to the development of waste, and (c) the transition to open-underground mining with laying of underground worked out area by overburden.

Problem solving of the existing disturbed land can be divided into the following options:
– use of disturbed lands for commercial purposes without land reclamation (waste storage, making of enterprises or recreation ares);
– conducting land reclamation in areas (biological, forestry, etc.);
– restoration of the broken massif properties and soil fertility to a natural state (Chetverik & Voron 2011; Bubnova 2011).

No less important resource, either directly or indirectly affected by mining is surface and groundwater.

Education and the presence within the natural environment disordered and anthropogenic environment has a negative impact on the natural hydrological conditions, which is manifested in the change of recharge area, movement and discharge of groundwater, formation and deformation of sufficiently large cone of influence. Violation of hydrological regime of territories, as opposed to violations of the rock mass massif and the land cover is a more dynamic factor that can dramatically increase the effects of other environmental factors (Chetverik & Bubnova 2004). As a result, disturbed land are generated implicitly (impounded, waterlogged, additional subsidense, change the properties of fertile rocks). This is due to the following reasons.

1 The main flows of groundwater and surface water intersect in the open-pit mining, which leads to disruption of hydrological regime (drain and draining aquifers, prevent the natural surface and underground run-off), and as a consequence, landslides and on ledges and pit (slag-heap).

2. Internal slag-heaps on existing technology forming have no hydraulic connection with the natural environment, and stacked rocks are detected. Thus, such a man-made environment, lying surrounded by natural geological environment is a barrier that prevents the movement of groundwater and surface water, which can lead to landslides and raise the water Table.

3. External slag-heaps affect the movement of surface water and groundwater, as often have in the gullies or ravines, i.e. in places where the discharge is usually ground and surface water, leading to flooding and landslides.

4. Sludge storage pits, tailing dumps, slime storages, etc. in most cases have in the gullies, ravines, etc., that at the very beginning of the operation leads to significant violations of the surface water and groundwater.

Currently, regulations of hydrodynamic groundwater regime in the areas of mining are laying different types of drainage systems to drain water from the producing fields. This technology allows a considerable distance to lower the water table and the aquifer water first through the development of depression cone radius of 1 km. However, the technology not only provides for the restoration of the natural state of the geological environment, but also introduces a significant negative impact on it, disrupting the natural balance and changing the hydrodynamic properties of the rocks drained array.

Therefore, it is proposed the concept of recovery of the hydrodynamic regime of underground waters in the mining pit areas, which is based on the following.

1. In the open cast method of mining is recommended as the main technical solutions for the conservation properties of the rocks in the environment adjacent to the breach, and to maintain the water balance of the territory to intercept groundwater flow before they reach the borders of the career and devote their corresponding aquifers outside of the career contour. The parameters of this technology depend on the career depth, quantity, capacity, and depth of the tapping of underground reservoir, the location of open-pit field relative to the direction of groundwater artery.

2. To preserve hydrologic regime in a man-made geological environment is proposed to intercept sur-face runoff drainage system with a branch in its natural place of discharge, which will not only preserve the water balance of the territory, but also will ensure safe use of man-made object.

3. During reclamation of man-made geological environment is proposed to restore the upper aquifers (Bubnova 2011).

4 CONCLUSIONS

Thus, it's necessary to develop and use resource saving technology for the rational use of natural resources. The technology of preliminary concentration of ore in quarries allows reducing losses and rock attenuation, removing balance reserves and increasing the production capacity for the ore pit.

Future research should be directed to the study of natural geological environment disturbances and man-made features of the formation, the laws of the formation of their new properties, as well as the scientific rationale and development of technical solutions that focused on the restoration of man-made and broken geological environment properties that correspond to the natural.

REFERENCES

Babiy, E.V. 2011. *Technology of preliminary concentration of iron ore in deep pits*. Kyiv: Naukova Dumka: 184.

Bubnova, E.A. 2011. *Restoration of properties damaged of land by mining*. Dnipropetrovs'k: IGTM NASU. Interdepartmental collection of scientific papers, 94: 17-23.

The branch instruction by definition, accounting and valuation losses of iron ore in the development, manganese and chromite deposits in the enterprises of the Ministry of Ferrous Metallurgy of the USSR. 1975. Belgorod: VIOGEM: 68.

Chetverik, M.S. & Voron O.A. 2011. *Patent for Utility Model 64879 Ukraine*, IPC class. A 01 C 7/00, E 21 Method of reclaiming lands affected by open cast mining to create potentially topsoil (UA).; Appl. 05.04.2011, publ. 25.11.2011, Bull. # 22.

Chetverik, M.S. & Bubnova, E.A. 2004. *The settling of the earth's surface during dewatering and sinking areas of mining regions*. Kryvyi Rig: KTU. Scientific and technical collection. Vol. 86: 31-36.

Method of calculation of the minimum pressure of hydro breaking of the coal layer

D. Vasilyev & Y. Polyakov
M.S. Polyakov Institute of Geotechnical Mechanics, Dnipropetrovs'k, Ukraine

A. Potapenko
POJSC "Krasnodonugol", Krasnodon, Ukraine

ABSTRACT: In the following article describes a method for calculating of the minimum pressure of hydro breaking of the coal seam. The method is based on the theory of cracks. It is assumed that the development gets an extreme crack in the top of which shear stress reach the maximum value. At determining the maximum value of shear stress we taken into account the hydraulic pressure of liquid, that operates on the inside of crack face, the vertical rock pressure effect, frictional force, the horizontal stress of backwater, the damping contact shear stresses from contact formation zones. Derive a set of equations to determine the minimum hydraulic pressure of hydro breaking of coal layer.

1 INTRODUCTION

By the Safety rules (Rules of mining works… 2005) to deal with sudden outbursts recommended static pressurization of liquid on coal layer two local methods: hydro spin cycle and hydro breaking of their bottom hole formation zone. The difference between these methods is mainly in the size of the filter section. The most widely used method is hydro breaking through of boreholes with length of 6-8 m at a depth of sealing in 4-6 m. Therefore there is a need to justify the minimum value of pumping liquid pressure.

2 CALCULATION OF THE MINIMUM PRESSURE

Imagine that there are randomly located cracks on inside of coal layer, but the development gets an extreme crack in the top of which Coulomb effective shear stress reach the maximum value according to expression.

$$\tau_e = \tau_{r\beta} - \mu\sigma_\alpha , \qquad (1)$$

where $\tau_{r\beta}$, σ_α – external shear and normal stresses; μ – coefficient of internal friction.

The layer is subjected to vertical rock pressure effect. At the contact plane between layer and country a frictional force is exerted that is directed against deformation. The contact shear stresses create lat-eral backup pressure to vertical ground pressure σ_y and are damped with distance from the contact zone according to the law.

$$\tau_{xy} = f_m\sigma_y\left(1 - \frac{2Y}{h}\right), \qquad (2)$$

where f_m – coefficient of contact friction, y – ordinate of tip crack, h – seam thickness

The stress at the tip of crack can be determined by known formulas from the theory of cracks (Panasyuk 1968) (Figure 1).

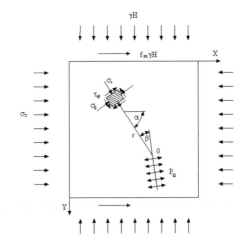

Figure 1. Scheme of loads at the tip of crack at the hydro breaking of coal layer.

The stress condition is accepted to represent according to (Rules of mining works... 2005) as the sum of the stress state in the solid part (no cracks) of infinite body that subjected to external stresses deformation and stress state with a crack ($-l \leq x < l$).

Taking into account the theory of friction, (1) can be written as

$$\tau_e = \sqrt{\frac{\pi E \gamma_n}{1-\upsilon^2}}\left(\sqrt{1+\mu^2}+\mu\right) = \left[\left(\frac{\sigma_y - \sigma_n}{2}\right)sin\,2\alpha + f_m\sigma_y\left(1-\frac{2y}{h}\right)cos\,2\alpha - \right.$$
$$\left. -\mu\left(\frac{\sigma_y+\sigma_n}{2}+\frac{\sigma_y-\sigma_n}{2}cos\,2\alpha - f_m\sigma_y\left(1-\frac{2y}{h}\right)cos\,2\alpha\right)\right]\sqrt{l}\,, \qquad (3)$$

where E – Young's modulus; γ_n – specific surface energy; υ – Poisson ratio; σ_n – horizontal stress of backwater; α – crack angle with respect to the plane of the crack; l – fracture half-length.

Now we must take into account the hydraulic pressure of liquid P, that operates on the inside of crack face. We introduce it in the negative part of

(3), as it decreases effect of the normal stress, that compresses a crack face.

The minimum hydraulic pressure of hydro breaking of seam with subject to damping contact shear stresses from contact formation zones is defined by the equation

$$P_e = \frac{1}{\mu}\sqrt{\frac{\pi E \gamma_n}{2(1-\upsilon^2)l}}\left(\sqrt{1+\mu^2}+\mu\right)+\frac{1}{\mu}\left[\left(\frac{\sigma_y-\sigma_n}{2}\right)sin\,2\alpha + f_m\sigma_y\left(1-\frac{2y}{h}\right)cos\,2\alpha\right]-$$
$$-\left[\frac{\sigma_y+\sigma_n}{2}+\frac{\sigma_y-\sigma_n}{2}cos\,2\alpha - f_m\sigma_y\left(1-\frac{2y}{h}\right)sin\,2\alpha\right]. \qquad (4)$$

Now we need to determine the minimum value of hydraulic pressure of hydro separation, and for this purpose we should differentiate (4) and set it equal to zero. Let's find the slope angle of extreme flat.

$$\alpha = -\frac{1}{2}arctg\left(\frac{-2\mu f_m\sigma_y\left(\frac{1-2Y}{h}\right)+\left(\sigma_y-\sigma_n\right)}{2f_m\sigma_y\left(\frac{1-2Y}{h}\right)+\mu\left(\sigma_y-\sigma_n\right)}\right). \quad (5)$$

The crack angle with respect to OX axis is defined by equation

$$\alpha_1 = \frac{\pi}{2}+\alpha\,. \qquad (6)$$

Based on this geometrical crack pattern, and equations (4-6) it is determined about the crack tip stress away from the rock face – the depth of sealing.

Now we must determine the horizontal stress σ_n, arising in consequence of the action of shear stresses from the contact friction in the direction from the bottom into the depth of massif. Use an expression received from the algebraic equation limit state (Vasilyev 2011)

$$\sigma_x = \frac{2(k+\mu\sigma_y)}{cos\,\rho}\left(sin\,\rho-\sqrt{1-b^2}\right)+\sigma_y\,, \qquad (7)$$

where $\rho = arctg\mu$ – angle of internal friction; k – coal shear strength; σ_y – normal stress at the crack tip.

We write this equation with the exponential increases of normal stresses σ_y and damped of shear stresses in the plane of the crack.

$$\sigma_x = \frac{2\left(k+\mu\sigma_0\,exp\left(\frac{f_m(1-2Y/h)(lg-x)}{h}\right)\right)}{cos\,\rho}\cdot\left(sin\,\rho-\sqrt{1-b^2}\right)+\sigma_{y0}\,exp\left(\frac{f_m(1-2Y/h)(lg-x)}{h}\right), \qquad (8)$$

where σ_{y0} – normal stresses at the top angular region of layer at the place of his outcrop; lg – the depth of sealing; x – current abscissa of the crack tip with respect to depth of sealing.

$$b = \frac{f_m\left(1-\dfrac{2y}{h}\right)\cdot\sigma_y}{k+\mu\sigma_y}.$$

Denote the expression $f_m\left(1-\dfrac{2y}{h}\right)$ through f.

To determine the backwater stress σ_n we should find the integral value of the stress σ_x (8) with respect to x

$$\sigma_n = \int_0^{x_1}\sigma_x d\left(l_g - x\right).\qquad(9)$$

Neglecting the value of k compared with the normal stress, the influence variation of the parameter b can be ignored. Then (9) takes the form

$$\sigma_n = \frac{2\left(\sin\rho - \sqrt{1-b^2}\right)}{\cos\rho}\cdot\left(k(l_2 - x) + \frac{\mu\cdot h\cdot\sigma_{y0}}{f}\left(\exp\left(\frac{f(lg-x)}{h}\right)-1\right)\right) + \frac{h}{f}\sigma_{y0}\left(\exp\left(\frac{f(lg-x)}{h}\right)-1\right).\qquad(10)$$

Thus, we have a set of equations to determine the minimum hydraulic pressure of hydro breaking of layer with respect to (4).

3 CONCLUSIONS

1. The equations of equilibrium forces that act on the body element, which is located at the crack tip, are composed and from which appears the opportunity to define minimum pressure of hydro breaking of coal seam.

2. The method is based on the theory of cracks, subject to internal and contact (external friction).

REFERENCES

Rules of mining works on the seams prone to gasdynamic phenomena. 2005. Standard of Ukraine Ministry of Coal Industry. Kyiv: Ukraine Ministry of Coal Industry: 224.

Panasyuk, V.V. 1968. *Limit equilibrium of brittle bodies with cracks.* Kyiv: Scientific Thought: 245.

Vasilyev, D.L 2011. *Regularities of the formation of horizontal normal stress in the rock mass. Geotechnical mechanics: collection of scientific works.* Dnipropetrovs'k: IGTM of Ukraine's NAS. Vol. 29: 17-21.

Mining of Mineral Deposits – Pivnyak, Bondarenko, Kovalevs'ka & Illiashov (eds)
© *2013 Taylor & Francis Group, London, ISBN: 978-1-138-00108-4*

Main directions and objectives of diversification processes in coal regions of Donbass

A. Petenko
Donets'k State University of Management, Donets'k, Ukraine

E. Nikolaenko
National Mining University, Dnipropetrovs'k, Ukraine

ABSTRACT: Restructing of Ukrainian mine pool has become a long-term task. That is why there is a need of coal enterprises closure mechanism improvement, which means improvement of techniques of working closure, building and surface construction removal, etc. On the other hand, new parallel technological and ecological solutions of overcoming of mine waters emission on surface, its overflow from closed mine to an operating one and ground impoundment are needed.

1 INTRODUCTION

It is proved that the value of demand is not absolute constraint in the problem of unprofitable mine liquidation because they can be closed also in terms of stable or even growing demand for coal (Salli 2009). The base for the problem solving with such statement is concept of retired capacity compensation and also possibility of mining production diversification, based on anthropogenic wastes recycling and closing mine management. Unfortunately, current economic mechanism do not provide real economic conditions for development and wide application of ecological innovations in production of finished coal products and diversification products of coal regions ecological situation improvement.

The leading hand in organization an management of ecological and innovative process should belong to the state (supporting of creation of ecological innovative infrastructure and organizational, juridical and informational base of ecological innovations) (Burkinskii 1999 & Veklich 2003). That is why state stimulation has to include both positive motivation stimuli, which promote development and implementation of ecological innovations and negative motivation ones, main task of which is reduction and closure of eco-dangerous enterprises (Figure 1). Domestic experience of coal industry restructing can be referred to negative ones: Attempts to close unprofitable coal mines without implementation of enough working places have created a situation of social stress and worsened the environment.

Meanwhile, ecological situation of Donbass is becoming more and more complicated. Liquidation of coal mines was and is still conducted without taking into account predictive estimates of ecological consequences and with often violations of environmental laws in terms of leftover principle financing. This leads to significant complication of situation in coal-mining regions turning them into unstable territories.

2 MAIN PART

Analysis of foreign experience of coal production diversification and based on it creation of new working places allow to make the following conclusions:

– social consequences of restructing can be softened as a result of well thought-out and efficiently-realized programs of available miners employment by means of new working places and production diversification creation;

– in order to realize these programs, one needs to create special budget support efficiently usage diversification companies;

– processes of further growth of coal mining are directly related with the development of "non-coal" sectors, which provides steadiness of enterprises and improves ecological situation in coal regions.

It is known that industrial waste recycling including mine tips reaches up to 70-80% in developed countries. In Ukraine and CIS countries this indicator is only 10-12%. Waste bank and tip rocks contain about 2.5% of sulfur and 3-20% of coal, as a result, they ignite spontaneously and burn for 7-12 years poisoning surface air of surrounding area with combustion products. In Donetsk coalfield basin there are 1185 acting and "retired" tips and waste banks, 400 of which burn every year and vent more than 500 th. tonnes of harmful gaseous substance, and rainwater, after getting on these tips, dissolves a fair amount of dangerous chemical elements and fill underground waters with them. Every year there are more than 35 th. tonnes of soil deflated from 1 hectare of medium-size waste bank and the biggest part of water-soluble salt is also washed out (Bardas 2010).

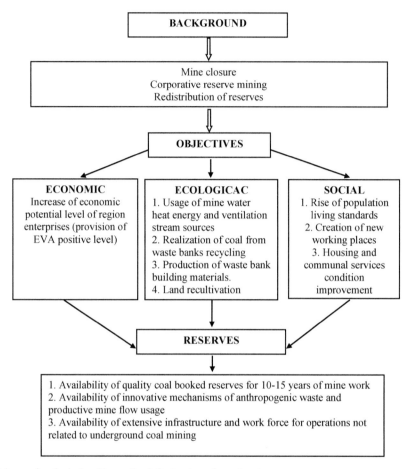

Figure 1. Scheme of ecological and innovative infrastructure of a coal region.

Commonly used (90% of mines) wet shutdown – is just flooding of the mine with the water which is constantly pumped out during its work. Because of its usage there is a number of ecogeological problems appeared. It is caused by working flooding which increases anthropogenic load on geological environment and hydrosphere. Over the entire history of the existence of Donbass as a mining region, about 1 thousand mines were built, that is why there is 3 hydraulically connected closed mines to 1 operating one for today. According to foreign experience, there is no demineralization of coal mine water in any coal extracting country of the world. One needs not less that 16 bln. hrn. of capital investments to realize countrywide mine water desalina-tion with 3.75 bln. UAH. annual exploitation costs. Mine water demineralization in state coal industry scale will lead not only to necessity of complicated engineering problems solving but also grand economic problem, which, obviously, can not be solved in the nearest future.

There is a typical situation relatively waste heap characteristics and useful components of Torezskoe-Stezhnyanskiy region mines. To avoid new sharp ecological and social conflicts in mining towns, it is immediately needed to start working on social and economic substantiation of directions and specific options of production diversification for at least a decade.

Table 1. Characteristic and useful components of mine waste heap in Torezskoe-Stezhnyanskiy region.

Mines	Height, m	Volume, mil. m³	Period of creation years	Coal content, %	Probable volume of hard coal extraction, th. tonnes	Probable volume of brick production, mil. units	Road building raw material, th. tonnes
"Lagutina"	71	4.06	50	21.3	864.7	262.9	6.89
"Miusskaya"	70	11.82	92	22	2600.4	116.5	19.9
"Kiseleva"	59	5.15	34	19	978.5	89.4	8.8
"Severnaya"	70	3.97	77	19.8	786.1	74.0	6.7
"Udarnik"	66	3.28	41	19.5	639.6	63.2	5.6
"Snezhnyanskaya"	61	2.86	77	23	657.8	89.5	4.8
"Vostok"	60	4.00	19	21	840.0	47.5	6.8
"Removskaya"	61	2.10	86	18.7	392.7	115.7	3.6
"Progress"	45	5.12	29	19.2	983.0	73.2	8.7
"Chervona Zirka"	71	3.29	86	21.8	717.2	96.7	5.5
"3-bis"	47	4.35	49	22.1	961.4	24.5	7.3
"Obyedenennaya"	70	1.11	63	23.4	259.7	33.0	1.8
"Yablunevka"	65	1.47	48	20.3	298.4	54.4	2.5
"Zarya"	65	2.43	45	20.6	500.6	29.2	4.1
"Lesnaya"	47	1.31	86	21.3	279.0	21.8	2.2
"Rassipnyanskaya"	50	0.97	29	19.6	190.1	90.7	1.6

The threat of non-renewable organic fuels exhaustion – is a vital present-day problem. That is why many scientific and technical programs are realized nowadays. Heat pump systems (HPS) will be used to substitute irreplaceable natural hydrocarbon. Topicality of this technology is growing, especially for mines with no sources of exhaust heat, hot-water supply potential and year-round need of hot water of which is satisfied by means of small coal boiler-houses, which combine low efficiency with high pollution in the environment. It should be also noticed that such type of systems are especially efficient, if geothermal source of heat and heated buildings are relatively not far. And it fully corresponds to conditions of the biggest part of the towns concetrated around Donbass mines. In terms of the majority of Donbass regions, private sector does not have centralized gas supply. During heating season (october – may), usage of heat, obtained from mine water low-temperature, is by 7 th. UAH more efficient than coal or electric heating usage. It is known that thermal pump usage is 1.2-1.5 times more profitable than the most effective gas boiler-house.

There is a lot of examples of testing of recycling innovative technologies of mine and processing plant anthropogenic wastes today. It is a question of coal extraction, production of aluminate cement, apogrite, claydite, building brick and extraction of rare earth, etc. It is proved that if organic matter content is not more than 10%, then solid flotation wastes can be used for brick production. Air brick is dried with heat of combustion gases that occur because of self-baking in stove ring kiln as a result coal-containing fuel burn-out. Firing temperature is 950-980 °C, firing cycle – 85 hours. Received from mien rock brick is lighter than the clay one and has better heat-insulating properties. Production technology of hyperpressed brick from burned out waste bank is worse studying (Nedodayeva 2006). The base of technological process of brick production without firing is hyperpressing method, which is based on the process of cold welding under high pressure with cementing components and water.

As a result, special development strategy has to be used in economic and ecological situation in coal regions with old and unprofitable state mines. Its essence can be represented by scheme of region development with complex diversification of production in the direction of environmental actions efficiency increase in terms of new work places creation, mine and processing plant anthropogenic waste recycling (Figure 2).

Process of diversification has always been a difficult managerial problem. It is explained by that fact that the enterprises which have made a decision of production diversification will be in state of continuous transformation. Every aspect of the work and life cycle of coal enterprise is checked for strategic.

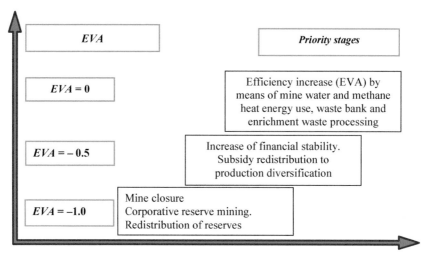

Figure 2. Priorities of diversification objectives.

Flexibility and strength, that means strategic decisions of managerial staff made to diversificate the production, aiming to provide balance between organization stability and changes of its state. With such strategic approach, coal mines should operate for advance and respond quickly to changing conditions, saving competitive advantages in production of diversification output.

3 CONCLUSIONS

1. For state coal mines with long lifetime in terms of unreality of extraction volume increase, it is possible to reorient part of production resources to waste recycling as a diversification direction.

2. Sector restructing direction includes not only closure of nonprofitable enterprises but also contemporaneous creation of enterprises complex of more quality extracted coal processing. Latter allows to increase efficiency of investments greatly and solve employment problem of cola regions. Creation of enterprises of high-quality, ecologically clean output production, which correspond international standard level can become one of such directions.

3. In the context of public management and nature management economics, process of diversification,

on a macro level, can be considered as capital migration from traditional and low-profit sector to the sphere of pressure on the environment reduction be means of the most perspective and outstanding technologies financing, development of professional managements to solve ecological and social problems.

REFERENCES

Salli, S.V., Bondarenko, Ya.P. & Tereshenko, M.K. 2009. *Management of technical and economic parameters of coal mines (NMU)*. Dnipropetrovs'k.: Gerda: 150.

Burkinskii, B.V. Stepanov, V.N. & Harichkov, S.K. 1999. *Nature management: basics of economic and ecological theory*. Odesa: IMPEER NASU: 350.

Veklich, O.O. 2003. *Economic mechanism of ecological management in Ukraine*. NSDCU Ukrainian institution of environment and resources research. Kyiv: 88.

Bardas, A.V. 2010. *Principles of ecological certification of coal processing enterprises of Ukraine in terms of sector restructing*. Dnipropetrovs'k: National mining university: 400.

Nedodayeva, N.L. 2006. *Ecological and economic politics of nature management in terms of mining specificity*. Donets'k: NASU. Institution of industrial economics: 356.

Mining of Mineral Deposits – Pivnyak, Bondarenko, Kovalevs'ka & Illiashov (eds)
© 2013 Taylor & Francis Group, London, ISBN: 978-1-138-00108-4

Complex use of coal of Northern part of Donbass

V. Savchuk, V. Prykhodchenko, V. Buzylo, D. Prykhodchenko & V. Tykhonenko
National Mining University, Dnipropetrovs'k, Ukraine

ABSTRACT: Data concerning composition and quality of coal of Northern part of Donbass are analyzed. Peculiarities of petrographic composition and chemical and technological properties of coal as well as peculiarities of their changes as for extent area of layers and in stratigraphic section are found out. Grade of working coal layers is determined; trends of their complex use are substantiated taking into account petrographic and technological peculiarities.

1 INTRODUCTION

Coal has always been the largest source of power and technological fuel. The analysis of grade composition of coal of middle Carboniferous period of Donetsk field shows that coal of Д, ДГ, Г grades prevail in its composition (Resources of hard fuel of Ukraine 2001). As a rule, coal is characterized by low combustion heat, higher humidity, ash content, and high content of sulphur and sodium and potassium oxides (Drozdnik & Shulga 2009). Current practice of using such coal shows that as a rule it is raw material basis for energetics. The problem of economic efficiency and environmental safety of its use will be solved together with the transfer from coal combustion in furnaces to the technologies of its complex advanced processing. There are considerable reserves of such slightly metamorphized coal in the north of Donetsk field within newly explored Starobelsk coalfield located to the north of Lugansk geological and industrial region. There are about 9 milliard tons of general reserves within this area. The coalfield covers Bogdanovka and Petrovskoye deposits as well as Starobelsk and Svatovka areas.

2 MAIN PART

Objective of this paper is to make complex estimation of the composition and quality of coal of Northern borders of Donbass with the following substantiation of its trends use.

To carry out such works within the areas of new coal zones it is important to analyze petrographic and chemical and technological properties of coal, to establish main area and stratigraphic regularities of its change, to delineate the extent area of layers considering their usability in these or those technological processes right at the very beginning of geological exploration.

Practical solution of this problem involves systematic approach to the study of coal composition and quality, using combination of geological methods. The problem of coal use means engaging wide range of parameters with the use of considerable number of regulatory documents. Systematic approach to the complex use of geological method of studying coal composition and quality means developing information and analytic database which would reproduce to the full extent all the data necessary for decision-making as for the trends of coal reserves use. *Surfer* was used for applied supporting of automated processing of task complex connected with mapping.

Technological value of coal and trends of its use are controlled by such geological factors as degree of metamorphism, petrographic composition, and degree of restorability (Yeriomin 1994).

According to the source material, coal of northern boarders of Donbass belongs to humus group. Quite rarely coal layers contain sapropelite-humites in the form of narrow bands. In general, coal is the complex mixture of macerals of the groups of vitrinite, inertinite, and leptynite. Each maceral is available according to the standard (Savchuk 2011).

Vitrinite group which amount within the area is 78.4% prevails in typical petrographic composition of coal layers. Its amount varies from 62.0 to 96.0% as for separate boreholes. Average values on the boreholes vary within 77.3-80.4%.

Content of semivitrinite group is negligible; it is about 0.2-0.9% at average values on separate boreholes within 0.1-6.5%. Average content within the area is 0.6%.

Inertinite group is the second as for its distribution. Average value within the area is 11.1%. Average values of inertinite group content on the separate boreholes vary within 2.0-23.0% at average values of 8.1-12.9% on the layer.

Leptynite group takes less percentage in petrographic coal composition. Its content is 9.9% on average. Leptynite group varies in its content on separate boreholes within 1.0-23.5% at average 8.7-11.3% values for layers.

According to petrographic classification of A.P. Karpinsky Russian Geological Research Institute (Yeriomin 1994), coal of northern borders of Donbass is represented by the class of heliolytes in which sub-class of hoelites (80.2%) exceeds by far the subclass of gelitites (19.8%). Sipoid-fusinite-hoelites prevails among petrographic types (64.9%) while lipoid-fusinite-geliolites accounts 15.3% and fusinite-hoelite accounts 12.6 (Savchuk 2011).

According to petrographic classification by Yu.A. Zhemchuzhnikov, coal is represented by the interstratified layers of clarain, duroclarain, and sometimes clarain-durainous coal (Savchuk 2011).

Genetic peculiarities of petrographic compositions of the region coal have been established, among which there are the following ones: prevailing of attrital ground mass (collinite) in the group of vitrinite at insignificant distribution of gelified fragments (telinite), available fragments of argillites different in conservation and oxidation level, character of ground mass, changeable colour of spores, and in some layer intersections their bedding is at an angle to stratification.

The layers coal is at the almost equal stage of carbonization. Average regional index of vitrinite reflection (R_0) is 0.47%, and standard deviation is 0.04%. Values vary along the separate boreholes within 0.38-0.62%. According to the current standards, the coal belongs to 03 class of metamorphism being at O_3 stage of metamorphism (Savchuk 2011; 2008; 2009; 2010). As for separate values of this parameter ($R_0 < 0.40\%$ and $R_0 > 0.50\%$) coal belongs to 02 class of metamorphism being at O_2 stage of metamorphism and 10 class of I stage of metamorphism correspondingly. It is established that standard deviations of values of index of vitrinite reflection are not more than 0.1% and they are within 0.03-0.05%. According to International Codification System (Savchuk 2012) reflectograms of the layers belong to 0 code.

Summary data considering chemical and technological properties of coal are given in Table 1.

Table 1. Chemical and technological parameters of coal on layers and areas of Starobelsk coal area.

Parameters from-to average	Sites								
	Svatovka Area		Starobelsk Area		Petrovskoye Deposit				Bogdanovka Deposit
	Synonymy of Layers								
	k_2^H	m_3	k_2^H	l_7	k_2^H	l_7	m_3	h_8	k_2^H
W_t^r , %	9.4-16.6 / 12.0	4.5-14.7 / 11.4	8.2-24.9 / 15.5	9.7-25.3 / 14.1	6.4-13.8 / 10.2	11.3-21.6 / 14.4	10.9-15.6 / 12.9	13.1-24.6 / 17.5	4.0-28.3 / 20.2
W_a , %	2.0-14.4 / 6.2	3.9-13.0 / 7.0	3.4-15.4 / 8.2	4.3-18.4 / 8.0	4.2-9.7 / 6.6	5.2-15.3 / 8.3	4.9-14.5 / 8.0	3.4-15.4 / 8.0	1.0-25.0 / 9.1
A_{seam}^d , %	9.8-43.8 / 29.1	7.5-24.9 / 14.8	7.1-43.7 / 18.9	7.2-28.2 / 13.9	4.7-44.1 / 14.3	4.1-23.2 / 12.8	7.5-34.3 / 14.5	6.2-41.4 / 15.7	3.2-44.9 / 17.3
$A_{c.p.}^d$, %	9.8-38.9 / 21.5	7.5-23.0 / 14.7	7.0-37.7 / 16.4	7.2-26.2 / 13.7	4.7-39.1 / 14.2	4.1-21.2 / 12.5	7.5-33.0 / 14.2	6.2-22.0 / 13.8	1.0-40.7 / 14.1
S_t^d , %	0.7-9.1 / 4.1	0.5-7.2 / 4.0	0.5-8.2 / 3.6	1.8-7.9 / 4.3	1.0-12.1 / 3.6	2.2-8.6 / 4.4	1.8-10.2 / 4.7	1.2-6.1 / 3.0	0.2-11.0 / 1.6
V^{daf} , %	37.7-50.5 / 44.0	39.9-48.4 / 42.9	33.4-54.3 / 45.2	38.5-44.8 / 41.0	42.0-48.0 / 45.5	40.2-44.3 / 42.6	39.7-45.3 / 43.6	32.0-50.7 / 43.9	32.0-58.9 / 42.9
Q_s^{daf} , MJ / kg	27.5	31.8	31.6	31.0	28.3	26.8	31.1	31.6	31.1
Q_i^r , MJ / kg	21.5	23.2	22.4	21.4	23.8	22.1	22.0	22.0	21.0
O_e	0.73	0.76	0.76	0.73	0.81	0.75	0.75	0.75	0.75

Average regional maximum water-absorbing ability (W_{max} , %) is 14.8%. Parameters on separate boreholes vary within 4.0-28.3%. Average regional moisture test value (W_a , %) is 7.9%. Parameters on separate layer intersection vary within 1.0-25.0%. Layer k_2^H is characterized by the highest variability of maximum water-absorbing ability and moisture test value. Parameters W_{max} and W_a of k_2^H layer occurring within the territory of Bogdanovka deposit is quite higher than the parameters of the layer in other areas of Northern Donbass.

Average regional ash content of the layer (A_{seam}^d, %), taking into account the contamination, is 15.9% while ash content of plies ($A_{c.p.}^d$, %) is 14.4%.

The following conclusions are possible to be drawn owing to the analysis of ash content of the region coal:

1. Ash content of k_2^H layer is characterized by the most variability.

2. Average ash content with the ash content parameter within 8-16% prevails on all the layers.

3. Coal of l_7 layer is characterized by minimum ash content while coal of k_2^H layer is the most high-ash according to average parameter.

Chemical composition of coal ash is represented by silicon and titanium dioxides, oxides of aluminum, iron, calcium, magnesium, potassium, natrium, phosphorus, and sulphur trioxide. Comparing to chemical composition of ash of coal of middle Carboniferous period of Donbass, in general chemical composition of considered area coal is characterized by higher content of Fe_2O_3, CaO, MgO, SO_3, Na_2O and lower content of SiO_2, Al_2O_3 and K_2O (Savchuk, Prykhodchenko & Kuz'menko 2011).

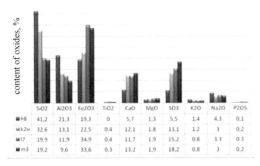

Figure 1. Chemical composition of coal ash of main coal layers in Starobelsk coal area.

As for the content of mass fraction of sulphur (S_t^d, %), coal belongs to polysulphide and higher sulphide ones. Its average regional content is 3.8%. Ratio of pyrites and organic sulphur does not corre-

sponds to the statistic relation determined for Donbass coal. Coal is characterized by higher content of organic sulphur (Savchuk 2012).

Average regional volatile content (V^{daf}, %) is 43.3%. Values vary on the separate boreholes within 32.0-58.9%. It is determined that the parameter has no areal changes. Coal of k_2^H layer is characterized by the highest changeability and highest average value.

The highest average specific heat of coal (Q_s^{daf}, MJ / kg) is 30.4; it varies depending upon layers from 23.6 to 34.0. The lowest specific heat (Q_i^r, MJ / kg) varies from 16.2 to 25.7 at the average value of 22.1.

Specific features have been found:
– average values for all the layers are within short interval and are characterized by close values;
– parameters of k_2^H and l_7 layers have the values lower than the average ones while the parameters of layers h_8 and m_3 have the values higher than the average ones;
– coal of k_2^H layer is characterized by the highest changeability of values of both highest and lowest specific heat.
– the highest specific heat of all the layers has inverse dependence upon the occurrence depth of a layer.
– the lowest specific heat of all the layers has inverse dependence upon the values of coal ash content.

Average values of element content of the coal are as follows: C^{daf} – 75.2%, H^{daf} – 5.2%, S^{daf} – 4.6%, $(N+O)^{daf}$ – 15.0%. Carbon content varies on the separate boreholes within 61.6-82.8%, hydrogen content – 2.6-8.9%, nitrogen and oxygen content – 5.0-25.2%.

Grade composition is the necessary coal characteristic which is taken into account while carrying out geological prospecting works as it defines the trends of its use. According to the current Ukrainian standard DSTU (Ukrainian national standardization system) 3472-96 coal of all the layers belongs to stone one of Д grade (Savchuk 2011; 2010; 2012; 2010).

Д grade covers a subgroup of long-flame vitrinite coal which is applied in CIS countries (GOST 25543-82 1983).

According to International Coal Codification the layer coal belongs to the middle rank (stone coal) (Savchuk 2012).

According to International Coal Classification (ISO 11760) (International codification system... 1988), coal of m_3 and l_7 layers belongs to subbituminous coal of low-rank A. Coal of k_2^u and h_8 layers is partially classified as subbituminous coal of low-rank A – 40 and 30%, correspondingly, and bituminous coal of medium rank D – 60 and 70%, correspondingly (Table 2). According to petrographic composition the layers coal belongs to the group with moderately high vitrinite content. According to ash content, the coal belongs to medium category (according to average values) (Kuz'menko 2010).

Table 2. Classification of coal of Starobelsk coal area according to ISO 11760.

Layer	Value of vitrinite reflection, R_0, %	Vitrinite content, %, per pure coal	Ash content, %	Classification
h_8	0.5	80.9	13.8	Low grade A/Medium grade D, moderately high vitrinite content, medium ash content
k_2^u	0.48	77.6	14.5	Low grade A/Medium grade D, moderately high vitrinite content, medium ash content
l_7	0.46	77.8	13.0	Low grade A, moderately high vitrinite content, medium ash content
m_3	0.41	75.9	14.3	Low grade A, moderately high vitrinite content, medium ash content

According to the current standards, the area coal belongs to Д grade of subgroup of vitrinite ДВ (Savchuk 2009, 2010). Such trends as power-generating and technological use as well as building materials production and hydrocarbonic adsorbents are possible for coal of this grade according to the standard (Yeriomin 1994 & Savchuk 2008).

Usability of coal in power generation is evaluated according to the current standard of Ukraine. It is determined that according to the value of lowest specific heat ($Q_{i_{av}}^r = 22.1\ MJ/kg$) coal of Starobelsk coal area belongs to the first quality category. Average values of power moisture content are more than the normative one (Table 3).

Table 3. Quality parameters of coal for layers of Starobelsk coal area and operating requirements according to DSTU 4083-2002.

Parameters	Operating requirements	Average values for layers of Starobelsk coal area				
		Area average	h_8	k_2^u	l_7	m_3
Q_i^r, MJ / KG	20.097	22.1	22.0	22.2	21.8	22.3
W_t^r, %	14.0	14.8	17.5	14.5	14.3	12.7

To identify zones of coal appropriate for combustion according to the requirements of DSTU 4083-2002, maps of Q_i^r and W_t^r content have been compiled. Different usability of the layers coal for the combustion has been determined. Coal of k_2^u layer is appropriate for combustion only within definite areas (Svatovka area, separate areas of Petrovskoye deposit) (Savchuk 2010, 2012, GOST 25543-82. 1983; International codification... 1988).

Problem of coal "mineralization" is quite important while estimating it as the raw material for the power generation. Coal is mineralized or "alkalined" when it has abnormally high content of alkaline metals, mostly natrium (Savchuk 2010). Main difficulties of the exploration of deposits and use of alkaline coal are due to its higher fluxing power and corrosive nature, negative environmental impact while its combustion or other thermal processing.

There are no common criteria of qualifying coal as "mineralized" today. Most scientists take content of sodium oxide Na_2O in coal or in its ash as the criterion of "mineralization".

Studies carried out in such countries as Germany, the USA, England, Poland, Australia, Ukraine and others have shown that specific sediments on working surfaces of potholes can be observed if ash contains more than 2% of Na_2O (Savchuk 2011 & Kuz'menko 2012).

Average value of sodium oxide content in coal

ash of Starobelsk coal area is 3.4%. Graph of Na_2O content distribution in coal ash of Starobelsk coal area is in Figure 2.

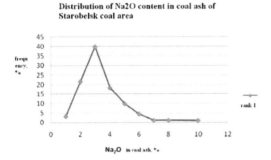

Distribution of Na2O content in coal ash of Starobelsk coal area

Figure 2. Distribution of Na_2O content in coal ash of Starobelsk coal area.

To identify distribution zones of coal appropriate for combustion as for this parameter, maps of Na_2O content are compiled for all the basic layers. Areal regularities of its content changes are studied using trend maps. Distribution zones of coal with Na_2O content less than 2% are qualified as appropriate for combustion.

Analysis of maps of Na_2O content and second order polynomials has allowed determining that in general, the layers coal is characterized by higher content of Na_2O. Only separate small areas with Na_2O content of up to 2% can be found. Thus, according to occurrence area of h_8 layer, Na_2O content is more than 2% and increases from north-west to south-east. As for k_2^H layer, Na_2O content is not more than 2% only in the west of Svatovka area and in the south of Starobelsk area; it regularly increases from south-west to north-east. As for l_7 layer, coal zones with Na_2O content of up to 2% can be observed in outermost southern and central parts of Starobelsk area. Petrovskoye deposit is characterized by its higher values. Its content increases from south-west to north-east. As for m_3, coal layer with Na_2O content of up to 2% can be observed in outermost part of Svatovka area and in the central part of Starobelsk area. Content increases from north-west to south-east.

Regulatory documents stipulate using coal with sulphur content within 1.0-1.5%; and 3% in some cases under tough conditions of limiting other parameters.

Sulphur content in coal of the region is higher; in average it is about 3.8%. For separate boreholes values vary from 0.2 to 12.1% at average values for layers (%): $h_8 - 3.0$, $k_2^H - 3.2$, $l_7 - 4.4$, $m_3 - 4.6$ (Savchuk 2012 & Kuz'menko 2010). Study of variation regularities of sulphur content over extent area of regional layers has found just small areas with sulphur content of < 3% all over the occurrence area (e.g. Bogdanovka deposit, north of Starobelsk and Svatovka area) (Savchuk 2012 & Kuz'menko 2010).

To find the possibility of reducing sulphur content in the coal we have performed a number of experiments as for their washability. It has been established that the layers coal is characterized by different degree of washability as for the sulphur (GOST 25543-82 1983). Coal of l_7 layer is characterized by the best washability while coal of m_3 layer is characterized by the worst one. It has been found that it is possible to reduce sulphur content using gravitational method just by 20-30%. Mostly strata and lenses containing secondary sulphur balls or concentration of fine grains both of primary and secondary origin will be separated into waste.

In general, coal of the region is characterized by changeable quality parameters and it is impossible to give unique estimate of its usability in power generation. Estimation of the layers coal according to the current Ukrainian standard DSTU 4083-2002 allows drawing conclusion that coal of m_3 layer meets the requirements for combustion best of all over its distribution area as well as coal of k_2^H layer on separate areas (Svatovka area, separate areas of Petrovskoye deposit).

As for Na_2O content in coal ash and sulphur content, coal can not be recommended unambiguously for combustion without its preliminary processing.

Solving the problem of economic efficiency and environmental safety of coal use is connected first of all with its deeper and more complex processing. Hydrogenation of coal is one of such ways to converse it completely.

Usability of coal of Starobelsk coal area for obtaining synthetic fuel is substantiated according to current methodologies being valid both in Ukraine and abroad (Savchuk 2010, 2010). To confirm the data, the results of laboratory research on coal hydrogenation by the Institute of Physical-organic Chemistry and Coal Chemistry were used.

According to methodology of I.V. Yeriomin next to all the coal of Starobelsk coal area except the part of k_2^H layer within Svatovka and Starobelsk areas meets the requirements to be used for obtaining synthetic liquid fuel. Coal of k_2^H layer within Svatovka and Starobelsk areas does not meet the requirements just in one parameter – ash content of plies which exceeds the acceptable requirements being more than 15%.

Evaluation of coal of Northern coal area according to the criteria by M.M. Ulanov has shown that the layers coal belongs to different groups as for its hydrogenization degree (Savchuk 2010; 2010). In general, coal belongs to the second group, i.e. to the coal appropriate for obtaining synthetic fuel. It is impossible to determine the difference of hydrogenization degree of separate layers of the region using this methodology.

The considered methodologies were developed mostly for coal of deposits located in Russia. Further there was proposed a methodology to identify Ukrainian coal usability for obtaining synthetic fuel right at the stage of geological prospecting. Grade composition of coal of Starobelsk coal area is quite stable and, according to current Ukrainian standards, is represented by Д grade. According to this methodology, coal of all the layers as for its hydrogenization degree belongs to the second group.

On average values of such parameters as R_0, V^{daf}, $V_t + L$, C^{daf}, coal mostly belongs to the group appropriate for liquefaction and as for the total of content of sodium and potassium oxides it belongs to the third group.

Additional maps were compiled for the parameters which belong to different groups as for hydrogenization usability and for the most informative parameters for classification on which zones of coal distribution according to hydrogenization usability groups were identified. According to vitrinite index of reflection coal of all layers belongs to the second group. Compiling map of vitrinite indexes of reflection of m_3 layer has allowed detecting distribution zone of the coal belonging to the first group.

On average values V^{daf}, coal of industrial layers of Starobelsk area mostly belongs to the second group. Compiling maps of V^{daf} has allowed identifying (according to distribution area of h_8 and k_2^H layers) distribution zones of the first and the second usability groups. It has been established that changeability of this parameter for l_7 and m_3 layers does not exceed the limits of usable group.

On average, content of leptynite group in petrographic composition of the regional coal varies from 6.5 to 12.8%. Its less values are typical for coal of l_7 and m_3 layers. In general, regional coal belongs to the second group according to average content of maceral leptynite group.

Ultimate composition is also the criterion for estimating coal as raw material for obtaining synthetic fuel. In particular, value of carbon content (C^{daf}, %) and hydrogen-carbon ration are used. The layers coal belongs to the second group according to average and specific parameters of carbon content. Coal of Northern region is characterized by high ratio values of atomic parameters of hydrogen and carbon content to be 0.80-0.83%. According to distribution area, coal of all the layers belongs to the group of both the most appropriate (first group) and usable one.

Similar to the previous classifications, ash content parameters correspond to coal group usable for coal hydrogenization. Small areas of k_2^H layer within Svatovka and Starobelsk areas are the exceptions. According to this parameter, coal belongs to the group of little use. Parameter values are as follows: module of coal ash varies in a wide range on both layers and its distribution area (Savchuk 2012; 2011; Kuz'menko 2010). Its values vary from the parameters characteristic for the group of the most usable coal to the parameters of the coal group of little use.

Maps are also compiled for sulphur content and zones of different usability degree for all the industrial layers are specified. Parameters of total content of sodium and potassium oxides content for h_8 layer vary within wide value range corresponding to the group of little use. Maps are compiled with singling out zones of different usability degree for k_2^H, l_7 and m_3 layers (Savchuk 2012; 2011 & Kuz'menko 2010).

Hence, according to the Ukrainian methodology, coal of Northern area is sure to belong to the second group on average values of quality parameters; coal of this group is usable for hydrogenization. Coal of the region belongs to the third group and is not usable for hydrogenization only according to values of parameters $Na_2O + K_2O$, ash content, and content of leptynite group. It should be added that content of leptynite group (9.9%) and amount of mineral impurities (15.3%) are quite close to limit values for coal of the second group and differ from them within the limits less than acceptable errors while determining these parameters according to the current standards.

Coal of all the areas is sure to be qualified as the second group. Fraction of parameter values according to which coal of the areas is unusable for hydrogenization varies within 15-27%; on average it is about 20%.

If the coal belongs to the second group it means that the degree of transformation of organic part of coal (OPC) should vary within 80-90%. According to the results by Institute of Physical-organic Chemistry and Coal Chemistry of the National Academy of Sciences of Ukraine in cooperation with Starobelsk Geological Prospecting Expedition it has been found that on average, value of OMC is more than 80%.

Thus, both according to the results of coal classification as for its quality parameter and as for the results of laboratory research, coal belongs to the second group being usable for obtaining liquid fuel.

3 CONCLUSIONS

1. Coal of Northern part of Donbass is characterized by considerable content and hard washability in sulphur, higher amount of sodium salts. It cannot be recommended unambiguously for combustion without preliminary processing.

2. Coal of m_3 layer best of all meets the combustion requirements (almost over its total distribution area), and coal of k_2^H layer is the best as for meeting the requirements on separate distribution area (Svatovka area and Petrovskoye deposit).

3. Taking into account petrogenetic and chemical and technological properties, the key trends of using coal of Northern areas are hydrogenization stipulating its most complete and complex use. It has been established that both according to values of separate parameters and results of laboratory research coal belongs to the second group and usable for obtaining liquid fuel.

REFERENCES

Resources of hard fuel of Ukraine. 2001. Kyiv.

Drozdnik, I. & Shulga, I. 2009. *On professional use of slightly metamorphized coal.* Mineral dressing. Issue 36(77) – 37(78).

Yeriomin, I. & Bronovets, T. 1994. *Grade coal composition and its reasonable use.* Moscow: Nedra.

Savchuk, V., Prykhodchenko, V., & Kuz'menko, O. 2012. *New data on composition and quality of coal of north-ern borders of western part of Donbass.* Lithology and geology of anthracides: inter-university scientific thematical collection of papers. VI (20). Yekaterinburg.

Savchuk, V. & Kuz'menko, O. 2011. *Petrographic peculiarities of the coal of Northern coal area of Donbass.* Dnipropetrovs'k: Scientific messenger of NMU, 5.

Savchuk, V. & Kuz'menko, O. 2008. *Grade composition and main trends of coal use of Svatovka perspective area.* Dnipropetrovs'k: Scientific messenger of National Mining University, 9.

Savchuk, V. & Kuz'menko, O. 2009. *Coal composition and quality of Starobilsk perspective area and main trends of its reasonable use.* Messenger of DNU. Geology. Geography. Issue 11. Dnipropetrovs'k.

Savchuk, V. & Kuz'menko, O. 2010. *Coal composition and quality of Bogdanovka deposit and main trends of its reasonable use.* Messenger of DNU. Geology. Geography. Issue 12. Dnipropetrovs'k.

Savchuk, V., Prykhodchenko, V. & Kuz'menko, O. 2011. *Chemical composition of coal ash of Northern coal area of Donbas.* Messenger of DNU. Geology. Geography Issue 13. Dnipropetrovs'k.

Savchuk, V., Prykhodchenko, V. & Kuz'menko, O. 2012. *Sulphur washability of coal of Petrovskoye deposit of Sterobelsk coal area.* Dnipropetrovs'k: Collection of scientific papers of NMU, 37.

DSTU 3472-96. 1997. Lignite, hard coal and anthracite. Classification. Derzhstandart of Ukraine.

GOST 25543-82. 1983. Lignite, hard coal and anthracite: Classification by genetic and technological parameters. Moscow: The USSR State committee on standards.

International codification system of coals of high and middle ranks. 1988. New-York. OON.

Kuz'menko, O. 2010. *Coal composition and quality of Petrovskoye deposit of Northern Donbas coal area and main trends of its reasonable use.* Dnipropetrovs'k: Scientific messenger of NMU, 9-10.

Kuz'menko, O. 2012. *Evaluation of coal of Sterobilsk coal area of Northern Donbas as energy raw material.* Collection Geotechnical mechanics. Dnipropetrovs'k: IGTM, Issue 102.

Savchuk, V., Prykhodchenko V., & Kuz'menko, O. 2010. *Methodical aspects of Ukrainian coal evaluation as for its usability for obtaining liquid fuel.* Miner's forum, papers of international conference. Dnipropetrovs'k.

Savchuk, V., Prykhodchenko V. & Kuz'menko, O. 2010. *Selection and substantiation of trends of professional use of Northern Donbas coal.* Region – 2010: Strategy of optimal development: papers of scientific and practical conference with international participation. Kharkiv.

Mining of Mineral Deposits – Pivnyak, Bondarenko, Kovalevs'ka & Illiashov (eds)
© 2013 Taylor & Francis Group, London, ISBN: 978-1-138-00108-4

On the limit angles of inclination of belt conveyors

V. Monastyrs'kyy, R. Kiriya, D. Nomerovs'kyy & N. Larionov
M.S. Polyakov Institute of Geotechnical Mechanics, Dnipropetrovs'k, Ukraine

ABSTRACT: The analysis of the high-angle conveyors and structures for each design are defined theoretically and experimentally, the limiting angles of inclination lump in transportation of bulk loads. A design of the vertical conveyor for lump goods and justified options.

1 THE PROBLEM AND ITS RELATION TO THE SCIENTIFIC AND PRACTICAL TASKS

In mining enterprises of the CIS to move the bulk of different particle size used means of continuous transport, different from other species (dump trucks, vibrators, rail transport) high performance, reliability and continuous flow of load to refineries.

According to (Sheshko 1996; Perten 1977 & Nikolaev 1999) to increase the efficiency of belt conveyors in mines may decrease the length of transportation through the use of high-angle conveyors (HAC) of various designs. Raising the limit angle conveyors achieved in practice (Perten 1977 & Nikolaev 1999) through:

– increase the adhesion forces between the bulk and weight belt with additional riffles, ledges on its surface;

– create additional pressure on the bulk load by changing the grooved belt, applying pressure tape and different partitions;

– the creation of special pockets on the belt or attached firmly to her blood vessels, allowing raising bulk load different lumpiness at 90° from the depth of 500 m.

2 STATEMENT OF THE PROBLEM

Questions to determine the maximum angle of transporting bulk goods of different granule composition were investigated in (Perten 1977; Nikolaev 1999; POCKETLIFT® 2004 & Monastyrs'kyy 2013).

Found that the bulk load is estimated ratio $\frac{a_{max}}{B_b}$,

where a_{max} – the maximum size of a piece, B_b – belt width (capacity of the vessel). On this basis

weights are divided: for small-sized $\frac{a_{max}}{B_b} < 0.1$, as-

sorted $\left(\frac{a_{max}}{B_b} < 0.2\right)$ and large-sized $\frac{a_{max}}{B_b} > 0.2$.

According to (Monastyrs'kyy 2013), a large piece, often have the shape of a parallelepiped, are oriented in the flow along the strip and has local projections, and after loading the conveyor belt can be at substrate without it. The results of theoretical and experimental studies of the stability of the load on the conveyor belt for general purposes. The limiting angles of transportation and small-sized lump of goods on the tape HAC with baffles and without. The following are, in abbreviated form, the results of studies (Monastyrs'kyy 2013).

For small-sized load addressed issues of stress-strain state of a bulk load, in which no relative sliding of particles relative to each other. For research in the load weight is highlighted with dimensions $d_x \cdot d_y \cdot 1$, inclined to the horizontal at an angle β_k. At each site there are a normal element (σ_x, σ_y) tangent (τ_{xy}) and the force of gravity $G = d_x \cdot d_y \cdot 1 \cdot \gamma$, which, depending on the angle β_k to be expanded to a normal ($G_m \cos \beta_k$) and tangential ($G_m S_m \beta_k$) components.

To determine the effective stress on each platform, inclined at an angle to the β_k to the x-axis shows the design scheme and the system of analytical expressions, the solution of which is performed under the condition $\tau_{ls} = \sigma tg\rho$, where ρ – the angle of repose of the material in motion, τ_{ls} – limiting shear stress on the faces of the element. An expression for the maximum angle of inclination of the pipeline:

$$tg\beta_k = tg\rho . \qquad (1)$$

Consequently, the critical angle of small-sized bulk load transportation on the tape HAC generally limited to an angle of repose. In the particular case of bulk load hold on the belt by friction ($F \leq Ntg\varphi$, where F – shear force; N – normal component of the force of gravity; φ – angle of internal friction, the limit value is equal to ρ). The expression for the maximum angle of inclination of the assembly line in this case is:

$$tg\beta_k = f,\qquad(2)$$

where f – the coefficient of sliding friction load on the tape.

For large pieces examined the cases of their movements on the tape at V_b. Given the interaction forces, inertia forces and a piece of the process of segregation. In general, the critical angle of transport is determined by the formula:

$$\beta_{ca} = \varphi_{ca} - (\beta_\kappa + \Delta\beta + \beta_i),\qquad(3)$$

where β_κ – angle conveyor; $\Delta\beta$ – yaw angle on the belt pulley; β_i – an additional piece of angle from the forces of inertia, β_{ca}, φ_{ca} – avenue respectively the critical angle of inclination of the conveyor and the angle at which the piece loses its stability, equal to $90°$.

The analytical expressions for each component of the maximum angle of CLS are given in (Monastyrs'kyy 2013). Parallelepiped to pieces and cubic (oval) shape proposed conditions:

$$\frac{a_{max}}{H} = \frac{2}{1 - \cos\alpha},\qquad(4)$$

where H – height partitions; α – angle, connecting the center of a piece with the top baffle (central angle).

Analysis of expression (4) indicates that the traffic load for steady tape regardless of the form piece, the ratio $\frac{a_{max}}{H}$ should not exceed 2.

Let us consider another factor that affects the efficiency of using pipes with baffles. According to (Monastyrs'kyy 2013), for transporting bulk goods on the conveyor lumps segregation occurs, at which the process of "floating" is as follows. When interacting with a piece of roll sets the rise of its center of gravity on the value of small-sized h_{sf} fraction fill the free space below the plane of the piece, and the piece continues to move relative to the roller substrate until crossing its center of gravity of the

vertical axis of the roller. Thus there is a stirring piece on bank filling of small-sized fractions and under the plane, on the other hand, also produces a free space where the flock small-sized fraction.

As shown filming the interaction of a piece with the roller carriage and the experimental determination of the dynamic characteristics (K_d), after repeated interactions piece on substrate value h_{sf} practically not increased and therefore the condition for sustainable movement piece on the tape is:

$$h \gg h_{sf}.\qquad(5)$$

The greater the strength of the interaction, the greater the h_{sf} and at the intersection of the vertical line (φ_{ca}) the last piece of the center of gravity becomes unstable if the condition. The value h_{sf} define, according to (Monastyrs'kyy 2013), the balance equation of rotational kinetic energy after interaction piece of work and the amount of potential and of the external forces:

$$h_{sf} = \frac{a^2}{2l_r}(K_d - 1) - \frac{J_i\omega^2}{2m_p g},\qquad(6)$$

where $J_i = m_o\left(\dfrac{a^2 + b^2}{3}\right)$ – the moment of inertia of a piece; l_p – respectively the length of a piece and step placement on the conveyor idlers; K_d – dynamic factor; m_p – mass of a piece; g – acceleration due to gravity; ω – angular velocity of the piece after the collision.

Analysis of the results showed (Monastyrs'kyy 2013) that the stability of the load on the belt HAC is determined taking into account the strength of interaction with large pieces of idlers, the inertial forces of pieces, the process of segregation and the ratio of the maximum size of the piece to the height of the partition. The limiting angle HAC obtained by calculation, in good agreement with the experimental data for bulk load size $a_{max} > 0.2B_b$; $a_{max} < 0.2B_b$, mm.

Therefore, the prior art limit detection angle HAC general purpose riffles, partitions during transportation of bulk load investigated adequately.

3 THE PRESENTATION MATERIALS AND RESULTS

In solving the problem of stability of the load on the tape by creating additional compressive forces on the tape with the load, preventing slipping down along the tape the following tasks: conveyor idlers equipped with deep-grooved step placement on the length l_r is loaded uniformly distributed load q_l. The conveyor belt moves at a speed V_b in the box with the raised grooved formed idlers.

On a plot of load carrying rollers between the force of gravity ($m_l g$), the tangential component ($F_m = m_l g \sin \beta_k$), the friction force of the load tape ($F_m = f m_l g \cos \beta_k$) compression force load on the rollers ($F_c = \sigma_c K(F - F_2)\omega_p ctg\alpha$) and the power of thrust load ($P_p = m_l g \cos \beta_k \cdot \sin \rho$), where m_l – load weight; g – acceleration of gravity; β_k – to the angle of inclination of the conveyor; σ_c – compressive stress; F, F_2 – respectively sectional area of the load on the tape before and after the passage of roller; K – coefficient taking into account the rigidity of the belt.

The process of load passage with a conveyor belt through a section of idlers can be drawn through the same section with a free (open) surface. Thus, in a very short time a change cross-sectional shapes of the load on the tape, the force of the actuator rod. After passing through the roller tape again assumes its original shape under the force of gravity, and in this case, there is the effect of freewheel: forward load strained on roller carriage under the force of traction, almost freely through its cross section, but during the return roller prevent its slipping down the tape, because no active force for change in section re-bulk load on roller carriages.

Equilibrium load area on the tape is determined from the condition that the goods in the form of trapezoidal shape, without losing its continuity, by the action of the tangential component (F_m) tends to move downward against the resistance of a deep grooved roller and head of load. The equilibrium equation is:

$$m_l q \sin \beta_k \leq \delta_c K(F - F_2)\omega_p ctg\alpha +$$

$$+ f m_l q \cos \beta_k + K_\partial m_l q \cos \beta_k \cos \rho . \quad (7)$$

If you submit a conveyor belt consists of 3 separate parts, based on the rollers roller, for practical purposes it can be assumed that in the middle of the span $\left(\dfrac{l_{sp}}{2}\right)$ value of the deflection ribbon is the same. In this case it is possible to adopt:

$$ctg\alpha = \frac{l_r 4 H_b}{2 m_l q \cos \beta_k l_r} = \frac{2 H_b}{m_l q \cos \beta_k} , \quad (8)$$

where H_b – belt tension.

Substituting (8) into (7) we get:

$$\sin \beta_k = \frac{2\sigma_c K(F - F_2)\omega_p H_b}{2 m_l^2 g^2 \cos \beta_k} +$$

$$+ (f + K_\partial \sin \rho)\cos \beta_k . \quad (9)$$

Optimizing the expression (9) with respect to the angle, after transformations we obtain:

$$\beta_k = arcsin \frac{m_l^2 g^2}{4\sigma_c K(F - F_2)\omega_p H_b} . \quad (10)$$

Analysis of the results showed that with increasing belt tension limit angle conveyor varies from 85 to 15°; shipping weight reduction from 500 to 100 kg reduces limit angle, the value of which is asymptotically approaches 0.

In IGTM NAS created HAC lumpy bulk load containing two traction round link chains, sprockets envelopes and tension drums that are installed on individual frames. The two chains are connected to each other by rigid triangular cross-section cross members between which the rubber-fabric conveyor belt, based on the cross with sagging in each bay. Along the perimeter of the deflection ribbon on both sides are fixed to the ends of its elastic corrugated board, the edges of which are in the form of parabola branches of different lengths. The excess of the branches relative to each other by the angle of the slope and shape of the HAC education tanks located between the tape and boards. Collars can be made from pre-stressed composite insert. HAC bulk load loading occurs continuously on a conveyor belt using the bootable device. Thus, each cross covered conveyor belt which presses them firmly without damaging the tape. To prevent spillage of load loading it on the HAC on the boot device has side rails made of thick rubber, and in the place of loading large pieces of devices are installed with shock absorbers. Unloading HAC is in its upper portion through the drive pulley by gravity from each container into the hopper.

On the basis of the research suggest the following conclusions:

1. Recommendations for determining the limit angles in the design for HAC transport conditions lumpy bulk.

2. Proposed construction of a new quality HAC, allowing transporting bulk load particle size of more than 300 mm at an angle of inclination of the conveyor from 0 to 85° and the method of calculation of their parameters.

REFERENCES

Sheshko, E.E. & Morozov, N.G. 1996 *Prospects steeply inclined lift in mines*. Leningrad, 6: 56-59.

Perth, Y.A. 1977. *Steeply inclined conveyors*. Leningrad: Mechanical Engineering: 216.

Nikolaev, E.D. & Dmitrin, V.N., Kasterin, A.N. & Fedorenko, A.I. 1999. *analysis of designs for high-angle conveyors of deep pits*. Dnipropetrovs'k: Mining journal, Ukraine, 11-12: 78-82.

POCKETLIFT ®. 2004. *An advanced high-angle and vertical transport to a height of 500 meters*. St. Petersburg: Metso minerals, 1809-05-04-WPC: 21.

Monastirs'kyy, V.F., Maksyutenko, V.Y., Kiriya, R.V. Braginec, D.D. & Nomerovs'kyy, D.A. 2013. *The influence of granule composition on the stability of bulk cargo transportation steeply inclined conveyors*. Dnipropetrovs'k: Proceedings of the School of underground mining: 210.

Mining of Mineral Deposits – Pivnyak, Bondarenko, Kovalevs'ka & Illiashov (eds)
© 2013 Taylor & Francis Group, London, ISBN: 978-1-138-00108-4

Investigation of the rock massif stress strain state in conditions of the drainage drift overworking

V. Sotskov
National Mining University, Dnipropetrovs'k, Ukraine

I. Saleev
Mining Machines, Donets'k, Ukraine

ABSTRACT: There was simulated conducting coal-face work on the drainage airway, which had previously been driven at the depth 8-9 meters from excavated coal seam C_5 of "Samars'ka" mine. There was researched the stress strain state of the coal-bearing massif near the drainage airway and approaching working face. The character of stress redistribution while changing of the distance from the working face to drift is described. The regions of the rock mass softening are defined.

1 INTRODUCTION

The most important component in the energy balance of Ukraine is the coals for coking and energy needs, they are making the coal industry priority in the economic development of the country. The stable operation of the coal mine is largely determined by the underground mine roadways states, among which in-seam workings are exploited in the most difficult conditions and situated in close technological communication with extraction. There is explanation and adoption of rational technical solutions while operation of underground mining roadways network is one of the components of the effective functioning of the coal mine (Bondarenko, Kovalevs'ka, Symanovych, Fomychov, Martovytskyi & Kopylov 2010). These tasks take the special urgency in the difficult geological conditions of thin coal seams extraction, particularly in the layered massif of weak water-bearing rocks of the Western Donbass. In this regard, the particular attention is given to in-seam drifts for two reasons: first, this group of mines is characterized by the non-uniform structure and properties of the host rock, and secondly, their state largely determines the stability and reliability of the mine (Bondarenko, Kovalevs'ka, Symanovych & Fomychov 2011).

2 FORMULATION OF THE PROBLEM

Conditions of the working out of seam C_5 from "Samars'ka" mine are characterized as complex due to the presence in the roof a sufficient number of watered shallow coal seams and interlayers, watered sandstone and intensely fractured mudstone and siltstone layers with almost no cohesion between the layers. The main roof is unstable with the formation during landing the quite extensive area of displacement and formation the heavy loads on the powered roof supports, mounting and security systems of drifts. There is added hydrodynamic factor to geostatic rock pressure: water inflow into the lava reaches 52 m^3 / h during the landing of the main roof. In order to drainage the mine water there was made the technical solution to construct the drainage ventilation roadway in the soil of coal seam C_5 at the depth of 8-9 m. In this regard, there is current problem of estimating the condition of rock massif in the vicinity of the drainage drift in the duration of the clearing works on the layer C_5.

3 MATERIALS UNDER ANALYSIS

To implement the task three-dimensional model of layered rock mass was built with the physical and mechanical properties of rocks according to the geological survey around the drainage drift on the "Samars'ka" mine (Bondarenko, Kovalevs'ka, Symanovych & Fomychov 2010). The simulated structure of the rock mass is presented in Figure 1.

The characteristic feature of this area is conducting of the clearing works above the drainage drift, which had previously been passed at a depth of 8-9 m meters of excavated coal seam C_5. The roadway is conducted in dangerous areas and the collapse of the roof rocks, which are characterized by weak intra-and interstratal clutch, so the change of the lithology of the roof may collapse.

As a result, the design model consists of 25 rock layers: on the fall / uprising – 55 m, height – 48 m and length –1 m, the angle of dip – 3 degrees. Each layer is modeled as a separate part of required size with the relevant mechanical properties. Feature of the calculation is the change of position of working face relative to the drainage ventilation roadway.

At a distance of 15 m from the left edge of the model and the depth of 9 m from excavated coal seam is the drainage drift, which section was modeled under the arched roof supports KSHPU-11.7. It is used to support the frame support resin-grouted anchors. In cross section the frame support maden from profile SVP-22 is inscribed. Step installation of frame support – 1m. All elements of the roof supports were modeled as separate parts with relevant mechanical properties. System of anchoring consists from the nine resin-grouted anchors, of which seven are set 2.4 m on the top of the drift and two length 2 m – on each side of the roadway. The diameter of the bearing element – 22 mm, installation step – 1 m.

Figure 1. Dimensional model of the rock massif.

Powered support for reducing the dimension of the task was modeled as a solid rectangular block with the reaction of the Resistance of the pressure corresponding to the bearing capacity of roof supports KD-80. Dimensions of the height and depth match is extraction height, the block length is equal to the length of the powered support section.

4 ANALYSIS OF THE COMPUTER SIMULATION RESULTS

For conducting numerical experiment there were built three models with a different working face arrangement. In the first stage this distance was– 14 m, in the second – 7 m, and in the third stage the working face caught up with the vertical axis of symmetry of the roadway. By the results of calculations there were constructed diagrams of the component stresses and displacements. For pair wise analysis shows diagrams of the first and last stages.

4.1 Vertical stresses σ_y

The first step of calculation was carried out at the location of the working face at the distance 14 m from the drainage roadway, and the developed space was set up to 15 m width of the model. If we analyze the resulting diagram of vertical stresses (Figure 2a) by complex, then you can see that almost completely massif is compressed at 8-9 MPa, with the exception of ground rocks of roadway, areas of the rocks softening around the developed space, as well as focal tensile stresses in the place of installation fastening constructions. The main concentration of compressive stresses is concentrated in the support area of the working face, due to the lowering of the roof rocks in the developed space. Continuity disturbances of the rock massif at the process of coal seam extraction provokes the stresses redistribution in the neighboring layers. The area of developed space provides the greatest impact. Lowering the roof is as long as there is no limit exceeding of the rock strength in compression, after that the destruction will occur and caving takes place in the developed space. Cracks and watering of massif can significantly speed up the natural process of collapse. As a result of this discontinuity rock layers in the conduct of clearing works over mechanized complex there is formed a zone of layers inflection, which provokes a sharp increase of pressure on the roof support and the formation a zone of bearing pressure with high compressive stresses about 20-30 MPa and size along the rise – 5-7 m and the height of 10-13 m.

Developed space region, formed directly behind the working face is characterized by a large area of softening roof rocks, which greatly reflected on the distribution of stresses. The compressive stresses are less intense than in the surrounding layers. This area is spread over 4-6 m in the roof, and 12-15 m in the ground from the developed space and generally takes to the rise 15 m and height of 20-25 m. In this case the compressive stresses ranging of 0.5-1.5 MPa. Further there is a gradual increase of stresses until reaching the initial state of the massif.

Around the drainage drift is sufficiently predictable distribution of the stresses is occurs, its associated with the effect of maintaining roadway, including alternation of the frame support and anchoring to the surrounding massif. At the roof and ground of the roadway there is forming the discharge area with reduction of the compressive stresses σ_y up to 1-3 MPa, which is associated with swelling of ground rocks (dimensions: width 1.5-2 m, depth 2-

4 m) and formation the fixed plate by rock anchors in the roof. In this case in the sides of the roadway, by contrast, there is an increase of the stresses up to 11-14 MPa (dimensions: height 4-5 m, width 3-4 m) in connection with the violation of the initial state of the rock massif as a result of the drift excavation.

(a)   (b)

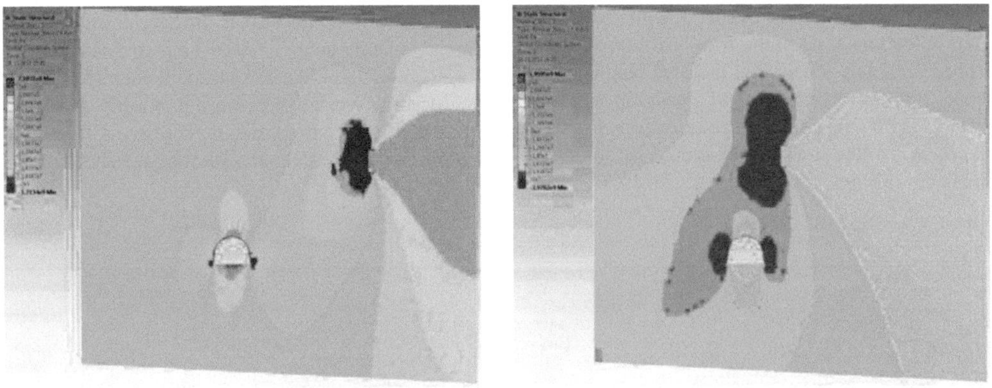

Figure 2. Diagrams of the vertical stresses σ_y distribution: (a) at a distance of 14m from the working face to the roadway, and (b) working face sharing it with the vertical axis of symmetry of the drift.

The characteristic feature of this numerical experiment is overworking of the drainage airway drift that at the first stage of the calculation yielded a whole new stresses region, which combined the bearing pressure zone ahead the working face area with high compressive stresses in the side of roadway. Due to the sufficient remoteness of the working face at the first stage of the calculation, this area does not have a significant impact on the roadway, but now it is possible to assume that with clearing works redistribution of loads in the buffer area between the bearing pressure zone and roadway will have some impact on the frame and the anchor fastening which installed in the drift. At this stage, the overall dimension of this area to the rise reaches 12-14 m and height 27-30 m, with compressive stresses at 12-16 MPa.

The distribution of the vertical stresses σ_y in the rock massif (Figure 2b) in the last stage of the calculation has undergone significant qualitative and quantitative changes. This is due with the position of the working face above the drainage drift that causes the formation of maximum influence on the roadway.

The bearing pressure zone that formed around the working face, reached a maximum size compared to the previous stages of the calculation. First, this is

due to an increase of the console hanging of the main roof, caving rock away from the face, as well as displacement the main part of the overlying layers, which provokes a displacement of rock layers not only over the extraction area, but also over mechanized complex. Pinch of large amounts of rock massif and formed at the same time mould with the increase in the area begins to exert increasing pressure on the supports. Second, in the domain of influence of the bearing pressure zone is adjudged the roadway around which is also take place the typical stress distribution. Due to the relatively small distance between the excavate coal seam and drift, the compressive stresses are superimposed from the bearing pressure zone in the roadway sides and around the mechanized complex. It leads to the formation the unified field of high compressive loads, which occupies the entire space between roadway and working face, as well as including all bearing pressure zones.

The highest concentration of compressive stress formed around mechanized complex with dimensions along the rise of 6-8 m and 13 16 m in height and also in the roadway sides with the dimensions: width 1-2 m and a height 3-4 m where the stresses exceeding 20 MPa. Considering the low strength characteristics of the Western Donbass rocks, such loads indicate their weakening. As described above, there is formed a single area between these zones with slightly lower stresses about 15-17 MPa, which also indicates the softening of rocks and the appearance of the cracks. Also the formation of this area around the roadway is uneven. From the working face advance of its dimensions reaches the 4-5 m in width and 9-11 m in height, while at the opposite side of roadway this area is much larger and up to 10-12 m in width and 16-18 m in height. Due to results of the calculations it connects with the formation of unloading zone in the rocks of the soil under the open area. Obviously that the area has some influence on the distribution of loads around the roadway. Also, the unloading zone formed in the roof and soil of the roadway. It should be noted that decreasing of the compressive stresses occurs gradually, therefore there is formed the sizeable area with loads in the range of 10-15 MPa around the described zones. It is indicates the occurrence of deformation processes at a considerable distance from the roadways.

4.2 *The intensity of the stresses* σ

Around the working face a zone of the bearing pressure with impressive size (Figure 3a), about 7-10 m along the rise and 16-20 m in height was formed. Feature of this area is rather high stresses of about 16-20 MPa, given the low strength characteristics of the rocks in Western Donbass, it is fairly safe to assume that in this area there is an active formation of fractures, including the destruction of rocks.

(a)

(b)

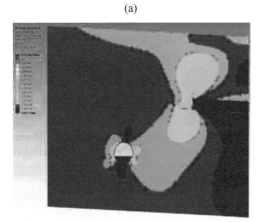 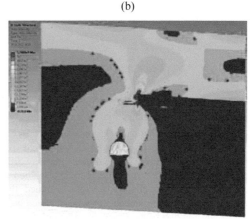

Figure 3. Diagrams of the stresses intensity distribution: (a) at a distance of 14 m from the working face to roadway, and (b) there working face sharing it with the vertical axis of symmetry of the roadway.

Over the mined-out space a zone of unloading is made, which naturally, given the collapse the roof rocks in mined-out space. In this case, in the soil here was a sharp decrease of pressure, to lows of up to 1.5 MPa.

Around the drainage drift is observed an active redistribution of stress intensity. In the soil of the roadway is unloaded, which is associated with the process of heaving rocks inside the section. On the sides, conversely, the stresses are high enough to reach the 10-13MPa for sizes 5-7 m in height and 2-3 m in width. At the top there is a slight unload, associated with the effect of anchoring.

Considering features of the experiments, special attention should be paid to the relationship of the stresses due sweeping the working face and the availability of drainage drift. At a distance from lava to drainage drift 14 m, we can already see the connection stresses of the two zones that form a common area of the stresses distribution of considerable size. By the uprising a new area is about 15-20 m and height 18-22 m. With concentration of the stresses is at a level of 8 MPa. Analyzing the location of this area, and the stresses size can be assumed that at this stage of calculation is increased of loads on the drainage airway, but this effect is still low.

The redistribution of the intensity of the stresses in the rock massif (Figure 3b) mainly due to the increasing influence of clearing works on the roadway, which is quite natural, given the location of the working face perpendicular to the drainage airway.

The most important factor of the last stage of calculation there is impact of the bearing pressure zone of the cleaning work on the roadway. At this stage, the drift, as well as the surrounding rock massif are completely under the influence of the increased stresses σ. In the layers that lie between the drift and the lava and are roof of the roadway, the loads reaches 15-20 MPa. There is roof of the roadway is unloaded, which is associated with the work of the frame and anchoring support. However, the high intensity of the stresses is concentrated in the roadway sides, exposing of the lateral pressure the rack of the frame support. In the soil of the drift occurs the regular unloading with the flow of the process of rocks swelling. It should be noted that the stresses are distributed on a substantial distance around the the drift, with values of about 10 MPa, which is slightly greater than the initial stress state of the massif, but it does not lead to intensive cracking and softening of the rock layers. There is soil under the developed area, is unloaded and at some distance completely stop the affect of the clearing works.

5 CONCLUSIONS

As a result of numerical experiment there was found that with decreasing distance from the working face to drainage drift, the character of the stress distribution changes. In the first stage the influence on the drift from the clearing works was minimal, but with the approaching of clearing works increasing the influence takes place and leads to significant changes in the stress-strain state of rock massif. There are bearing pressure zones ahead the mechanized complex and they are connected in the sides of the drift,

which leads to a redistribution of stresses throughout the massif between the coal seam and the roadway. There are patterns of changes in the stress strain state that mean that more active processes occur of weakening of the soft rocks that extend for considerable distances in the roof, the soil and the sides of drainage drift. In this regard, it is possible to predict increasing of the vertical and lateral loads on the fastening system and intensification of swelling processes of the soil. The aim of the numerical experiment results is strengthening of the mounting system by strengthening the deep rock roof and more active opposition to lateral forces.

REFERENCES

Bondarenko, V., Kovalevs'ka, I., Symanovych, G., Fomychov, V., Martovytskyi, A. & Kopylov A. 2010. *Methods of calculation the displacements and strengthening of the border rock drifts in the Western Donbass mines.* Monograph. Dnipropetrovs'k.

Bondarenko, V., Kovalevs'ka, I., Symanovych, G. & Fomychov, V. 2011. *The influence of the heterogeneity characteristics of the thin-layer rock massif on the numerical experiment results at the stability of excavation mining workings.* Materials from V International conference "Mining school underground": 10-18.

Bondarenko, V., Kovalevs'ka, I., Symanovych, G. & Fomychov, V. 2010. *The methodology and the results of computer modeling of the stress-strain state of the system "layered massif-support" of the development roadway.* Mining magazine. Special issue: 62-66.

Mining of Mineral Deposits – Pivnyak, Bondarenko, Kovalevs'ka & Illiashov (eds)
© 2013 Taylor & Francis Group, London, ISBN: 978-1-138-00108-4

Methodology of gas hydrates formation from gaseous mixtures of various compositions

M. Ovchynnikov, K. Ganushevych & K. Sai
National Mining University, Dnipropetrovs'k, Ukraine

ABSTRACT: The given paper is devoted to mine methane utilization method development. The method consists in creation of gas hydrates from air-gas mixture exiting the degassing wells of coal mines. Some results showing formation parameters of mixed gas hydrates are presented in the paper. Description of laboratory studies on gas hydrate creation using definite parameters is given.

1 INTRODUCTION

The analysis of coal deposits development under modern conditions shows the necessity of new solutions for a line of problems to provide safety of mines exploitation, complex development of mineral resources and protection of the environment. One of these problems is a mine methane utilization recovered to the surface with various degassing methods, and methane taken onto the surface by ventilation current.

A known fact is that gas hydrates form and steadily exist within a wide range of temperatures and pressures. However, each separate gas is characterized with strictly defined parameters of pressures and temperatures of hydrates stable existence. Hydrates formation process is defined with the gas composition, water state and its mineral content, external pressure and temperature.

Natural gases consisting of various components mixtures form mixed gas hydrates. At this, crystals characteristic to both methane and ethane, and other gases are formed simultaneously. That is, for conditions of gas-air mixtures of degassing wells it is necessary to use the notion of mixed gas hydrates.

2 RESEARCH OF GAS HYDRATES FORMATION PROCESS KINETICS

Based on the existing data about the composition of researched gaseous mixtures shown on Table 1, the equilibrium parameters of gas hydrates formation were determined (Figure 1) (Makogon 1985).

The gas hydrate formation process occurs at a boundary of "water-gas" contact under conditions of gas full saturation with moisture. Thus, it is important to know moisture content in gas and its change

under various thermodynamic conditions. Hence, moisture content of the natural gas rises with temperature increase, reduces with pressure increase and salt content that is necessary to be taken into account when substantiating parameters of mixed gas hydrate formation from various gases.

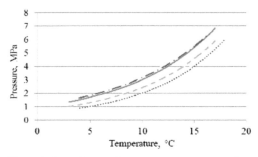

Figure 1. Dependences of pressure on temperature of gas hydrate formation for various gaseous mixtures (the composition is given in Table 1): — ·— ·- mixture 1; —— mixture 2; - - - - - mixture 3; ·············· mixture 4.

There are two principle trends in experimental practice to research gas hydrate formation kinetics: the first – researches without implementing mechanical stirring, that is hydrate-forming components supply is carried out due to diffusion, and recovery of the heat releasing during phase transition is implemented by the heat conductivity laws; the second trend (dynamic mode) – use of mechanical stirring.

An advantage of the first method is the possibility of gas hydrates process formation research at the stage of the crystallization centers nucleation, and also the possibility of visual observations conduction and kinetics-morphology researches.

Table 1. Compositions of researched gaseous mixtures, (molar fractions).

Number of mixture	Methane CH_4	Ethane C_2H_5	Propane C_3H_8	Isobutane C_4H_{10}	Carbon dioxide CO_2	Nitrogen N_2
1	0.8641	0.0647	0.0357	0.0099	–	0.0064
2	0.7516	0.0595	0.0333	0.005	0.002	0.143
3	0.932	0.0425	0.0161	–	0.0053	0.0043
4	0.674	0.037	0.019	0.006	0.008	0.25

If the gas hydrate formation kinetics is defined by the diffusion rate, it can be stated that:
– nucleation centers occur at the ice surface at the presence of gas and then convert into gas hydrate until the entire ice vapors and converts into the hydrate;
– after gas hydrate nucleation centers formation at the ice surface the further ice conversion into the hydrate takes place as a result of the gas diffusion through the formed gas hydrate layer.

3 LABORATORY RESEARCHES

At present, scholars of the department have been conducting researches on artificial gas hydrates on a new laboratory unit NPO-5 (Figure 2).

The experiment conduction is carried out by the following way: special vessel of gas hydrates formation (24) is placed in the cooling chamber (23). The special vessel volume can vary from 0.05 to 2 liters. Unit for water pressurized injection (25) and tank with methane (1) are installed outside the cooling chamber. At the beginning of the experiment the shaft (11) is inserted into the special vessel (24) and fixed with capscrews with nuts.

At the beginning of the experiment the shaft (11) is inserted into the vessel (24) and fixed with coupling bolts with screws. A definite water volume is injected into the shaft (16) which is inserted into the cylinder (17) on the body of hydraulic jack (21).

Cooling chamber (24) is set in a working mode with setting the temperature up to a necessary value and LED cluster (10) is plugged into the electrical network. Necessary pressure is created in the cylinder (17). The tap for water injection control (15) is opened and simultaneously to it the tank tap (2) is opened, that is water and gas injection into the unit (25) is performed simultaneously.

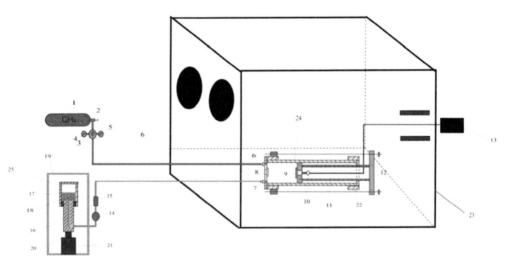

Figure 2. Laboratory unit NPO-5 for gas hydrate receive, where: 1 – tank with methane; 2 – tank tap; 3 – reduction gear; 4 – high-pressure manometer; 5 – low-pressure manometer; 6 – fitting for methane injection; 7 – fitting for water injection; 8 – cylinder transparent glass; 9 – shaft transparent glass; 10 – LED cluster; 11 – shaft; 12 – coupling bolts with screws; 13 – battery; 14 – water pressure manometer; 15 – tap for water injection control; 16 – shaft of water pressure unit; 17 – cylinder of water pressure unit; 18 – water; 19 – rigid frame; 20 – shaft of hydraulic jack; 21 – body of hydraulic jack with pressure up to 5 MPa; 22 – guidance flange; 23 – climatic thermal unit; 24 – vessel of gas hydrates formation; 25 – unit for water pressurized injection.

Water pressure is controlled with the pull-out value of the hydraulic jack shaft (20) and gas pressure – with reduction gear (3). In order measure the pressure in gas and water hoses there are manometers installed: (14) – for water, (4) and (5) – for gas.

Gas and water are simultaneously injected into the cooling chamber through the fittings (7) and (6) moving the shaft (11) back till the position where it is regulated with coupling bolts with screws (12). Water and gas form a mixture filling the shaft (11) area that leads to formation of a gas hydrate. At this, maximal effect of gas hydrates formation rate is achieved since water and gas molecules interaction activity value reaches its peak.

The readings are written down in a special table showing the following parameters: gas hydrate formation starting time, end time of the process, temperature change, water and gas pressure change with time. The readings are taken every 30 minutes. The computer takes pictures of the process taking place in the unit.

Table 2. Equilibrium parameters (P-T) of methane hydrate.

Temperature, °C	Pressure, MPa
-10	1.8
-7	2.1
-2	2.4
0	2.7
+5	4.8
+9	6.8
+13	9.8

Unit NPO-5 serves as a laboratory base of gas hydrates creation and research of their stable existence under certain values of pressure and temperature. Formation conditions of methane hydrates and equilibrium parameters of their stable existence received by experimental way under are presented in Table 2 and on Figure 3.

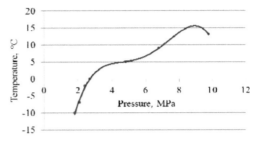

Figure 3. Pressure and temperature correlation of gas hydrate formation.

4 CONCLUSIONS

1. Formation conditions of mixed gas hydrates from air-gas mixtures of various composition is presented within the range of temperatures from -10 to +17 °C and pressures from 1 to 10 MPa.

2. Description is given to the working scheme of the gas hydrate receiving unit using simultaneous injection of water and gas within the temperature range from -30 up to +100 °C.

3. Further direction of the researches is focused on the establishment of dependence of various mineral content of water on thermobaric conditions for gas hydrate formation.

REFERENCES

Makogon, Y.F. 2007. *Gas hydrates: prevention of their formation and use.* Moscow: Nedra: 232.

Mining of Mineral Deposits – Pivnyak, Bondarenko, Kovalevs'ka & Illiashov (eds)
© 2013 Taylor & Francis Group, London, ISBN: 978-1-138-00108-4

The development of methodology for assessment of environmental risk degree in mining regions

A. Gorova, A. Pavlychenko & O. Borysovs'ka
National Mining University, Dnipropetrovs'k, Ukraine

L. Krups'ka
Pacific National University, Khabarovs'k, Russia

ABSTRACT: Sources of environmental risks in the mining industry are considered. The necessity of developing the methodology for assessment of the ecological risks degree for the environment and people in mining regions is validated.

Ukraine is a country with developed and powerful mining complex, which activity leads to the development of degradation processes in nature and society. Contamination of the environment as a negative by-product of human activities is a factor that limits the sustainable development of society, and its importance is growing every day (Rudko 2006 & Grebenkin 2009).

Mining industry prevails among other industries by the degree of environmental impact, adversely affecting surface and ground water, soil, subsoil and landscape. It yields up to chemical, petrochemical, metallurgical and energy industries only by the effect on air basin. Long-term use of mineral resources in mining regions resulted in significant changes in the environment and emergency situations appearance (Rudko 2006).

It is impossible today to plan and implement further development of industry and agriculture without considering the atmosphere, natural waters and soils contamination, as well as its impact on human health and welfare and the state of ecological systems in general. In this connection the receiving objective information about the level of environment contamination and response of biological objects, including humans takes on special significance.

For that reason there is a need for development and implementation of methodology to determine the environmental risk degree caused by industrial activity mining complexes, as well as ways to reduce these risks for environment and public health (Gorova 2012; 2009 & Koskov 1999).

Therefore the aim of this work is to study the characteristics of influence of mineral extraction processes on the state of environmental components,

biological systems, and population health for early warning of environmental risks.

To create a methodology to determine the degree of environmental risk in mining complexes at different stages of their operation (from designing to the date of liquidation) the following tasks must be performed:

– to generalize and systematize published data and materials of patent search on the issue of environmental risk management in mining regions;

– to identify technological and environmental risks and conditions for their occurrence in mining complex;

– to examine the characteristics and regularities of formation of natural and mining systems and their impact on the environment and public health;

– to create a methodology for assessment the degree of environmental risks in the territories of various mineral resources development;

– to develop a complex of measures to reduce environmental risks in the mining regions.

Methodology for defining the environmental risks in the mining regions should take into account the peculiarities of their formation in underground and opencast development of mineral resources, as well as in areas of mining waste disposal.

In addition, the establishment of scientific bases of environmental risk management includes the development of evaluation criteria and risks valuation for the environment and the population in areas of mining regions.

To investigate the ecological state of environmental objects it is recommended to use not only physical and chemical methods, but also bioindication methods which allow determining the toxicity and mutagenicity level of contaminated environmental objects as well as to assess their risk degree to biota.

Using the set of methods, including highly sensitive bioindicaton methods, allows to consider the overall negative impact of anthropogenic factors and to determine their complex effects on different biological systems and also to manage environmental risks in mining enterprises for their environmental and social security. It will allow to prevent crisis situations in the mining areas in time.

The studies of ecological risks levels for the environment and population health state were conducted on the territory of Dnipropetrovs'k region. Towns with different type of mining industry were selected for studies: Vilnogirs'k (extraction of polymetallic ores), Zhovti Vody (uranium ore), Pavlograd (coal mining) and Nikopol (mining and metallurgical industry). The territory of the resort "Medical treatment and recreation complex "Solonyi Lyman' of Novomoskovs'k district of Dnipropetrovs'k region

was used as a control. There are no mining companies on the control area and it undergoes minimal anthropogenic impact. From two to four test polygons, covering both industrial and residential areas were selected on the territory of each town. This range of test polygons was caused by the need to assess and analyze the ecological state of territories with different types of mining. From 4 to 8 monitoring points were allocated at each test polygon and the sampling of soils and plants was conducted there. Fifty two monitoring points was examined in total.

Integrated environmental assessment of the environment was carried out by the method (Gorova 2007; 2009). State of the atmosphere and soil by toxic and mutagenic background was assessed by cytogenetic biotests. Assessment of population health was based on statistics analysis of the prevalence of all diseases classes.

The results of calculations are given in Table 1.

Table 1. The assessment of environmental and ecological and genetic danger to humans and biota in areas with different kind of mining.

Town	Ecological and genetic danger	Danger level	Environmental danger	Danger level
Nikopol'	0.434	average	0.461	above average
Zhovti Vody	0.637	high	0.613	high
Vilnogirs'k	0.497	above average	0.515	above average
Pavlograd	0.487	above average	0.493	above average
Solonyi Lyman	0.182	below average	0.192	below average

Analysis of the data given in table 1 indicates a "high" level of environmental risk to biota in the territory of Zhovti Vody town. Environmental danger has "above average" level in urban areas of Vilnogirs'k, Pavlograd and Nikopol'. "Below average" level of environmental risk was found on the control area. It should be noted that a similar situation is observed with the level environmental and genetic danger, the exception is the Nikopol town, where "average" level of environmental and genetic danger was found.

Thus, the exceeded levels of environmental danger (in 2.4-3.5 times in comparison with the control area) and environmental and genetic danger (in 2.4-3.2 times) were elicited in the mining towns. By the level of environmental danger studied towns can be ranged as follows: Zhovti Vody > Vilnogirs'k > Pavlograd > Nikopol' > Solonyi Lyman.

After years of research at the territories of various minerals extraction the information database on ecological state of environment and population health was created. This database is constantly supple-

mented with the results of new research.

As a result of researches at the territories of mining centers the following outcomes were obtained:
– significant theoretical and practical experience was gained in conducting research of ecological condition of the environment objects and public health in areas of development of different mineral types;
– environmental characteristics of natural systems state by the degree of anthropogenic disturbances in the process of mining are formed;
– regularities of changes in the ecological state of the environment and biological systems depending on the method of mining are established;
– theoretical, practical and information base of methodology for determining the degree of environmental risk caused by production activities of environmentally hazardous mining complexes is created;
– modern information and mapping environmental data base is formed, on this base it is planned to develop environmental management activities.

The developed methodology for determining the degree of environmental risk can also be used to adjust the value of land, to calculate insurance rates and compensation by people living and working in unfavourable environmental conditions etc.

Implementation of the methodology in areas of mining enterprises operation will prevent the unwanted environmental effects and will allow creating favourable conditions for human health and the preservation and reproduction of the environment.

REFERENCES

Rudko, G.I. & Goshovskiy, S.V. 2006. Environmental safety of technic and natural geosystems (scientific and methodological basis): Monograph. Kyiv. Nichlava: 464.

Grebenkin, S.S., Kostenko, V.K. & Matlak E.S. 2009. *Conservation of the natural environment in the mining operations:* Monograph. Donets'k. VIK: 505.

Gorova, A., Pavlychenko, A. & Kulyna, S. 2012. *Ecological problems of post-industrial mining areas. Geomechanical processes during underground mining.* Leiden. The Netherlands: CRC Press/Balkema: 35-40.

Koskov, I.G., Dokukin, O.S. & Kononenko, N.A. 1999. *Conceptual Foundations of Ecological Security in the Regions of Mines Closure.* The Coal of Ukraine, 2: 15-18.

Gorova, A. & Pavlychenko, A. 2009. *Integral assessment of the socio-ecological state of mining regions of Ukraine.* Mining Journal. 5: 49-52.

Gorova, A.I., Ryzhenko, S.A. & Skvortsova, T.V. 2007. *Guidelines 2.2.12-141-2007. Survey and Zoning by the Degree of Influence of Anthropogenic Factors on the State of Environmental Objects with the Application of Integrated Cytogenetic Assessment Methods.* Kyiv: Polimed: 35.

Bolt support application peculiarities during support of development workings in weakly metamorphosed rocks

V. Lapko, V. Fomychov & V. Pochepov
National Mining University, Dnipropetrovs'k, Ukraine

ABSTRACT: Practice of drivage and support of galleries requires reorientation to new, more progressive types of support yielding to mechanization, less labor and metal intensive. Analysis is conducted and researches are generalized for development workings support interaction processes with rock massif. Rational systems selection methods are considered together with parameters of near-the-contour rocks strengthening by roof bolts. The received results point at the fact that use of existing theories of roof bolts application possesses a row of peculiarities meeting specific mining-geological conditions. At this, each theory realizes its working mechanism of bolt support and does not allow to carry out a forecast of bolts installation optimal configuration.

1 INTRODUCTION

Coal deposits with weakly metamorphosed rocks have their distinctive features from viewpoint of support and protection of development mine workings. First of all, conditions characterized as "high depths" in deposits with weakly metamorphosed rocks are bedded on significantly shallower depths than in other deposits. Authors (Usachenko 1990) state that if in Central Donbass such conditions begin from depth of 700-800 m, then in Western Donbass corresponding manifestations of rock pressure are observed at absolute depths of 200-300 m.

Researchers (Braginets 2004 & Pirs'kyi 1990) have observed manifestations of significant absolute value of rock pressure on mines developing coal seams in weakly metamorphosed rocks. At this, galleries contours dislocations reach dozens of centimeters and have practically continuous behavior. As a result, there is no possibility of galleries maintenance without repairing.

Modern development of Ukrainian coal industry has served as a rapid increase of drivage works and, correspondingly, expenses for supporting underground mine workings. It is obvious that during this period active application of bolt support that had a whole row of advantages compared to other types of support was introduced.

Long-term experience of bolt support application has shown that this support can be applied in wide range of geomechanical characteristics of rocks both independently and in combination with frame supports, sprayed concrete and bulk concrete (Korchak 2001).

Annually, there are more than 500 million roof bolts installed in world coal industry. Besides, 70-90% of total volume of driven galleries are supported by bolts in the most developed coal countries.

Under conditions of Western Donbass in 2006 all driven galleries were supported by frame support, in 2008, 64583 linear meters of galleries were supported by bolts, from them – 55287 linear meters by frame-bolt support, 9296 linear meters – by bolts, that allowed to reduce labor intensiveness and end operations time (Vivcharenko & Liadets'kyi 2009). In crisis 2009 year 58995 linear meters were driven and in 2010 production volume returned to previously gained results – 63074 linear meters.

2 FORMULATING THE PROBLEM

When armoring a roof by bolts the function of bolt support comes down to binding unstable part of massif to stable one, or stitching roof layers to "bearing" beam, or to carrying out two functions especially at the beginning period of galleries exploitation.

Performance of binding and supporting functions at the final stage of exploitation depends on the shape and dimensions of contour of massif unstable part around the gallery, properties of bolted rocks, scheme of bolting. Choice of bolting scheme influences level of drilling means development, elementary basis of bolt support, knowledge of host rocks properties, shapes and qualities of rock pressure manifestations depending on quality of near-the-contour massif armoring by bolts.

3 STATEMENT OF BASIC RESEARCH MATERIAL

Bolt support can be applied (Vysots'kyi 2005): (a) independently – in haulage and ventilation drifts, crosscuts, shaft station, brake slopes, inclines and man ways, intermediate galleries, ventilations linkages, ramps and development workings; (b) as an additional (strengthening) support in combination with prop support (so called combined support) – in capital galleries at an increased rock pressure and in development galleries located in stoping works influence zone; (c) as a temporary support – in long galleries, abutments of mine workings, chambers , etc., with successive support using props; (d) as means against bottom heaving in unwatered rocks of capital and development workings.

Two schemes of bolting are used depending on bolting depth: single-level, with bolts of 2.6 m in length; multiple-level with bolts of standard length (up to 2.6 m) and bolts of deep installation with length more than 2.6 m (2.6-10 m and more).

By rigidity of load-bearing rod the bolts are differentiated as: rigid (steel, wooden, fiberglass); flexible (rope, bunch-type).

Boreholes for bolting have various shape and cross-section depending on the applied bolting scheme, type of bolts, drilling means with rigid base of drilling pipes or with flexible drilling pipe.

Bolts by interaction type with blast-hole walls are divided into the following classes: lock, lock-free, combined.

Raisin-grouted bolts of ampule fastening cannot be applied in flooded galleries and also in open-pits that use mass explosion technologies. When an explosion takes place chemical-mechanical links of polymer plug with rocks and load-bearing rod of bolt get broken after that raisin-grouted bolts do not carry out their function of binding and supporting. In wet semi-stable rocks of mudstone type the raisin-grouted bolts during tests after 4-5 hours had insignificant bearing capacity (60-80 kN) and to increase it – there is a necessity to significantly deepen the bolt behind unstable contour of the massif. Raisin is washed away at flooded seams.

The most widespread bolts are raisin-grouted ones with application index equal to more than 60% because of polyester raisin used in ampules: high mechanical compression strength $(80 \, N / mm^2)$, bending strength $(15 \, N / mm^2)$, shear strength $(3.6\text{-}4.2 \, N / mm^2)$, elasticity modulus $(15 \, N / mm^2)$

provide high bearing capacity of bolts; high adhesive strength with rocks and metals provide formation of a "compression cone" in unstable rocks; ampules elasticity allow to use them for direct and deviated boreholes; little volume and packing create comfort in transportation and in handling; significant (up to 6 months) storage time does not interfere in supply; possibility of ampules selection with precise choice of setting-time allow to implement active method of bolts fastening; raisin plasticity provides sealing that prevents rock weathering, metal corrosion, rock shearing deformation, decrease loads on support elements (Remezov 2003).

This property of raisin allows to install bolts of various length and rigidity using similar technique and technology including bolts of deep installation (steel ones connected with coupling, rope one with length of 3-10 m) and flexible fiberglass, coal-plastic bolts of ordinary length. Possibility of double-level support scheme realization is created under complex conditions (Figure 1).

Except for mentioned factors, the preference is given to rope bolts also because of the fact that jointing bolts do not have residual strength whereas rope bolt get broken along different strands. After one strand gets broken (in total there are 7 strands) residual strength makes up approximately 80% of maximal. Next strand gets broken after additional dislocations at level of residual strength and so on.

Examining theories of bolts and rock massif interaction the authors (Shyrokov 1983) should be mentioned. They state that during intermediate rock binding to more strong rocks by bolts in development working the bolts length is defined based on the condition of their locks fastening behind possible zone of caving. At the same time, density of bolts installation should correspond to conditions at which their summary bearing capacity exceeds weight of supporting by them rocks.

Basic calculation value to determine parameters of bolt support is roof area for one bolt and defined as function of its lock bearing capacity or preliminary initial tension of bolt. Theory's essence (Kovalevs'ka 2011) lies in the fact that during bolt support application the rocks are exposed to artificial strengthening. And the load-bearing construction is formed in the massif analogical to composite beam, plate, vault or arc. Bound rocks are exposed in this case mainly to compression and the bolts receive extension forces.

Bolts length is calculated based on calculation methods of this theory (Fomychov 2012). Roof area depends on stress (bearing capacity) of bolts or strength properties of materials from which they are made.

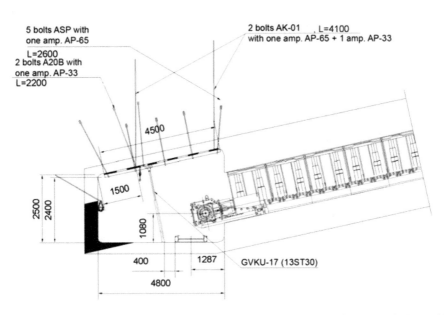

5 bolts ASP with
one amp. AP-65
L=2600
2 bolts A20B with
one amp. AP-33
L=2200

2 bolts AK-01   L=4100
with one amp. AP-65 + 1 amp. AP-33

4500

1500

2500
2400

1080

400   1287   GVKU-17 (13ST30)

4800

Figure 1. Fragment of extraction district pattern with double level bolting scheme in trapezoidal shape of mine working.

Supporters of the compression theory of supported rocks assume that bolt support work comes down to compression of bound rock strata and prevention of extension stresses existing in its lower layers. If to install bolt support right after rock mass excavation having compressed the rocks by force equal to preliminary tension of bolts the development of stresses will halt and rocks destruction will not occur. If the gallery stays unsupported during some time then the roof is exposed to bending and extension stresses developing in lower layers will lead to dangerous deformations and extensions of roof (Bondarenko 2012). Roof area allocated for one bolt is defined by authors of this theory depending on the value of initial tension of bolts.

Considered theories of bolt support work conform with practical data in those cases for which they are recommended. However, proposed calculation methods are based on simplification considered phenomena and possess a row of assumptions decreasing reliability and practical value of received results during the calculations.

Thus, authors consider interaction of bolt with its supported rock as static task, although it is known, that after support installation the rocks continue dislocating towards gallery during a definite time.

As a basic factor, defining density of bolts installation is accepted as bearing capacity of bolts locks or their initial tension. It is considered that locks bearing capacity value and bolts tension during support working process stay unchangeable although practical and special researches data show that bolts

tension significantly change. In connection with that, last years are dedicated to research of bolts work under conditions of their long-time exploitation.

Interaction mechanism of bolt with supported rocks is explained by theory of mutual work of bolt with rock and shows that bolt support being yieldable cannot fully prevent process of inelastic dislocation of roof rocks, it can only restrict rate of deformation and stop it under definite conditions.

At the first moment of bolts installation their pressure on rock over the holding clips is defined as initial tension. Then under influence of rock pressure forces the support reactive resistance rises, and bolts, as a result of elastic extension of metal and lock sliding along the borehole, dislocate towards gallery together with its contour. During dislocation, the value of surrounding rocks stresses decreases and ability of bolts to receive load rises.

This process will be developing until balance state establishment in the system support-rock, i.e. until reactive resistance of bolts will be equal to rock pressure forces influencing them. If bolt will be installed following roof rocks exposure and create necessary stress in it, then rock layers will be held by natural force and their lamination will be stopped.

This also raises bearing capacity of rocks. Thanks to bolt support work the rock layers sliding reduces. With help of bolt support interaction of separate rock layers in roof is provided. It means that separate layers of rocks in zone of stresses reduction are bound by bolts, hence vertical extension and compression influence is excluded, and roof bearing capacity is provided.

That is why in thin-layered rocks it is possible to bind roof rocks with help of bolts and, hence, prevent rock layers delamination from massif, mutual dislocation and bending into mine working. Roof stability will depend on bending strength limit and number of bound rocks.

Supporters of energy theory of rock massif interaction with bolt support count that potential energy releases during mine working drivage, value of which is defined by initial stress state. Given approach is universal but its practical application is complicated by initial data determination.

Mine working can be supported by bolt and when the roof is not plane but arc-shaped then the rock is partially destroyed or various characters of layers bedding is presumed. At arc-shaped gallery, an additional resistance to side dislocations of rock layers is provided.

When a gallery is of an arc shape the bolts are installed in rock in which there is no equilibrium stress state or where the stresses did not exceed rock elasticity limit.

The bolts should be installed with some angle to the plane of cracks in order to counteract the extension stresses, i.e. in roof – to lamination planes, and in walls – to longitudinal cracks planes. In rock layers with various dip angles the bolts should be installed in a fan-shaped manner (Figure 2) considering dip angle or rocks continuity breakage.

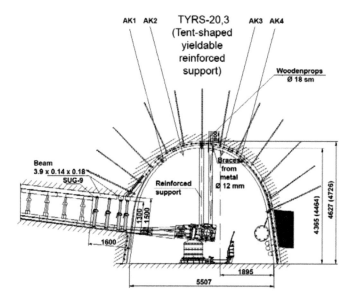

Figure 2. Fragment of extraction district pattern with double bolt scheme at arc shape of a mine working.

Calculation model of bolt-frame support interaction with near-the-contour rock massif for conditions of layered mine working of "Yubileynaya" mine has shown that distribution of stresses in bolts and surrounding their rocks depends on current mining-technical conditions to a certain degree.

Figure 3 shows that mechanism of bolts work can significantly vary within the range of single calculation model. Bolts in their upper part are squeezed by strong siltstone but stresses distribution in lower part of bolts is substantiated by not only strength characteristics of surrounding less strong mudstones but also by their location. Bolt distanced from the central one receives significant stresses in the zone located close to a gallery contour. Such picture is not uncharacteristic and caused by concrete combination of bolts location and mining-geological characteristics of rock massif.

Characteristics of bolt model influences calculations. Hence, calculation experiment was conducted to analyze influence of bolt tension degree and value of supporting plate area on stresses distribution in rock. The task was being solved for elastic-plastic state for 25 combinations of variable values.

When analyzing entire range of calculations the following feature of full dislocations on value of support spacer area. Deviation values from maximal dislocations from initial ones made up not more than 5% that is comparable to calculation error, and after transfer to over-the-limit state – deviations rose and made up from 25 to 55% depending on value of initial bolt tension.

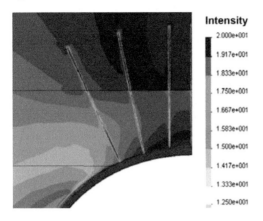

Intensity

2.000e+001
1.917e+001
1.833e+001
1.750e+001
1.667e+001
1.583e+001
1.500e+001
1.417e+001
1.333e+001
1.250e+001

Figure 3. Fragment of diagram of bolt-frame support stresses and surrounding rock massif.

4 CONCLUSIONS AND PERSPECTIVES OF FOLLOWING REEARCHES

Modeling results of bolt interaction with weakly metamorphosed rocks point at strength increase of "bolt-rock block" system relatively to ordinary rock block during its elements transfer to over-the-limit state. Hence, all world experience of mining industry development, conditions coal deposits of Ukraine mining and specifically quite complex mining-geological situation of Western Donbass mine underlines high currency (both at present and in perspective) of bolt support application for effective and resource-saving support of mine workings. As bolt support application experience shows, multiple-level supporting schemes are more reliable when bolts of deep installation are fastened behind the contour of massif disintegration zones.

As a result of above mentioned factors, when ropes bolts are preferred to be used when bolting depth exceed gallery height. Although when mastering technology of rope bolts application the calculation methods of their parameters, technological

rules, technological parameters depending on supporting method and fastening material application. In order to achieve it, further development of design and calculation of rational support for concrete conditions are needed.

REFERENCES

Usachenko, B.M., Cherednichenko, V.P. & Golovchan-s'kyi, I.E. 1990. *Geomechanics of galleries protection in weakly metamorphosed rocks.* Kiev: Naukova dumka: 144.

Braginets, I.D. 2004. *Methods of bottom rock heaving prevention of transport mine workings.* Sc. herald NMU of Ukraine, 2: 13-18.

Pirs'kyi, A.A. & Stovpnik, S.N. 1990. *Experimental-industrial tests of rocks strengthening method for prevention of bottom heaving.* Coal of Ukraine, 4: 9-11.

Korchak, A.V. 2001. *Methodology of underground structures design.* Moskow: Nedra communications LTD: 416.

Vivcharenko, A.V. & Lyadets'kyi, A.N. 2009. *Intensive mining of thin coal seams under mining-geological conditions of Western Donbass.* Dnipropetrovs'k: School of underground mining: LizunoffPress: 18-22.

Vysots'kyi, D.V. 2005. *Use of bolt support in mining.* Col. of sc. works "Scientific-technical problems of mineral deposits development, mine and underground building": Shakhtinskiy institute (branch) SRSTU (South Russian State technical university): 181-186.

Remezov, A.V., Kharitonov, V.G. & Zharov, A.I. 2003. *Improvement of supporting methods and means of stopes abutments with galleries.* Kemerovo: Kuzbass-vuzizdat: 167.

Shyrokov, A.P. & Gorbunov, V.F. 1983. *Increase of rocks stability.* Novosibirsk: Science: 170.

Kovalevs'ka, I., Vivcharenko O. & Fomychov V. 2011. *Optimization of frame-bolt support in the development workings, using computer modeling method.* Istanbul: XXII World mining congress & Expo. Volume I: 267-278.

Fomychov, V.V. 2012. *Bases of calculation models plotting of bolt-frame support considering non-linear characteristics of physical environment behavior.* Scientific herald of NMU, 4: 54-58.

Bondarenko, V., Kovalevs'ka, I. & Fomychov, V. 2012. *Features of carrying out experiment using finite-element method at multivariate calculation of «mine massif – combined support» system.* Geomechanical Processes During Underground Mining: AK Leiden (The Netherlands): CRC Press/Balkema: 7-13.

Mining of Mineral Deposits – Pivnyak, Bondarenko, Kovalevs'ka & Illiashov (eds)
© 2013 Taylor & Francis Group, London, ISBN: 978-1-138-00108-4

Pressure variation of caved rocks in mined-out area of face

O. Dotsenko
Donbas State Technical University, Alchevs'k, Ukraine

ABSTRACT: The article gives the results of approximate the experimental data on worn row ground pressure variation in time by modified form logistic correlation of Pearl-Reed and was obtained correlation describing the process of rock pressure increasing in the mined-out area in the development at the depth of 240 meters.

Deteriorating of coal-mining conditions at great depths because of intense manifestation of mining pressure contributes to extension of workings which have poor operational conditions. It is possible to solve this problem by maintaining preparatory mine workings conducted after stope (Babiyuk 2011) in caved and compacted rocks of mined-out area. In order to establish the parameters of this method it is necessary to know the law of the worn row ground pressure variation while face progressing ahead.

Usually, the pressure is studied by putting dynamometers into mined-out area after stope (Gapanovich 1974 & Zborshik 1978) and measuring their pounding in time while face progressing ahead. In (Gapanovich 1974) the results of pressure variation in mined-out area of faces where the developments are conducted at a speed of 1.25 m 24 hours and at 240 m depth in conditions of Chelyabinsk coal basin are given. Dynamometer readings are shown in Figure 1 as the points in a coordinate system $t - P$, where t – time of measurement after face passing, 24 hours, P – pressure, fixed by appliance, MPa.

Duration of pressure stabilization of rocks at the level γH coincides with the time of rock subsidence t_{act} (Gapanovich 1974) and equal to 73 days after face passing. During this period the pressure varies in 3 stages: moderate growth, intense growth and stabilization.

In order to determine the pressure at each time point the author (Gapanovich 1974) suggests to use a piecewise approximation where for each of the stages of pressure variation a separate correlation is proposed. Suggested approach is not a convenient one from a practical point of view, due to the lack of precise information on the distance where one or another stage of pressure variation of each face begins, and what is the weight of the rocks, not participating in this or that period.

The purpose of the research is to approximate the experimental data on worn row ground pressure variation in time by one correlation and determine the possibility of its extrapolation to the depth of the development of more than 240 m.

The basic requirement in the construction of the model is sufficient accuracy and simplicity of description of the basic curve showing the process of rock pressure growth in mined-out area for the whole distance of its stabilization, proceeding in three stages. Before the main roof caving (spacing of roof caving $l_n = 10$ m) the pressure is zero, since the rock fall (on the 7th day after passing mined-out area) a stage of moderate growth begins. Caving down with fissure, rocks form an arch. Forming the arch of caved rocks pressure increases intensively in mined-out area, and taking the weight of superstrata rocks the arch becomes stabilized.

Simple exponential $P(t) = \dfrac{t}{e^{a_0 - a_1 \cdot t} - 1}$, hyperbolic $P(t) = \dfrac{t}{a_0 - a_1 \cdot t}$ and logistic functions were considered. In the best possible way the data of correlation field (Figure 1) can be described by S-shaped curve, called the logistic. The use of logistic function in a "closed system" is justified, since we have "condition of the ceiling" – the value of γH, beyond which the pressure increasing in mined-out area, is impossible.

Let's consider logistic correlation of Pearl-Reed (Kinin & Priven 2011), which reflects a slow increase from the asymptotic minimum and turning into multiple rapid increases at the stage of intensive growth, at the end of which stabilization stage begins, and traced in slowing of pressure growth and in achieving the maximum of asymptotic:

$$P(t) = \frac{A}{1 + a \cdot e^{-b \cdot t}}, \tag{2}$$

where A – limit value of pressure, MPa; t – time of process, 24 hours; a, b – parameters of logistic curve.

To describe the process regarded, let's consider the correlation (2) in the modified form:

$$P(t) = \frac{\gamma H \cdot 10^{-3}}{1 + a \cdot e^{-b\left(t + \frac{l_n}{v} - t_{act}\right)}}, \qquad (3)$$

where P_t – pressure value in mined-out area, MPa; γ – bulk weight of massive, H / m³; H – mining depth, m; t – time after passing the face, 24 hours; l_n – spacing of roof caving, m; v – average speed of heading advance, m / 24 hours; t_{act} – duration of rock pressure stabilization, 24 hours.

The ratio l_n / v transposes the curve point to the point corresponding to the time of main roof caving, and from which the pressure starts to increase moderately. Parameter a specifies the location of a curve on the time axis. Its change shifts the curve to the right or to the left. On the steepness of the middle part of the curve influences the value of parameter b.

At the preliminary stage of processing the data have been tested on abnormality by Irwin's criterion λ_p:

$$\lambda_p = \frac{\left|P_t - P_{t-1}\right|}{\sigma_p}, \qquad (4)$$

where σ_p – mean-square deviation of the resulting feature (P).

Table value $\lambda_{tabl} = 1.1$ at a significance level $\alpha = 0.05$ and the number of observations $n = 47$ exceeds the calculated values, abnormal data were not found out.

To find the parameters of regression equation (3) assumption formula rearranges to the following form:

$$\frac{\gamma H \cdot 10^{-3}}{P(t)} - 1 = a \cdot e^{-b\left(t + \frac{l_n}{v} - t_{act}\right)}. \qquad (5)$$

Denoting the left side of the equation by Y, let's take the logarithm of the equation (5):

$$ln(Y) = ln(a) - b\left(t + \frac{l_n}{v} - t_{act}\right). \qquad (6)$$

Replacing again $ln(Y) = D$, $ln(a) = c$, we shall obtain a linear equation:

$$D = c - b\left(t + \frac{l_n}{v} - t_{act}\right). \qquad (7)$$

Finding the parameters of expression (7) is reduced to solving a system of normal equations obtained by the least-squares method:

$$\begin{cases} nc - b \cdot \Sigma\left(t + \frac{l_n}{v} - t_{act}\right) = \Sigma D; \\ c \cdot \Sigma\left(t + \frac{l_n}{v} - t_{act}\right) - b \cdot \Sigma\left(t + \frac{l_n}{v} - t_{act}\right)^2 = \Sigma D\left(t + \frac{l_n}{v} - t_{act}\right). \end{cases} \qquad (8)$$

By solving the system (7), the parameters: $b = 0.1109$ and $c = -3.6136$ are obtained. Since $c = ln(a)$, then $a = e^{-3.6136} = 0.027$.

The relation of the pressure variation in the mined-out space according to the data of the research (Gapanovich 1974) has the form:

$$P(t) = \frac{5.886}{1 + 0.027 \cdot e^{-0.1109\left(t + \frac{10}{1.25} - 73\right)}}. \qquad (9)$$

Figure 1 except for the actual measurement data shows logistic curve, which approximates them.

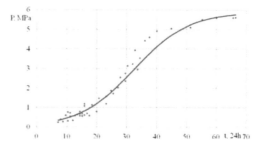

Figure 1. The actual measurement of pressure variation data in the mined-out space and logistic curve approximating them.

Determination coefficient of regression equation (9) is equal to $R^2 = 0.966$, which confirms a very high interrelation between pressure value and time after the face passing. Essentiality of the equation as a whole is verified by F-Fisher' criterion at a significance level $\alpha = 0.95$. The calculated value of F – criterion ($F_p = 26.7$) is higher than its table value ($F_{tabl} = 1.46$), therefore, the equation is statistically significant. The average error of approximation is equal to 19.26%, so the quality of the equation is good.

CONCLUSIONS

The correlation describing the process of rock pressure increasing in the mined-out area in the development at the depth of 240 m is obtained in this paper. Resulting logistic relation is limited by the values of the maximum pressure γH and duration of the pressure stabilization period t_{act}. Knowing the value of t_{act} it seems possible to extrapolate the obtained correlation (9) to the depth of the developments typical for Donbas region.

REFERENCES

Babiyuk, G.V, Dotcenko, O. G., Averin, G.A. & Brajinsky, S.S. 2011. *Patent of Ukraine 60791Mwthod of developing flat-lying seam by long wall face*. E21C41/16. Publish 25.06.2011. Bul. #12

Gapanovich, L.N., Gaydukov, U.G. & Lavruhin, V.N. 1974. *Research regularity of demonstration rock pressure in mined-out area of face under mining layer coal in the Chelyabinsk's basin*. Coal, 2: 13-18.

Zborshik, M.P. 1978. *Heading protection on deep mines in mined-out area of face*. Kyiv: Technique: 176.

Kinin, A.T & Priven, A.I. 2011. *Elementary models of developing technique system* [Electronic resource]. Metodolog. Electronic data. [St.Petersburg?]. – access mode: http://www.metodolog.ru/node/940, free – title from screen.

Mining of Mineral Deposits – Pivnyak, Bondarenko, Kovalevs'ka & Illiashov (eds)
© 2013 Taylor & Francis Group, London, ISBN: 978-1-138-00108-4

Magnetic stimulation of transformations in coal

V. Soboliev, N. Bilan & D. Samovik
National Mining University, Dnipropetrovs'k, Ukraine

ABSTRACT: Results of experimental studies shows that the chemical reactions in the coal substance under the influence of weak magnetic fields are essentially directed at formation of stable gas molecules due to re-combination of free radicals, which also enter into chemical reactions with components of the coal organic mass and as a result increasing the crystalline constituent of the structure. The action of the external magnetic field alters the direction of the magnetic moments (spins) of electrons and stimulates the formation (or extension) of crystalline phases of carbon and hydrocarbon chains, two-dimensional carbon structures. It is assumed, that magnetic fields of weak strengths could be applied for the creation of a stable coal-gas system.

1 INTRODUCTION

The unstable state of coal nanostructure occurs as a result of phase transitions and chemical reactions excited by mechanical, electrical, thermal and other impacts, is forced to rebuild, thus providing the formation of stable phases and as low as practicable value of the stored proper energy.

The behavior of carbon nanostructures, the variability of the physicochemical state of its macro-volume has a direct action on the stability of the system coal-gas. As a rule, the system degradation starts to break some chemical bonds in coal (Soboliev et al. 2009) and result in the creation of conditions for instability in the limited volumes of coal-bed with the risk of gas-dynamic phenomena origin. Thus, the formed system with the new state can contribute markedly to the distribution of stresses in the rock mass around the excavation.

Investigations of the character of gas-dynamic phenomena should focus on the factors and physicochemical mechanisms of changes in the degree of stability both of separate nanoscale components of coal and general coal nanostructures. The relevance of these studies is due to the need to establish the unified system of physical understanding of sudden outbursts and effective ways to suppress them.

The purpose is to establish from experiments that the possibility of solid phase chemical reactions formed a thermodynamically stable solid phases and stable gas molecules under the influence of a weak magnetic field on coal.

2 MATERIALS AND RESULTS

In the experiments we use bituminous coal with the following characteristics: the carbon content of 86.6%, hydrogen 5.7%, vitrinite reflectance $R_0 \geq 1.03\%$, $Y = 18$ mm, $W = 1.1\%$, $V^{daf} = 33.2\%$, heat of combustion $Q^{daf} = 35860$ kJ / kg. Samples were prepared from coal powdered in a porcelain mortar and sieved to fractions 200 / 100 µm (according to a laser light scattering analyzer the size of the initial coal particles was 214.5-111.7 m). The average weight of the sample was ~ 1.28 g. Previous studying coal was dried at a temperature of 35 °C for 24 hours. The maximum heating temperature in magnetic treatments did not exceed 300 K. Current passing through the heating coil generated the magnetic field in the sample (Soboliev 2008). The maximum intensity of the pulsed magnetic field generated by the coil was not more than 240 A / m. During treatment in magnetic field coal samples were compressed in a container with steel electrodes. The treatment time for each sample did not exceed 4 hours.

X-ray diffraction studies of coal were carried out on X-ray diffractometer DRON-3M, which has an attachment for results output to a PC. Survey was made by the Debye-Sherrer method. Monochromatic CuK_α radiation was used. Spectrometer ВИГТ.421.410.001 (Russia) was used for the study of electron paramagnetic resonance. Comprehensive studies of the physical and chemical characteristics were carried out in the laboratories of the State Enterprise Research-Industrial Complex "Pavlograd Chemical Plant" using the following equipment: the thermogravimetric analysis apparatus (TGA) and differential scanning calorimetry (DSC) METTLER

TOLEDO (Switzerland), the optical microscope LEICA DM ILM (Germany), laser light scattering particle size analyzer SHIMADZU SALD-301V (Japan), and the calorimeter C-2000 IKA (Germany). The infrared spectra of coal were recorded on the Fourier transform infrared spectrometer ФСМ-1201 (Russia) with transmission in the spectral range of 400-5000 cm^{-1}.

The regularities of coal particles size distribution depending on the type of treatment are shown in Figure 1.

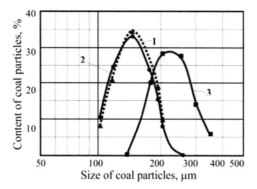

Figure 1. The pattern of coal particles sizes distribution: 1 – the initial sample; 2 – after treatment in the electric field ($E = 107$ V / cm, $T \approx 304$ K); 3 – after treatment in the magnetic field.

3 DISCUSSION

After treatment in the magnetic field (MF) the range of particles' sizes increased and reached 350-132 µm in relation to the initial sample and treated one in the electric field (EF).

Increase in particles size is probably due to the spin-dependent reaction (Buchachenko et al. 1978), running between the movable radicals and active sites on the solid surfaces. There is a clear proportionality between the concentration of surface active sites and the rate of a chemical reaction. Minimum activation energy is sufficient either for chemical reaction involving surface atoms or for the formation of stable gas molecules in the system of mechano-activated coal-gas.

The use of magnetic field as a way of energy stimulating of chemical reactions between particles in gases and liquids (Buchachenko et al. 1978) influences on properties of many non-magnetic crystals (Belyavsky et al. 2006), which energy of structural change is about 1 eV (phase transformation takes more energy). If we consider the Zeeman energy of the magnetic field with the order about 10^{-5} eV, which is much smaller than the thermal

energy kT of the particles, it becomes clear why the effects of magnetic fields (especially weak ones) seriously was not considered as an additional energy of reacting atoms, molecules or radicals even in the most intense magnetic field is negligible quantity compared to the energy of the thermal motion. However, it is known (Buchachenko et al. 1978) that magnetic treatment in weak fields of all others stimulates chemical reactions between radicals, transiting radical pair from the triplet spin state to the singlet one, thus increasing the probability of radical pairs recombination.

Two highly diffuse maximums, which correspond to the angles 2θ – 24 and 43 degrees, are showed up in all diffractograms of bituminous coal, Figure 2 (data on difractorgamms are given in Table 1).

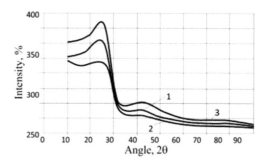

Figure 2. Diffractograms of coal under investigation: 1 – the initial sample of coal; 2 – coal after treatment in the electric field; 3 – coal treated in the magnetic field.

The position of the first maximum fluctuates from sample to sample slightly; half-width almost does not vary. The intensity of coal lines decreased after the treatment in the electric field. Growth of the half-width indicates the decrease of the particles dispersion, in particular, and the increase of the "amorphism" degree, as a whole.

Lines of the crystalline phase for the sample treated in the MF are significantly more intense with respect to the initial sample and treated in the EF. The values of the interplanar distance of these lines (in nm): 0.455; 0.424; 0.403 – weak lines 0.371 and 0.338. The last of these corresponds to graphite. These lines fall into in the region of the first maximum. Lines are arranged about the second maximum in diffractograms of the initial samples. These lines correspond to the crystalline phase having the interplanar distance $d = 0.199$-0.200 nm, which is close in value to the line of graphite – the second one on intensity $d = 0.202$ nm. This line is present in all samples, but the most intense in samples treated in the magnetic field.

Table 1. Manipulation of data on diffractograms.

Sample name	The position of the first maximum, degrees	The position of the second maximum, degrees	The interplanar distance, nm		The half-width of the first maximum, degrees
			The first maximum	The second maximum	
Initial	25.0	42.7	0.355	0.280	0.18
Treated in the EF	23.6	42.4	0.361	0.213	0.20
Treated in the MF	24.6	42.2	0.377	0.214	0.16

According to Figures 2 and 3, Table 2, coal treated in the pulsed magnetic field exposes the greatest stability in relation to the initial sample and treated one in the electric field. Most of the weight loss (7.3%) of the initial sample of coal is attributable to the preliminary grinding (the effect of mechanochemical activation), i.e. the mass of liberated water is mixed with the mass of volatiles formed additionally during coal grinding. If the pre-ground coal is treated further in the electric or magnetic field, the characteristics of these coals are markedly different from of the initial mechano-activated coal.

This is evidenced by the results of electron paramagnetic resonance analyzes the studies of chemical composition, the nature of the diffractograms and the interplanar distance, particles geometry, especially their distribution by size and other physical parameters.

Table 2 shows that the heat of combustion of the sample Q_2 differs from Q_1 only about 0.3% (error in measurement is 0.1%). Q_3 increased in relation to initial one to 0.9%.

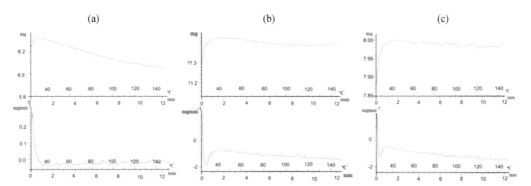

Figure 3. The results of the thermogravimetric analysis of coal: (a) the initial sample, (b) after treatment in the electric field; (c) after treatment in the magnetic field.

Table 2. Basic data on colorimetric analysis, TGA and DSC.

Sample No		1	2	3
The name of the sample		Initial	Treated in the EF	Treated in the MF
Sample weight, mg		6.240	11.300	7.880
Coal particles size, μm		214-112	252-112	350-131
The heat of combustion, Q, kJ / kg		35860	35720	36210
Coal mass loss upon heating to 120 °C, Figure 3	%	7.30	2.12	1.50
	mg	0.46	0.24	0.12

There are following main regions of the absorption bands on the infrared (IR) spectra of coal (Figure 4). The region (3800-2500 cm^{-1}) of stretching vibrations of ordinary links: O–H, N–H, C–H, S–H. The region (2500-1500 cm-1) of stretching vibrations of multiple bonds (unsaturated fragments): C=C, C=O, C=N, C≡C, C≡N. This region of the spectrum allows to determine the aromatic and heteroaromatic nucleus. The region (1500-500 cm^{-1})

of stretching vibrations of ordinary links C–C, C–N, C–O allows to identify uniquely deformation vibrations of ordinary links O–H, N–H, C–H and S–H as well as various types of carbon-hydrogen bonds: $C(sp3)$–H, $C(sp2)$–H, $C(sp)$–H, $(O=)C$–H (aldehyde). The position and intensity of the absorption bands in this range are very individual for each coal rank.

In case of full agreement of frequencies and intensities of lines in this region of the IR spectrum, we may prove the identity of the compared coal samples (initial ones and treated in EF and MF). Regions 2500-1500 cm^{-1} and 4000-2500 cm^{-1} are most informative when interpreting IR-spectra.

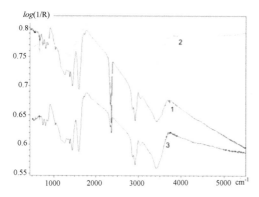

Figure 4. IR spectra of coal: 1 – the initial sample; 2 – after treatment in the electric field; 3 – after treatment in the magnetic field.

An increase in the intensity of absorption is observed in the region of oscillations of hydrogen-bonded OH-group, band 3300-3700 cm^{-1}, after the treatment in the magnetic field relative to the initial coal and treated one in the electric field. This range of the natural frequencies corresponds to C–H bonds in the primary and secondary amines, in CH_3 groups, in C_2H_2 molecules, HCN, $\equiv C – H$-groups in alkenes. The destruction of bridge aliphatic chains in relation to the initial coal barely noticeable (as determined by a slight decrease in the optical density of the bands 2920 and 2860 cm^{-1}), and corresponds to the stretching and deformation vibrations of the C–H bonds in the structures that contain the CH_3-, CH_2- and CH-groups. In this case, intensity and frequency are the same to the initial samples of coal and treated in the magnetic field. In addition, there is a slight decrease of bands 3000-3100 cm^{-1} in aromatic hydrocarbons.

Intensity in the region of stretching vibrations 2300-2400 cm^{-1} is far less. The band 2360-2325 cm^{-1} corresponds to the CO_2 molecule vibrations, including as impurities provided from the atmosphere. The intensity of the absorption spectrum increases for coal treated in the EF and reduces several times after the treatment in the magnetic field. In this region, the absorption of impurities (inorganic macrocomponents) may mask the bands of the analyzed coal samples. Thus, organophosphorous compounds (P – H valence bonds) correspond to the frequency range 2440-2300 cm^{-1}, nitrogen-containing compounds are identified also by the presence of intensive transmission in the region of stretching vibrations of the specified frequency range.

The band 1570-1620 cm^{-1} is formed by the peaks of the stretching vibrations of C_{ar}–C_{ar}-bonds in the aromatic rings. The intensity of this band is enhanced by the presence of aromatic structures of the quinoid carbonyl groups C=O, nitrogen of the pyridinic type and phenolic hydroxyls. During the treatment in the MF band intensity decreases relative to the initial coal and becomes almost twice higher than treated in the EF. The intensity of the lines of 1670-1620 cm^{-1} increased mainly due to the formation of new valence bonds C=C.

The region of vibration frequencies 970-1880 cm^{-1} includes the following sections, where peaks 1050 cm^{-1} (C–O group in aliphatic structures) dominate (intensity of the bands slightly decreased after treatment in the MF). Here it is possible overlapping of bands from vibrations of silicate groups of mineral part at 1040 cm^{-1}. Intensity of the peak declines with increasing metamorphic grade, which is consistent with a decrease in the proportion of oxygen-containing compounds in the coal. Section with the band 1150-1350 cm^{-1} (maximum is 1280 cm^{-1}) is caused by oscillations of oxygen-containing groups C – OR, C_{ar}–OH and R–C=O. Less intense peaks of deformation oscillations in aliphatic structures (1380 cm^{-1}) – methylene CH_3-group, and (1450 cm^{-1}) – methyl and methylene groups. The frequency range of absorption bands of 1450-1410 cm^{-1} is also consistent with fluctuations in ion-carbonate CO_3. Fluctuations in the C–O-bonds in the ester groups, displaced under the influence of neighboring alkenyl and aromatic groups, is observed at the section 1035 cm^{-1}.

Weak vibrations of sulfur-containing groups correspond to the section of bands 1200-1250 cm^{-1}, data on initial samples of coal treated in the magnetic field are almost the same, and although for coal treated in the MF intensity is weaker than for the initial ones. There are bands oscillations of several groups with aromatic, aliphatic and naphthenic structures in the region to 745-890 cm^{-1}. These are bands' groups of the out-of-plane deformation oscillations of the four types of C_{ar}–H-groups.

Ratio of amorphous phase fractions of different types has changed highly compared with the data of the initial sample after treatment in the magnetic field. The treatment reduced the amount of hydrocarbons and increased graphite phases (probably graphene). In comparison with the initial structure it is observed not only the ordering of the particles' periodic behavior, but also the formation of new crystalline structures, as evidenced by the appearance of new lines and the increase of intensities of the main maximum in the diffractograms.

The concentration of paramagnetic centers – radicals, such as OOH, $COOH$, CH_2, C_6H_5, CH_3, OH, CH_2O etc. increases two to five times in coal due to mechanical impacts or treatment in the electric field. According to the results of physical studies, the influence of the external magnetic field may change the direction of the magnetic moments (spins) of the electrons and stimulate the formation (or completion) of crystalline phases, carbon and hydrocarbon chains, two-dimensional carbon structures.

4 CONCLUSIONS

The influence of magnetic field on the system "coal organic mass – radicals", besides stimulation of the magnetic scenario of radical-based reactions, leads to the stabilization and growth of carbon structures with a regular arrangement of atoms. The effect could be used in the development of physical and chemical models of the process of carbonization and the formation of solid carbon phases.

Studies have shown that the chemical reactions occurring in the coal substance under the influence of weak magnetic fields, mainly directed to the formation of stable gas molecules because of the recombination of free radicals. Moreover, free radicals react with the components of the coal organic mass, thus increasing the crystal part of the structure.

Issues of coal formation and suppression of gas-dynamic phenomena in mines may be considered tentatively as one physical-chemical task related to the stability of the "coal-gas" system, one of the solutions is an imitation of coalification process. It could be used magnetic fields with low intensities to create coal-gas system to a greater degree of stability.

Currently, it was proved theoretically and experimentally that the magnetic field is a stimulating factor for the development of chemical reactions between free radicals in the gas phase and the liquid medium. Moreover, the kinetics of these reactions in liquids essentially depends on the dynamics of molecular motion, which is determined by the liquid's structure and radical pairs' properties.

To date, any reports of spin effects in solid-phase reactions are not encountered in the literature. There are also no theoretical developments aimed at identifying the effect of weak magnetic fields on chemical reactions in the solid phase. The analysis produced by us over many years of experimental results on the thermomagnetic treatment of solid-phase systems (minerals, rock, coal) could be interpreted from the point of view of the appearance of spin effects. Almost all of the studied materials were porous or used as pressed powder. Therefore most likely, that the chemical reactions take place on the surfaces of the powder particles, pores, cracks and grain boundaries. Such a conclusion is, first, from the considerations of the necessity of conditions mobility (even if very limited) of reaction complexes.

The practical application of the results obtained by the magnetic treatment of coal may be related directly to the creation of a new method of suppressing the outburst in coal.

REFERENCES

Soboliev, V.V., Chernay, A.V., Bilan, N.V. & Filippov, A.O. 2009. *The formation of gas as a result of mechanical degradation of the coal organic mass.* Proceedings of International Conference. Forum miners-2009. Dnipropertrov'sk: National Mining University: 186-191.

Soboliev, V.V. 2008. *Formation of New Phases in Ground Calcite with Added Silicon Under Heating and Current Passing Through.* Mineralogical Magazine, 4: 25-32.

Buchachenko, A.L., Sagdeev, R.Z. & Salikhov, K.M. 1978. *Magnetic and spin effects in chemical reactions.* Novosybirs'k: Nauka: 296.

Belyavsky, V.I., Ivanov, J.V. & Levine, M.N. 2006. *Magnon mechanism of reactions defects in solids.* Physics of Solid Fuels, 48. No. 7: 1255-1259.

Mining of Mineral Deposits – Pivnyak, Bondarenko, Kovalevs'ka & Illiashov (eds)
© 2013 Taylor & Francis Group, London, ISBN: 978-1-138-00108-4

Receipt of coagulant of water treatment from radio-active elements

O. Svetkina
National Mining University, Dnipropetrovs'k, Ukraine

ABSTRACT: The method of vibroloading obtains mixed coagulants of clearing of water. The advantages of usage share mechanochemical of activation $Al(OH)_3$ and iron ore are rotined. Is clarified, that at activation there is a formation of padding fissile centers, accountable for process of coprecipitation of radioelements from sewages.

1 INTRODUCTION

The application of the mixed coagulants it is known that a greatly effect at water treatment being mixture of salts of aluminum and iron. In this case the area of rational values of pH broadens considerably due to the variety of products of hydrolysis and physical and chemical properties of the by latter.

During water treatment by the mixed coagulant even at a low temperature fallouts of oxide of iron are not seen caused by the formation and besieging of flakes before the filters. Flakes are besieged more evenly than in the case of the application of coagulants separately, and more complete lighting up is reached in settlers that allows consider ably to decrease loading on filters. The effect of treatment of water at the temperature of 20 °C by the mixed coagulant is near to the effect of the coagulation by the sulfate of iron at 50 °C and sulfate of aluminum at 80 °C. In this work authors offered a coagulant , the technology of preparation of which consists of mixing in certain correlation of $FeCl_3$ and unrefined sulfate of aluminum. The correlation of initial matters can be changed depending on the terms. In this case the size of it is 2:1 in a count on waterless salts is maximal. The indicated reagents can be utilized both as solutions and given separately as salts in the refined water.

The mixed coagulant can be also got from the sulfates of aluminum and iron at the correlation of Al_2O_3 / Fe_2O_3 equal 1/(0.5÷3). This coagulant is got by the use of high-ferrous gibbsitic bauxites. A bauxite is processed by 60% sulphuric acid in an amount 90% of stoichiometrical. A process was conducted at 100-150 °C during 1-2 hours. The got mash is exposed to the granulation drying at 150-200 °C or crystallizations with partial dehydration on tableskristollizatory with the reception of lump product. At processing of high-ferrous bauxites with maintenance 40.8% Al_2O_3, 27% Fe_2O_3 and 8.7% SiO_2 extraction of aluminum and iron in solution as sulfate salts was 90%. The got coagulant contained water conducting 15.7% Al_2O_3 and 10.4% Fe_2O_3, and also 0.3% of free sulphuric acid and 5.5% insoluble remain.

The application of this method of receipting of the mixed coagulant allows, from one side, to simplify the process of processing of bauxits, because the operations of defending of sulfate asid mashes, evaporation and dehydrations of sulfate asid solutions, crushing of product, are eliminated, but from the other side – content of insolubles is increased. As an operation of division of liquid and hard phases are absent in a method, with the purpose of diminishing of the content of insoluble remain in a product it is expedient to process bauxits with the small content of silica and aluminosilicate.

In the case of the processing of bauxites with increased content of these admixtures and at a necessity the reception of the cleared coagulant it is necessary to add more operations as to dilution of sulphate add of dissection, division of liquid and hard phases and washing of latter.

2 FORMULATING THE PROBLEM

The described technologies include in the technological process the use of plenty of water for washing of the prepared product, which contain insolubles. These coagulants are ineffective for cleaning of radio-active waters. In this connection there is a necessity of production of new coagulants for cleaning of mine waters.

One of asks of creation of new effective coagulants is a reception them on the basis of utilization of wastes of metallurgical, chemical, and also mining industry. We offered the coagulant, consisting of gidrargilit and tails of the oxidized quartzites which are the wastes the process. The technology of reception of the coagulator is based on the conducting of the joint mechanochemical activating of iron-ores and trihydrate of aluminum (gidrargilit). Afterwards will designate this mixture (I).

3 RESULTS OF ANALYSIS

Mechanochemical of reaction, mainly, depend on power effort of vehicle for grinding down. Changing the correlation of different types of affecting the ground material blow and abrasion, it is possible in a different degree to change the inner structure of particles. Mechanochemical activating (I) was conducted in a vertical oscillation mill basic advantage of which is the vibroshock affecting the destroyed material. Preliminary experiments on grinding down of different materials showed the high degree of mechanoactivation them as compared to the results of grinding in the vehicles of other types.

The number of experiments on grinding was conducted on a laboratory vertical mill with the use of the different technological modes the row (I). Components were mixed up and jointly ground in the correlations of $Fe_2O_3 : Al_2O_3 = 1{:}8, \ 1{:}1, \ 8{:}1$. A grinding chamber was filled with grinding bodies such as balls from steel of different diameters. Material was skipped through a grinding chamber certain amount of times for the getting of general way of grinding 1, 2 and 3 m. Thus time of being it in the working organ of mill was regulated. Initial, the intermediate and eventual products of grinding were exposed to the analysis. At the same time power descriptions of the activated surface were measured by the method of pH-titration.

The degree of disactivating is determined by isotopic composition of radionuclidess and their condition of in the solution. If radionuclidess are adsorbed on dispersible admixtures or are in the colloid-dispersible state, steady decontamination of water is achieved at 97-99%. Thus, the degree of decontamination depends in this case on the degree of lighting up of water.

Tests of activated (I) were conducted in the process of cleaning of water on such parameter as turbidity of water and isotopic composition of radionuclidess. The rate of turbidity depends on intensity of light diffusion and is proportional to the concentration of the self-weighted matters. Determination of

the content of the self-weighted matters was conducted by weighing of dry sediment after coagulation of water.

As is known, process of lighting up of water, from formation of micelle and concluding their besieging , can divide of into a few stages. On the first stage, after introduction of coagulants into the cleared water, there is a hydrolysis of it with formation of micelle and their subsequent agregirovanie in bigger spherical particles zolua (about 0.01-0.1 μ). Ospalescenciya appears. This period is called of the hidden coagulation. The second stage is formation of chain structures and tiniest flakes which agreriryutcua in bigger ones. The third stage is connected with segmentation , i.e. settling under the action of gravity flakes of certain sizes . Often these stages do not follow in sequence, but recovered, complicating the process of lighting up.

On a Figure 1 curves characterizing a change turbidity of water in the process of coagulation at a standard coagulant and activated under various conditions (I) are presented. From experimental information it is evident, that the process of lighting take place in of presence activated (I) more quick and practically at once there is the stage of flocculation passing the stage of the hidden coagulation. Rational is the correlation of components 8:1.

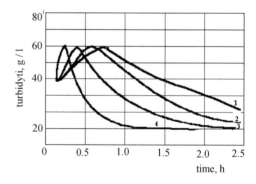

Figure 1. Change turbidity of warter in hroctss of coagulating: 1 – industrial cjfgulant; 2 – activated 1:8; 3 – activated 1:1; 4 – activated 8:1.

The substantial disadvantage of the use of salts of Fe^{2+} as a coagulant are corrosive activity of solutions, large expense of chlorine and necessity of careful dosage of the applied reagents. Insignificant rejections in dosages result in substantial violation of the technological mode, to caused by the under-oxidation of iron, and, as a result, to the incomplete process of hydrolysis. As a result of it there is appearance of Fe^{2+}, due to what water has unpleasant taste, the its coloured and turbidity increase. At application as a coagulant salts of iron it is necessary to give a preference to salts of Fe^{3+}. The process of oxidization intensively take place at pH 8.With this purpose before addition of green vitriol or simultaneously with it in more frequent than all dead lime lye is put in water. Oxidization of Fe^{2+} is more effective in Fe^{3+} passes at joint treatment of water a green vitriol and chlorine. The use of additional reagent results in limitation of application of green vitriol for lighting up and discolouring of water. However in the case of simultaneous lime-soda made soft of water he is an extraordinarily useful reagent.

As a result mechanical activation components takes place before oxidization of iron and its transition of Fe^{2+} in Fe^{3+} about what testify spectrums, presented on a Figure 2.

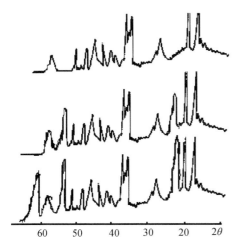

Figure 2. X-ray tables of activated components: 1 – initial mixture; 2 – activated 1:1; 3 – activated 8:1.

From a Figure 2 it is evident, that lances, proper formation of trivalent iron, appear at mechanical activation As a result of activating there is formation of hydrokco-combinenation of trivalent iron, type $\alpha - FeO(OH)$, which is the product of hydrolysis of trivalent iron. Therefore the finished product of hydrolysis $\beta - FeO(OH)$ appears mush more quick, than at the hydrolysis of chlorides of iron (contain of iron coagulants).

During the mechanochemical activating of ores in a vertical oscillation mill there is formation of clusters in the continuous mode, that is obviously evident from radio structural spectrums.

Thus, in a coagulant, got at a vibroladening, iron additionally is not needed to oxi and it forms chains

of few atoms, thus accelerates the process of coagulation.

The application of coagulants is very effective with enhanceable basicity. They are polymeric hydrocomplex and require considerably less alkaline reserve. As a result mechanical processing (I) ions of iron of hydrolysises in a greater rate and are the centers of coagulation, and as a result of, it large and durable enough flakes. Appear durability of which is determined by the presence of hydrocomplexs aluminum. Exactly these complexes lug away the admixtures of radio-active elements.

At the application of the mixed coagulants, got by the method of oscillation ladening, there is stabilizing of pH environment. After activating (I) such important factor for coagulation, as pH environment was measured.It appeared to be equal 8. The application of this coagulant results in diminishing of salts of radio-active ruthenium.

During the mechanochemical activating there is an increase of total adsorption potential. It results in that besides molecular and electrostatic forces the structural factor of aggregate stability of the dispersible systems appears addition.

Because of presence on-the-surface of particles of active centers there are electrostatic forces of pushing away, which cause an antihunt action. Hydrate shells a round particles promote stabilizing of the dispersible system. At the taking away particles on large as compared to their sizes distance of cooperation between them does not take a place. As a result of brownian motion of the positively charged particles they are drawn together and electrostatic forces pushing away appear, which are added up with molecular attractive powers. With diminishing of distance between particles the resulting action of these opposite forces results in predominating of pushing away. At further rapprochement of force of pushing away diminish and attractive powers begin to predominate. In order that coagulation takes place, particles must overcome forces of pushing away (so-called "power barrier"), that can happen in the case of sufficient large energy of motion of particles or decline of height of barrier. The higher this barrier and the less than energy of motion of particles, the less than probability of their clinging and slower the process of coagulation is absent at all.

4 CONCLUSIONS

As a result mechanochemical treatments there is a change of power descriptions of the ground up matter, that results in structural changes in a superficial layer (I). Appearny power barrier layers are cause intensification of process of coagulation.

The addition of clean hydracids of aluminum as a coagulant results in formation of flakes with the relative density of equal 1.001-1.003 kg / l. It is related to the fact that the volume of particulate matters in unit of volume flakes is small and is measured as the tenths of percent. Thus, appearing loose and easy flakes can not fully clean the colored little turbid waters. The application of the mixed coagulants, got by the method of vibroladening results in formation of durable smaller crystals flakes not depending upon the cleared water. It can be explained that the activated mixed coagulant contains hydro complexes of aluminum and iron simultaneously.

As a result mehanoactivethion (I) there is formation of active centers on the atoms of iron and aluminum. Co-ordinating areas of cations of metals of besiege are little soluble products of reagentnogo soft of water, thus there is not only chemical co-operation, but process of physical adsorption. As many isotopes form little soluble hydrates in an alkaline environment, these matters are separated together coagulant weigh in sediment.

The analysis of experimental data is shows the expedience of the application of these coagulants. On their basis compositions can be also offered and for ionochange resins, applied in industry for cleaning of sewer and industrial waters.

Blasting works technology to decrease an emission of harmful matters into the mine atmosphere

O. Khomenko, M. Kononenko & I. Myronova
National Mining University, Dnipropetrovs'k, Ukraine

ABSTRACT: It is shown how use of explosives increases safety during explosion works conduction and decreases emission of explosion products due to change of physical-chemical properties of explosives. Existing methodology of drill and blast works parameters calculation is improved and ore breakage technology in extraction chambers is developed that foresees application of emulsion explosives under conditions of CJSCK "ZZhRK".

1 INTRODUCTION

Mining industry of Ukraine is one the most polluting industries by level of formation and emission of harmful matters into the atmosphere. In general view, anthropogenic influence on air basin as a result of coal and ore enterprises is authorized and unauthorized emissions of harmful matters into the atmosphere that leads to dust pollution and air pollution in working zone and at bordering territories.

Extraction of iron ores is predominantly executed with help of room and pillar mine methods (up to 85%). Mining operations for mine workings drivage and stoping are carried out by drill and blast method (up to 90%). Worked off air current from ore mines is emitted into the atmosphere through air shafts without purification. At present, there are no effective cleaning facilities for capturing and cleaning mine gases that come into the atmosphere in significant volumes. Thus, harmful matters volumes reduction in mine atmosphere has current meaning for ecological situation in Ukraine.

2 ANALYSIS OF EXPLOSIVES BY THEIR NOXIOUSNESS

Results of noxious gases concentration determination in worked out air current of main fans channels have shown that mine air is saturated with noxious gases formed as a result of underground operations conduction and causes harmful influence on the environment. It was stated by CJSC "Zporozhskiy iron-ore plant" with help of physical-chemical method.

Further researches have allowed to establish that harmful matters concentration in worked-out air current depend only on annual specific consumption of explosives and change based on linear dependence (Gorova & Myronova 2011). Conducted researches substantiate necessity in implementing ecologically clean emulsion explosives and development of safe technology of explosion works conduction both during mine workings drivage and during stoping.

Iron-ore extraction by underground method is predominantly connected with conduction of blasting works that define efficiency of deposits exploitation. Analysis of production situation on mining operation conduction has shown that annually at CJSC "ZZhRK" there are about 2700-3100 thous. kg explosives are spent. Considering high cost of industrial of TNT-bearing explosives (grannulotols, grammonite, aquatols), their danger during transportation especially in large volumes and perspectives of coal enterprises development, it is expedient to use explosives created directly at places of blasting works conduction. It is connected with not only safety of blasting works conduction but with lower volumes of explosion products emission (Kuprin 2012).

As an example we will consider chemical decomposition of emulsion explosive of domestic production of "Ukrainit-PP-2B" type.

$$1.509Ca(NO_3)_2 + 6.435NH_4NO_3 +$$

$$+ 0.148C_{37}H_{46}O_2N + 8.761H_2O^l +$$

$$+ 0.024H_2O_2 = 1.509CaO + 8.018N_2 +$$

$$+ 25.059H_2O^v + 5.42CO_2 + 0.056CO .$$

When 1 kg of emulsion explosive of "Ukrainit-PP-2B" types blasts there is 0.056 moles of CO or 0.056 moles 22.4 1/ moles = 1.25 liters of carbon oxide. Emission of nitrogen oxide thermodinamically is unlikely. In addition, presence of *CaO* in blast products provides absorption of nitrogen oxides that can form during violation of stochiometric ratio of components or incomplete reaction behavior of explosion conversion of charges.

Based of compound composition and ration of these components, it can be stated, that in blast products of "Ukrainit-PP-2B" there are no toxic nitrogen oxides "Ukrainit-PP-2B". Emission of *CO* in amount up to 1.25 liters per 1 kg of emulsion explosive is 2 times lower than when using TNT-based explosives. From practice of blasting works conduction and analysis of scientific-technical data it is known that qualitative and quantitative content of noxious gases and solid products of reaction of charge blasting conversion depends on both the explosive type and blasting conditions (chemical composition of an explosive, physical-mechanical properties of rock massif, blasting works technology). It is known, that during blast of 1 kg of grannulotol – 250-300 liters of carbon oxide are released into atmosphere, grammonite 79/21, acvatol GLT-20 – 84-150 liters (Gushchin 1990). Despite the fact that compound composition of these explosives is balanced on zero oxigen balance, they are factually the sources of emission of carbon and nitrogen oxides in large quantities.

3 ANALYSIS OF STOPING TECHNOLOGY

Based on bedding conditions and geometrical parameters of Yuzhno-Belozerskiy deposit, (SROMI) scientific-research ore mining institute of State University of Krivyy Rig in 2001 has developed room and pillar mining method for levels of 640-940 m at CJSC "ZZhrk" with following backfill of goaf with solidifying mixtures. Based on existing stoping technology, ore deposits mining at levels of 640 m is executed with primary and secondary chambers. Chambers of hanging wall along the strike represent rectangular – rhomboid shape and across the strike – chamber rocks have inclined bottom towards hanging wall rocks. Chambers of footwall along the strike have identical shape as well as chambers of hanging wall, and across the strike represent by themselves a shape of rectangular triangle.

The following complex of working processes during ore mining in chambers are included in production stage of stoping: undercutting, cutting and breaking the chamber's reserves, undercutting of chamber's reserves is carried out by blasting wedges turning. To turn wedges from undercut drift the circular rising blasthole ring is drilled that is charged with an explosive and blasted towards preliminary driven cut raise. Cutting of the chamber reserves is

carried out by the following way. Descending parralel bunch holes are drilled at each sublevel horizon along the bottom of shrink drift. To form vertical cuttoff stope along the entire height of the chamber, the bunch of parallel holes are blasted towards cut raise. Ore breakage in chambers is implemented by the following way. Vertical rising blashole ring is drilled from sublevel cross cut along the entire length of chambers and parallely to cuttoff stope. Separation and crashing the ore is carried out by way of blasthole ring blasting towards preliminary formed cuttoff stope. Broken from chambers ore is discharged and loaded in wagons at haulage horizon with help of VVDR-5PS.

Having conducted the analysis of stoping technology in chamber it can be concluded that ore breakage technology with help of room-and-pillar mining method foresees usage of significant volumes of TNT-containing explosives. Thus, to improve stoping technology in chambers it is proposed to change technological parameters of drill and blast works. Specifically to improve existing methodology of drill and blast works parameters calculation adapting them to emulsion explosives use and to change hole drilling direction in rings that will allow to use emulsion explosives.

4 STOPING TECHNOLOGY IMPROVEMENT

To define parameters of drill and blast works the methodology shown in the work (Instructive-methodological... 1977) is used at present. Stoping parameters calculation comes down to determination of explosives rated consumption, line of minimal resistance of charges and distance between holes faces in the ring.

Rated consumption of explosives for breakage at ring-type location of holes

$$q = 0.1f \frac{\Delta q}{\Delta b}, \qquad (1)$$

where f – coefficient of ore strength according to Protodyakonov scale; $\Delta q = \sqrt[3]{d/0.085}$ – coefficient of explosives distribution evenness in mining massif; Δb – coefficient of an explosive rated capacity that is equal to 1 for ammonite #6 ZhV, for rock ammonite – $\Delta b = 2.04 - 0.58 \cdot d$, for grammonite 79/21 $\Delta b = 0.71 + 0.16 \cdot d$; d – borehole diameter.

Value of minimal resistance line for charges during the ore vertical layers breakage with deep holes at ring-type location is defined as

$$W = 114 \cdot K \cdot d \cdot \sqrt{\frac{\delta \cdot \Delta b}{f \cdot \Delta \cdot q \cdot m}}, \qquad (2)$$

where K – correction coefficient depending on breakage direction and rocks strength that is equal to 1 during vertical layers with $f > 10$ breakage and at ore with $f < 10 - 0.9$; δ – charging density (explosives quantity in the hole volume along the charge length); Δ – an explosive density; m – coefficient of charges convergence, equal to $0.8 - 1.2$.

Distance between holes faces

$$a = m \cdot W. \qquad (3)$$

The presented methodology of drill and blast works parameters calculation during stoping foresees application of TNT-containing explosive the ore breakage. Hence, to use ecologically clean and safe emulsion explosives it is necessary to improve the used calculation methodology. Coefficient of an explosive relative capacity for "Ukrainit PP-2B" has been established under industrial conditions at various hole diameters. Based on gained data the graph of an explosive relative capacity coefficient was plotted depending on holes diameter (Figure 1). The implemented analysis of an explosive relative capacity coefficient values has allowed to establish that as the hole diameter increases this coefficient decreases and changes by linear dependence.

Having conducted an approximation of maximal values with help of Microsoft Excel, empiric dependence equation of an explosive relative capacity coefficient for "Ukrainit PP-2B" on the hole diameter

$$\Delta b = 1.385 - 0.23 \cdot d, \qquad (4)$$

where d – hole diameter.

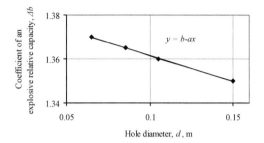

Figure 1. Graph for dependence of an explosive relative capacity coefficient for "Ukrainit PP-2B" on the hole diameter.

The improved methodologies of drill and blast works parameters calculation has contributed to the development of the new technology of stoping in chambers that is protected by useful model patent (Kononenko, Khomenko & Myronova 2011).

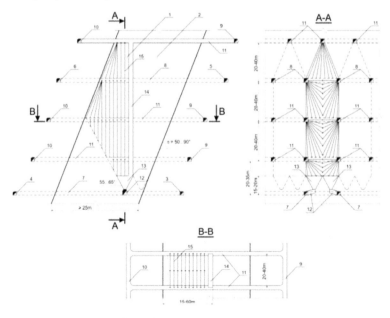

Figure 2. Proposed stoping technology in primary chambers using emulsion explosives: 1 – first order chamber; 2 – second order chamber; 3 – haulage drift of footwall; 4 – haulage drift of hanging wall; 6 – ventilation drift of hanging wall; 7 – haulage cross cut; 8 – ventilation drilling cross cut; 9 – sub-level drift of hanging wall; 10 – sub-level drift of hanging; 11 – drilling cross cut; 12 – access under vibrational feeder; 13 – receiving wedge; 14 – cutoff stope; 15 – ring of descending holes.

233

The model is based on the improvement of the known mineral deposits extraction technology with drill and blast method. It describes introduction of new technological operations and parameters to use emulsion explosives during mineral deposits extraction by underground mining.

This promotes general safety level increase during blasting works conduction and decrease of environment pollution by blast products. Due to this, ecologically clean and economically effective extraction of mineral deposits is achieved that provides resource saving and rational use of minerals. Figure 2 presents proposed technology of stoping in chambers of first order.

The proposed technology is realized by the following method. The deposit within the level is mined with help of stoping chambers of first and second orders of mining. The chambers are prepared by haulage drifts drivage of foot 3 and hanging 4 walls, and cross cuts 7. Ventilation drifts 5 and 6 and ventilation drilling cross cut 8 were driven earlier during development of above-situated level, they served as haulage ones.

Further, sublevel drifts 9 and 10 are driven, and drilling cross cuts 11. Access under vibrational feeders 12 are driven from haulage cross cuts 7 for their following dismantling.

After drivage of all development-cutting galleries the drilling of exploitation boreholes are carried out together with undercutting area formation 13, cutoff stope 14. Ore drilling-out from each chamber of cross cut 11 and ventilation cross cut 8 is implemented before charging by holes ring 15 towards underlying sublevel. After that, separate components of blast-proof emulsion explosive are transported to charging machine and hole rings. The machine prepares the mixture and charges the holes of the corresponding ring 15 to form emulsion explosive in them. According to the proposed scheme it is possible to charge the holes in the ring 15 with emulsion explosives up to set level excluding explosive losses and possibility to control the breakage process. Ore reserves extraction is carried out in the ring 15 towards cutoff stope 14 until the stoping chamber 1 of first order takes its project contours. Broken ore goes into receiving wedges 13 and with help of vibration units is loaded into transport units located in haulage cross cuts 7 and transported along drifts 3 and 4, and cross-cut to tippers after which it is lifted to the surface.

Further by analogical way the second order chambers are developed (Figure 3).

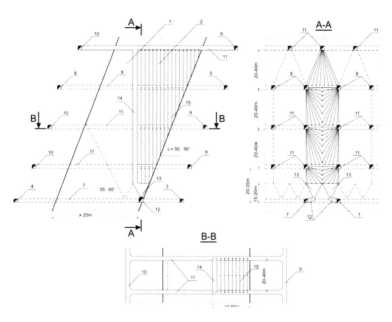

Figure 3. The proposed stoping technology in secondary chambers using emulsion explosives: 1 – first order chamber; 2 – second order chamber; 3 – haulage drift of foot wall; 4 – haulage drift of hanging wall; 7 – haulage cross cut; 8 – ventilation cross cut; 9 – sublevel drift of foot wall; 10 – sublevel drift of hanging wall; 11 – drilling cross cut; 12 – access under vibration feeder; 13 – receiving wedge; 14 – cutoff stope; 15 – ring of descending holes.

5 CONCLUSIONS

Substantiation of emulsion explosives use during stoping works will lead to decrease of blast products emission by 50%. It is connected with mixture content of an explosive and components ration in it. Use of emulsion explosive of "Ukrainit-PP-2B" type during stoping works conduction in chambers leads to decrease of carbon monoxide emission that is 3 times lower than when using TNT-containing explosives and also to absence of toxic nitrogen oxides.

The proposed technology of ore breakage during stoping of the chambers reserves is implemented using the emulsion explosive "Ukrainit-PP-2B" allows to increase safety of blast works that are in the first place connected with usage of emulsion explosives and also to decrease environment pollution by blast products and provision of more ecologically clean extraction of ore.

REFERENCES

Gorova, A.I. & Myronova, I.G. 2011. *Determination of harmful matters concentration in upcast air.* Dnipropetrovs'k: Collection of scientific works of NMU, 36. Vol. 2: 192-200.

Kuprin, V.P. 2012. *Development and introduction of emulsion explosives at open pits of Ukraine.* Dnipropetrovs'k: NHEI UDHTU: 243.

Gushchin, V.I. 1990. Book of problems on blasting works. Moscow: Nedra: 174.

Instructive-methodological directions on blast works rational parameters selection during underground mining at mines of Krivorozhsk basin and ZZhRK-1. 1977. Krivyy Rig: 54.

Kononenko, M.M., Khomenko, O.E. & Myronova, I.G. 2011. *Useful model patent #64145 Ukraine E21C 41/16. Method of mineral deposits extraction by blasting method*; applicant and owner is NMU. – # u201105285; app. 26.04.2011; publ. 25.10.2011, Bul. #20.

Mining of Mineral Deposits – Pivnyak, Bondarenko, Kovalevs'ka & Illiashov (eds)
© 2013 Taylor & Francis Group, London, ISBN: 978-1-138-00108-4

The investigation of rock dumps influence
to the levels of heavy metals contamination of soil

A. Pavlychenko & A. Kovalenko
National Mining University, Dnipropetrovs'k, Ukraine

ABSTRACT: The features of the impact of coal mine waste dumps on the ecological state of the surrounding areas are analyzed. The regularities of migration of heavy metals from waste dumps are fixed.

Intensive coal mining in Ukraine has led to significant changes in natural ecosystems. Coal mine wastes in the form of overburden and mine rocks are one of the factors of an intense effect on the environment. Large areas of the earth's surface are withdrawn to organize dumps. In addition, the risk of stockpiles is increasing by leaching and acid flow of aggressive compounds on the surface of rocks, by filtering of mixtures in soil horizons and further into the ground water, by the release of large amounts of toxic dust and poisonous gases, ten times higher than permissible limits. Intensive coal mining has led to a tense environment situation in the mining regions of Ukraine (Grebenkin 2012 & Gorova 2012).

Waste dumps have constant negative impact on the ecological status of the environmental components at all stages of the transformation of rocks, from the moment of their delivery to the surface until the extinction of internal and external physical, chemical, biological and other processes (Petrova 2002 & Zborshchik 1990).

The character and intensity of the negative impact of the stockpiles to the environment directly depends on the intensity of deflation and carbonation, oxidation, combustion, secondary mineralization, which are the main stages of the overburden rocks transformation. There is a weathering of rock mass that falls within the new physic and chemical conditions: oxygen-rich environment, atmospheric precipitation, temperature, pressure, humidity, etc. All these phenomena contribute to the removal of particulate matter from the body of dump, to the formation of acid flow of easily soluble chemical compounds, to the oxidation and decomposition of various mineral complexes. These processes are accompanied by the release of energy which is transformed into the heat that warms up the rock mass from the inside, creating, thus, a kind of uncontrolled reactor cycle. Products of physical and

chemical transformations in the form of acid and alkaline solutions of salts of heavy metals, newly organomineral complexes are drained into the soil, surface water and groundwater, contributing to the migration of toxic substances, and depressing the overall ecological condition of coal-mining regions.

It should be noted that the majority of scientific studies is aimed at studying the environmental problems arising from the combustion of waste dumps because the significant pollution of the environment requires the development of measures to extinguish the combustion sources (Zborshchik 1989; 1990).

However, each stage of internal and external processes in the dumps is characterized by complex interactions and deserves the special attention. Therefore, the aim of this work is to establish the characteristics of the contaminants migration from the coal mines waste dumps being on different stages of the flow of internal and external processes.

The investigations were conducted in areas adjacent to the waste dumps placed in the Lugans'k region, where intensive commercial production of coal is led from the late XIX century. The studied rock dumps were conventionally divided into groups, describing their state, which is the result of a spontaneous chemical, physical, biological and other processes, as well as carrying out certain operations.

Three conditional statuses of waste dumps were identified for the survey:
– at the stage of attenuation of internal and external physicochemical changes;
– at the stage of decay of physical and chemical transformations at infringement of the dump body integrity and its inner rock outcrop;
– at the stage of decay of physical and chemical transformations at fresh overburden entering.

Areas with active processes of leaching and acid corrosive fluids flow along the surface of the slopes have been identified as a result of visual inspection of waste dumps. Dusting of dumps and high content of suspended particles in the air surrounding areas testified about wind erosion. Combustion of mass was manifested with elevated temperature of rocks in places of smoke and gases exit to the surface which have sharply expressed smell of ammonia and sulfur.

Forty eight rock samples from the body dumps, the soil at their feet and at a distance of 100 m from it in four world directions were selected to study the migration features of heavy metals from waste dumps in the surrounding areas. The determination of mobile forms of heavy metals (Co, Cu, Pb, Cd, Fe, Mn, Zn) is done in ammonium acetate buffer extract with pH 4.8 by atomic absorption spectrophotometry.

The analysis of the received data allowed determining the characteristics of migration and distribution of heavy metals in the waste dumps and soils of investigated territories. High activity of physical and chemical processes and intensive migration of the majority of heavy metals from the rock mass were revealed, as well as their accumulation in the environmental objects of surrounding areas.

Migration of the mobile forms of manganese can be one of the examples of the features above. It is known that reduction of manganese content occurs during burnout of rocks, which may be connect with its high activity in an acid and oxygen enriched environment (Zborshchik 1989). To identify the characteristics of migration manganese content in four world directions for three conditional states of waste dumps was analyzed.

The studies showed a tendency of gradual removal of manganese from the rock mass and its accumulation in the environment, in this case – in the soil at a distance of 100 m from the stockpile. Significant activation of chemical reactions and intensive removal of manganese in the soil of surrounding areas is observing in case of fresh rocks entering to the already burned-out and relatively stable rock mass. Although the concentrations of heavy metal do not exceed the maximum permissible concentration, waste dumps both at the stage of attenuation of internal and external physical and chemical processes and at fresh overburden entering, are the source of constant environmental pressure on the environment.

Because every stage of internal and external processes that occur in the dumps carries the complex physical, chemical, biological, mechanical and other interconnections, the choice of methods to minimize the negative impact of these man-made objects to the environment should take into account their characteristics at the time of activities implementation. It is necessary to determine characteristics of the overburden, to analyze their state, to reveal patterns of their behavior and to develop engineering measures to protect the environment in the surrounding areas on this basis. For example, to waste dumps in the decay stage of physical and chemical processes, the formation of trenches at the base as a way to change the migration routes of chemical compounds and their accumulation in designated facilities can be one of the remediation method. The use of drainage systems, reshaping the waste dumps slopes to minimize surface flows, terracing and planting of greenery in order to prevent further erosion, etc. is also possible.

The established patterns of migration of heavy metals indicate that the waste dumps at every stage of the internal and external processes are the constant source of additional load on the objects of the environment. It causes necessity of development of environmental technologies aimed at reducing the mobility of heavy metals and improving the ecological status of the studied areas.

REFERENCES

Grebenkin, S.S., Kostenko, V.K. & Matlak, E.S. 2009. *Conservation of the natural environment at the mining enterprises:* Monograph. Donets'k: VIK: 505.

Gorova, A., Pavlychenko, A. & Kulyna, S. 2012. *Ecological problems of post-industrial mining areas.* Geomechanical processes during underground mining. Leiden, The Netherlands: CRC Press / Balkema: 35-40.

Petrova, L.O. 2002. *The environmental impact of coal mining and coal waste processing.* Geological Journal, 2: 81-87.

Zborshchik, M.P. & Osokin, V.V. 1990. *Prevention of spontaneous combustion of rocks.* Kyiv: Technique: 176.

Zborshchik, M.P., Osokin, V.V. & Panov, B.S. 1989. *Mineralogical features of sedimentary rocks that are prone to spontaneous combustion.* Development of mineral deposits. Kyiv, 83: 92-98.

The magnetic susceptibility of granular manganese sludge of Nikopol'skyy Basin

A. Zubarev
National Mining University, Dnipropetrovs'k, Ukraine

ABSTRACT: The experimental research of specific magnetic susceptibility of magnetic crops of granular manganese sludges of one of Nikopols'kyy Basin sludge depositories has been done. The results of the research show that the magnetic crops studied have sufficient difference in magnetic susceptibility for crude sludge treatment with dry magnetic separation methods.

Nikopols'kyy Region concentrates comparatively large depositories of manganese ore dressing wastes, among which oxide and mixed manganese sludges stand out. The preliminary analysis shows that granular part makes up to 50% of the whole volume of stored sludges. Involvement into processing the given raw material with the dry magnetic separation technique, in prospect, will provide reduced cost of searching and developing new deposit, will allow growth of qualitative manganese concentrate, will clear the land occupied by sludge depositories for reclamation, will eliminate environmental pollution sources, will reduce the problems of draining and water consumption, will improve complicated ecological situation itself nearby operating ore dressing enterprises. The results obtained regarding experimental observation of magnetic susceptibility of manganese products will allow enlarging understanding of essential separation characteristics of devices and range of controlling technological parameters of dry magnetic ore dressing process.

Task definition. The objective of the research is experimental determination of magnetic susceptibility of manganese sludge as function magnetic field intensity.

Theoretical bases of determining magnetic susceptibility of manganese products. Magnetic force affecting a sample in the magnetic field is calculated according to the formula (Karmazin 1978 & Svoboda 2004):

$$F_m = \mu_0 \cdot \chi \cdot m \cdot H \cdot |gradH| , \qquad (1)$$

where μ_0 – magnetic constant, $4\pi \cdot 10^{-7}$, (H / m); χ – specific magnetic susceptibility, (m^3/ kg); m – weight of the sample, in suspension (kg), H – the magnetic field intensity (strength), (A / m);

$|gradH|$ – gradient of the magnetic field intensity, (A / m^2); x – a coordinate measured along the axis of the studied sample, (mm).

To carry out the present experiment, a few variants of magnetic systems were considered. According to the results of series of preliminary research we chose the variant, equipped with a magnetic coil with a magnetic conductor and metal core (Figure 1).

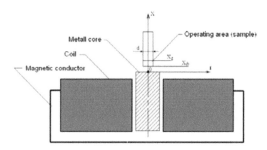

Figure 1. The variant of the magnetic system with the operating area to carry out the experiment.

The magnetic coil with the outer diameter 360 mm and the internal opening diameter 85 and 90 mm height is made of copper bus rectangular in shape, cross-section $1.5 \cdot 3.5$ mm^2. A metal core with a 60mm diameter is placed into the coil to increase the magnetic field intensity in the operating area. For this variant of the magnetic system the degree of the magnetic field nonuniformity was that which allowed neglecting the change of required magnetic susceptibility χ through the change of the magnetic field intensity within the operating area. The operating area implies that part of space in which the studied sample is placed, restricted by the surfaces: underneath and from above by horizontal planes removed from the top of the core for 4 and 10 mm correspondingly; from sides by a cylindrical plane, coaxial with a coil and, diameter, $d = 26$ mm.

Thereat, the given variant of the magnetic system ensures constancy of the gradient of the magnetic field intensity $|gradH|$ within the operating area and the sample volume correspondingly, and the average length value of the sample can be taken as H.

$$H = \frac{H_t + H_b}{2},$$

where H_t and H_b – vector projections of the magnetic field intensity onto axis x at top and bottom points of the operating area, correspondingly:

$$H_t = \frac{B_t}{\mu_0}; \quad H_b = \frac{B_b}{\mu_0},$$

where B_t and B_b – unit values of the magnetic field density at the top and bottom point of the operating area correspondingly which were measured with an electronic milliteslameter "Щ43.2214".

In case of a small change the magnetic field intensity in radial direction, the gradient of the magnetic field intensity $|gradH|$ can be accepted approximately:

$$|gradH| \approx \frac{dH}{dx} \approx \frac{dH_x}{dx} \approx \frac{H_t - H_b}{x_t - x_b}, \quad (2)$$

$$H_r \approx 0, \quad H = H_x, \quad F_{m,r} \approx 0, \quad F_m \approx F_{m,x}.$$

where x_t and x_b – coordinate x values in top and bottom parts of the operating area correspondingly.

The specific magnetic susceptibility of the sample is found from formula (1), taking into account the replacement (2):

$$\chi = \frac{F_m}{\mu_0 \cdot m \cdot H \cdot \frac{dH}{dx}}. \quad (3)$$

The minimal weight of Sample m, for which the filling coefficient does not change with the volume increase, should be no less than 2.9 g (Mladets'kyy 2011).

The experiment description and the results. The initial sample of the granular manganese sludges underwent magnetic fractionation on a laboratory magnetic roller separator with various current values in the coil. The result of the separation and chemical analysis of the products obtained are presented in Table 1.

Table 1. Characteristics of the presented magnetic crops of the manganese sludges.

Number of magnetic crops of the manganese sludge (of Sample)	The current value in the roller separator coil, A	Magnetic crop output, %	Manganese content, %
1*	0.5*	0.7*	11.7*
2	1.0	19.9	42.8
3	1.5	10.6	40.6
4	2.0	4.2	41.7
5	2.5	1.9	36
6	4.0	1.3	29.6

*The preliminary mineralogical analysis of the crops obtained, showed the occurrence of ferromagnetic phase in terms of concretion and ferruginization of manganese minerals (up to 5%) in Sample 1. Taking into consideration the low output of the product, determining its specific magnetic susceptibility is not depicted in the article, and these values are presented in the conclusions.

To measure magnetic susceptibility of the presented probes a certain stand has been prepared equipped with the specified magnetic system, weight measuring device removed from the effect of the magnetic field, a rigid lifting bar for a test tube with the sample, direct current source which provides electricity supply to the stand with a magnetic coil up to values of 20 A with voltage of 110 V.

According to the research methods the magnetic crop sample was loaded into a cylindrical container with the inner diameter 26mm, and consolidated to the required level. The container was secured on a lifting bar and moved into the operating area coaxial to the coil and core in such a way that the flat bottom of the sample (the internal container bottom) was at level $x_b = 4$ mm from the top of the coil, and the top of the sample was at level $x_t = 10$ mm. Further, the sample was weighed without magnetic field and with different values of the current in the coil correspondingly. According to the results of weighing, the magnetic force affecting Sample F_m

was determined. The weighing was done with an electronic weighting unit of RADWAG WPS 210/C/2 type.

The results of measuring the magnetic force F_m and conditions under which the experiment is conducted (the sample number, weight, current force, induction in the operating area) as well as the results of χ according to formula (3) are given in Table 2.

Table 2. The results of measuring the magnetic force F_m, experiment conditions and calculation data χ.

No. of an experiment	No. of tests	I, A	m, g	F_m, mN	H, kA/m	$\dfrac{dH}{dx}$, kAm²	χ, 10^{-6}, m³/kg
1	2	6		0.706	147.8	1856.8	0.377
2	2	8		1.088	188.8	2520.0	0.335
3	2	12	5.43	1.853	249.1	3183.1	0.343
4	2	16		2.554	305.2	3713.6	0.330
5	2	20		3.791	359.3	5039.9	0.307
6	3	6		0.657	147.8	1856.8	0.335
7	3	8		0.961	188.8	2520.0	0.283
8	3	12	5.68	1.588	249.1	3183.1	0.281
9	3	16		2.206	305.2	3713.6	0.273
10	3	20		3.339	359.3	5039.9	0.258
11	4	6		0.569	147.8	1856.8	0.297
12	4	8		0.833	188.8	2520.0	0.251
13	4	12	5.55	1.460	249.1	3183.1	0.264
14	4	16		1.988	305.2	3713.6	0.252
15	4	20		3.022	359.3	5039.9	0.239
16	5	6		0.549	147.8	1856.8	0.272
17	5	8		0.814	188.8	2520.0	0.233
18	5	12	5.85	1.471	249.1	3183.1	0.252
19	5	16		2.010	305.2	3713.6	0.241
20	5	20		3.098	359.3	5039.9	0.233
21	6	6		0.373	147.8	1856.8	0.187
22	6	8		0.559	188.8	2520.0	0.162
23	6	12	5.77	0.951	249.1	3183.1	0.165
24	6	16		1.356	305.2	3713.6	0.165
25	6	20		1.925	359.3	5039.9	0.147

According to the results we have constructed dependency diagrams of specific magnetic susceptibility χ of different manganese sludge crops as a function of the magnetic field intensity (strength) H (Figure 2).

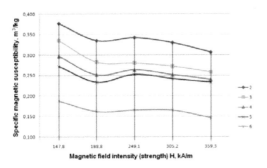

Figure 2. Dependency of specific magnetic susceptibility of different manganese sludge crops (coarseness 0.15-2.0 mm) upon the magnetic field intensity (strength).

The *fractional accuracy* of the experimental determination of the specific magnetic susceptibility can be found according to the following formula:

$$\Delta = \delta_1 + \delta_2 + \delta_3 + \delta_4 + \delta_5,$$

where δ_1 – fractional accuracy of determining the weight of the sample; δ_2 – fractional accuracy of determining the magnetic force affecting the sample; δ_3 – fractional accuracy of determining the magnetic field density at specified points of the operating area; δ_4 – fractional accuracy of determining $\dfrac{dH}{dx}$; δ_5 – fractional accuracy of determining approximate of the magnetic field in the operating area.

In all the experiments the sample weight changed within 5.43-5.85 g, fractional accuracy of a weight measuring device is 0.001 g, thus: $\delta_{1max} = 0.00018$;

$\delta_{1min} = 0.00017$.

The magnetic force F_m affecting the samples changed within 0.037-0.35 g fractional accuracy of a weight measuring device is 0.001 g, therefore: $\delta_{2max} = 0.027$; $\delta_{2min} = 0.0028$.

The magnetic field density at specified points of the operating area (x_b, x_t) changed within 0.178-0.469 mTl, milliteslametre deviation is 0.001 Tl, thus: $\delta_{3max} = 0.0056$; $\delta_{3min} = 0.0021$.

Fractional accuracy of $\dfrac{dH}{dx}$ is defined as a sum of fractional accuracies of density determining (the same fractional accuracy occurs while determining intensity H) and fractional accuracy of determining coordinates of the operating area points, hence:

$\delta_{4max} = 0.0056 + 0.05 = 0.0556$;

$\delta_{4min} = 0.0021 + 0.02 = 0.0221$.

The same deviation will be observed while determining the averaged density of the magnetic field in the operating area:

$\delta_{5max} = 0.0056 + 0.05 = 0.0556$;

$\delta_{5min} = 0.0021 + 0.02 = 0.0221$.

Summing up the deviations, we get the fractional accuracy of the whole experiment.

Moreover, the method deviation which is connected with the allowance made in formula (2) remains unaccounted. However, this deviation can be estimated by comparing the results of calculating χ using formula (3) an formula

$$\chi = \frac{F_m}{\mu_0 \cdot Q}, \qquad (4)$$

where $Q = \rho \int_V H |gradH| dV$; ρ – apparent density of the sample.

In addition, under the integral sign in formula (4) we applied the disposal $H_{(x,r)}$ in the operating area found by experiment.

The deviation which occurred while calculating according to formulas (3) and (4) made up as maximum as 9%.

CONCLUSIONS

It has been determined that the manganese sludge samples are classified as feebly magnetic material. The magnetic susceptibility of the first sample ranges from 359 to 9 10^{-9} m^3/kg while the magnetic field intensity varies from 64 to 262 kA/m. Meanwhile considerably increased magnetic susceptibility and its further reduction while increasing the magnetic field intensity are conditioned by ferromagnetic phase available in it. All this considered, while choosing magnetic separators and recovering iron-containing components it is necessary to make choice of separators with possibly lower level of magnetic density. The results of studying magnetic properties of magnetic crops of the material under research show their considerable difference in magnetic susceptibility for dressing initial sludges by dry magnetic separation techniques.

REFERENCES

Karmazin, V.I. & Karmazin, V.V. 1978. *Magnetic dressing techniques*. Moscow, 23.
Svoboda, Jan. 2004. *Magnetic Techniques for the Treatment of Materials*. Kluwler Academic Publishers: 5-10.
Mladets'kyy, I.K., Kuvayev, Ya.G., Lysenko, A.A. & Pavlenko, A.A. 2011. *Minimal sample mass for raw quality parameter analysis*.
Karmazin, V.I., Mostyka, Yu.S., Shutov, V.Yu. and others. 2000. *The analysis of the influence of magnetic saturation of highly gradient magnetic separation matrix and reducing magnetic susceptibility of extracted particles by means of the magnetic field intensity growth on magnetic separation efficiency*, Moscow: 221-223.
Vonsovskyy, S.V. 1971. *Magnetizm*. Moscow.

Mining of Mineral Deposits – Pivnyak, Bondarenko, Kovalevs'ka & Illiashov (eds)
© 2013 Taylor & Francis Group, London, ISBN: 978-1-138-00108-4

Inner potential of technological networks of coal mines

S. Salli & O. Mamaykin
National Mining University, Dnipropetrovs'k, Ukraine

S. Smolanov
Paramilitary Mine Rescue Service of Ukraine, Donets'k, Ukraine

ABSTRACT: Over the past century 9.4 billion tons of coal were extracted in Ukraine, that is almost one third of the existing reserves. It is quite natural that layers with the most favorable conditions were processed. Almost all reserves of hard coal in Donetsk and greatly in Lugans'k regions were fully extracted; extraction of scarce crozzling coal was also reduced. For this reason the quality of extracted coal will change for the worse in the nearest future. However, the main problem is still mine pool condition. The mines will be more aged and in 20-30 years they will turn into quite complicated enterprises, with low efficiency and more difficult working conditions. Donbass is greatly loosing its potentials (Pivnyak 2004).

1 INTRODUCTION

Donbass coal industry regeneration has started right after its territory had been freed from the occupation forces. However, in terms of continuation of military action the mine activity was restored in the former, pre-war type. Only separate units were reconstructed. Naturally, such engineering policy of restoration was in some way justified by rigid conditions of wartime and the urgent necessity to supply the economy with energy resources. However, hastily restoration has abandoned outdated technology, equipment, methods of work organization, making ground for further economic and social disorder in the basin.

In other words, for today Ukraine inherits not only explored coal reserves but also a mine pool with pre-revolutionary, pre- and post-war enterprises. Large new-built mines are very rarely. The evidence – average production capacity of Donbass mines of 500 th. tonnes / year. Donbass is confronted with an alternative to close the majority of unprofitable mines which has almost exhausted their booked reserves. Meanwhile, harmful consequences of age-old coal underground extraction with storing rocks on the surface, unsystematic fault of highly mineralized mine water and other aspects of irresponsible activity have greatly effected the environment. "Region problems" and "unstable territories" more and more often occur as a regular final of mass closure of mines.

We will consider two alternatives in terms of the region with quite aged mine pool and significant value of anthropogenic waste. One of the latter includes wide-ranging works of mine and processing plants' waste recycling, and the second one – continuation of unprofitable mine work for which possibility of closure is considered. Each of these alternatives requires approximately equal assets. Economic effect on the unit of final output will differ, according to the grade of left reserves and anthropogenic area reserves quality. This is fundamental statement of the question.

2 MAIN PART

In synergetics processes of system evolution are reflected with help of bifurcational diagrams (Figure 1), where y – characteristic system parameter (level of sophistication, organization, differentiation, etc.); x – time; x_1, x_2, x_3, x_4 – points of bifurcation; y_0 – value of parameter y in point x_2; solid line – stable solutions; cut-line – unstable solutions. In point x_4 the bottom bifurcational branch leads to system slip to the bottom branch of the previous bifurcation, points. In general, it would be more correctly to talk not about bifurcation but polyfurcation, because there may be a lot of trajectories going out of branching point (Anishenko 1990).

Thus, bifurcation is a gain of a new property during dynamic system movement with minor scaling. For example, with $x*$ parameter value

$$x* = 0.64 \text{ (with } a = 2.8\text{)}; \quad x_1* = 0.48; \quad x_2* = 0.82$$

(with $a = 3.3$);

$$x_1^* = 0.38; \quad x_2^* = 0.50; \quad x_3^* = 0.83; \quad x_4^* = 0.88$$

(with $a = 3.5$)

Strive of mine's technological scheme in terms of cleaning-up of the remained reserves in bifurcational stability state is explained by attempts of left mine fields extraction or switch to unbooked reserves. That means that the condition of technological network has to be in terms of structural and dy-namical theory, from stable development and changes point of view. Thus, on the one hand, several system elements correspond to certain problem of mine working development planning solution, on the other hand, one element of technological scheme can provide solution of several problems. On this basis, the efficiency of task processing is determined not only by mines subsystem functioning efficiency but also by interaction inside the system and between tasks which need solution.

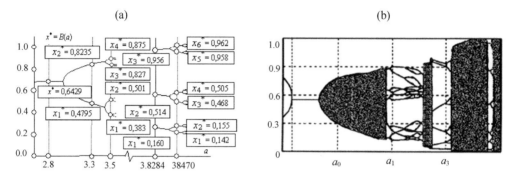

Figure 1. Bifuractional diagrams with numerical transformation of x^* parameter.

As a result, the formation of mine technological network potential is a symbiosis of level factor interaction of mining operation concentration, ventilation stability and impact on enterprises productive flow formation. Each of the factors is accordingly characterized by one of the indexes "technological units current capacity ratio", "ventilation factor power limit" and "productive flow density". Optimization of target "technical and economical stability of mine's technological network", which characterizes mine's potential to innovations is the result of these and second order factors interaction which determine formation of several key indexes. It can be claimed that stability of technological network is stability to keep its integrity and coal output supplier's mission, to operate according to required mode in terms of indeterminacy conditions of inner and outer factors, regulating its carrying capacity in simple and extended reproduction mode.

It is obvious that all the factors of technological scheme potential level interaction are synthetic, occurred as an interaction result of a number of natural component and production and economic activity of mine manifestation. Almost all the factors which characterize mining and technological conditions of extraction greatly depend on mine activity specificity and its sector profile. This determines the presence of not only functional connections between the factors in the process of potential formation but also analytic ones, which are quite significant. Such linkage allows to indicate technical potential inner components, which directly form the potential of mine's topological network. That is why it can be claimed that the technical potential of mine's technological network is the result of a set of concurrent and interrelated factors of the first, second, third and following orders.

In order to determine their influence on given parameter formation, it is reasonable to use statistic analysis method and reveal dependency between technical potential index and a line of independent indexes of production and economic activity of mine. Reception of analytical dependences which describe "technical potential" index is based on the research of dependencies between given index as a result one, mining and technological indexes, and factor ones. Dependencies research and reception of "technical potential" index's regression equation as integral estimate of mine's technological network innovative level is based on optimal programming usage and decision-making in terms of indeterminacy.

Following indexes from different groups which characterize production activity, mining and technological conditions of Donbass mines were accepted as background factors with the purpose to form "technical potential" parameter: level of mining operation concentration K, labor capacity of a worker extracting P, monthly displacement of longwalls V, prime cost of 1 tonne of coal S extraction.

"Technological scheme potential" category is mostly use during consideration of separate regions. Here, output increase and improvement of technical-economic indexes can be achieved using different ways: construction of the second order mines, reconstruction or even closure of operating mines. For the region the term "technological schemes potential" get a new meaning, and it is appropriate to talk about development, one element of which is modernization of separate mines and tightly coupled with it concentration and intensification of production.

It is commonly known that the main property of a coal mine which determines all the elements of its activity is spatial development. This property is objective as being caused by fundamental characteristic of coal its non-reproducibility. Rate of development is determined by human activity and it depends on many factors, level of scientific-and-technological progress in particular. But the development necessity is set by nature and can not be excluded or replaced with something, even if the technology of production process is changed.

First of all, the idea of "mine development" has to be determined. That means removal from building core along and transverse strike of layers and also vertical lowering from the first horizon from which mine exploitation has started. A mine is developing in three dimensions but not simultaneously. If we take a long enough period (one or several decades), the mine will develop in three dimensions, for a short period (one-two years), in general, the mine will develop along strike and for two other dimension there may be no changes.

In order to determine the inner potential of technological network, main factor features which can define formation of technical potential of mine's topological network as an integral estimate of mine's potential in innovation part were considered. As for labor capacity of an extracting worker, monthly displacement of longwalls and prime cost of 1 tonne of coal, their values were taken according to the actual data concerning activity of hard coal mines which are a part of SC "Sverdlovanthracyte", "Rovenkyanthracyte" and "Donbassanthracyte" for 2010-2012.

In order to derive multiple regression equation which describes "technical potential", step-by-step method of variables injection is used. As a result, the determination of maximum available value of economic additional charge, created by mine, is one of a variety of multicriterion problem with four criteria which needs to be reduced to singlectriterion with objective function (1).

$$k_k = -\alpha K + \beta P + \gamma V - \mu S \rightarrow max,$$

where k_k – free index of network potential; K – parameter which characterizes working and faces length; P – labor capacity of an extracting worker, tones / month; V – annual face movement, m; S – prime cost of 1 tonne of coal extraction.

As mine technological network technical potential formation being described by the equation (1), it is obvious that achievement of its maximum value depends, firstly, on relation between K, P, V and S. Besides, is should be considered that "echnical potential" index maximization is achieved in terms of limited power of ventilation factor and production flow density accordingly. Which means that maximization task of parameter k_k, which is the main meter of mine technological scheme potential, is turning into a compromise search between four main factor values.

The objective of the task is to determine optimal parameters joint influence of which allows to assign a limit of possible worsening of technical and economic indexes of mines.

From economics point of view, limiting state can be figuratively represented in three dimensions (X, Y, Z): where X – consequences of working depth increase manifestation; Y — worsening of technical and economic indexes; Z – expenses for extraction volume and quality maintenance (Figure 2).

As limiting state economically being function of time, the task has to be formulated in dynamic statement. With such graphic interpretation, economic and mathematical model is written in the following way: determine optimal values of objective function according to following criteria:

$$\left.\begin{array}{l} K(X,Y,Z) \rightarrow max; \\ P_i(X,Y,Z) \rightarrow max; \\ V_i(X,Y,Z) \rightarrow max; \\ S_i(X,Y,Z) \rightarrow min. \end{array}\right\} \quad (2)$$

The most suitable way to solve given problem is multicriteria Pareto method, used in such problems which are quite original: extremum concept can not be built for them but concept of improving situation ("Pareto optimum") can be put in (Podinovskyi & Nogin 1982).

During the process of impact priority setting on strategic parameters of production and economic mine activity, their actual values need to be compared with optimal ones, defined from solution of equation system included in model (1)-(2). Achievment of optimal values if technological scheme parameters means total realization of mine's economic potential, in other words, limiting achievable (reference) level.

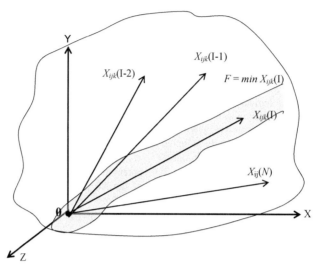

Figure 2. Geometrical interpretation of the determination task of technological scheme parameter limiting values.

3 CONCLUSIONS

1. There is practically no system of quantitative estimation of technological scheme state for coal mines. And existing separate scheme characteristics have principle errors, they prefer extensive reproduction to the detriment of innovative preferences.

2. Inner potential of technological networks is one of the most important parameters of coal mine state estimation. Its formation is influence result of many factors, which determine efficiency of underground mining and, first of all, property of a coal mine – spatial development. This property is objective and its rate of development is determined by human activity, and it depends on many factors, level of scientific-and-technological progress in particular.

But the development necessity is set by nature and can not be excluded or replaced with something, even if the technology of production process is changed.

REFERENCES

Pivnyak, G.G., Amosha, A.I., Yashenko, Yu.P. and other. 2004. *Mine pool reproduction and investment processes in coal industry of Ukraine*. Kyiv: Naukova dumka: 312.
Anishenko, V.S. 1990. *Compound tones in simple systems*. Moscow: Nauka: 312.
Podinovskyi, V.V. & Nogin, V.D. 1982. *Pareto – optimal solving of multicriteria problems*. Moscow: Nauka.

Mining of Mineral Deposits – Pivnyak, Bondarenko, Kovalevs'ka & Illiashov (eds)
© 2013 Taylor & Francis Group, London, ISBN: 978-1-138-00108-4

On parameters influence evaluating method application in some geotechnical tasks

G. Larionov, R. Kirija & D. Braginec
M.S. Polyakov Institute of Geotechnical Mechanics, Dnipropetrovs'k, Ukraine

ABSTRACT: The paper is devoted to sequence approximation method application for influence parameters evaluating in some geotechnical tasks. The proposed method efficiency was showed on the transformation solving into univariable function product for dynamic elastically deformed medium under borehole pressure impulse. Another task is devoted to loading hopper for averaging rock mass problem. Influence parameters evaluating consist of univariable function powers comparisons in point vicinity representation as univariable power function product.

1 INTRODUCTION

The reconstruction of the analytical form function is made with mesh value function methods as usual. When the function value obtaining time expense and number of design parameters are small the approximation methods are used. However in case of complicated and multivariable task when using the methods is time expensive, it is hard enough and impossible to do reconstruction even. Using the analytical reconstructed form function to design parameter influence evaluating has two demands. These demands are opposite. From the one hand, the reconstruction function is to be closed to the mesh value function data, from another to supply possibility for the design task parameters evaluating.

2 FORMULATING THE PROBLEM

The new sequence approximation method (Larionov 2011) is proposed for solving the problem. The method allow to approximate reconstruction in analytical form function when it exist in table form in vicinity point $M(X_0)$, $X_0 = X\left(x_1^0, x_2^0, ..., x_n^0\right)$, $X_0 \in \overline{D}$, where $\overline{D}$ – close domain. There are theorem have formulated and top limit error estimation for analytical form function representation as univariable function product obtained.

Function representation is proposed as: $F(x_1, x_2, x, ..., x_n) \approx \alpha_n g_1(x_1) g_2(x_2) ... g_n(x_n)$, where $g_1(x_1), g_2(x_2), ..., g_n(x_n)$ – approximation functions of the lines which caused by crossing the function surface by planes parallel to co-ordinate ones, and

which pass through $M(X_0) \in \overline{D}$, α_n – approximation factor. The demands to the functions $g_1(x_1), g_2(x_2), ..., g_n(x_n)$ are continuity and the first partial derivatives continuity. So, as evident, it is enough wide class of functions. But all work, as you see, have done for function representation in point vicinity. What to do for expand the results in domain? As a using experience demonstrated, solutions of real tasks allow expand representation in point vicinity in whole domain. In spite of the fact that the errors of the function representation raised up to domain boundary errors do not exceed 5-7%, what is sufficient for numerous geotechnical tasks. Thus influence parameter evaluating method consists of approximate solution reconstruction (when it exists in number table form) in form of exponent functions product and its index exponent comparisons. Influence parameter estimation procedure formulated as: the greater index value, the stronger function parameter influence.

It is known, that in points vicinity a function can be represented in a number of different ways. However different natural function presentations with unidimension functions are very useful in engineering applications (Larionov 2012).

Existence proof function notation theorem in univariable functions product in point's vicinity and its upper error limit is presented below (Larionov 2012).

Let a function $F(X) = F(x_1, x_2, ..., x_n)$ – the continuous function of n independent variables $x_1, x_2, ..., x_n$, defined in domain D. We consider that function F has a first order partial derivative, limited in a domain D. Assume that $X_0 = \left(x_1^0, x_2^0, ..., x_n^0\right)$ –

certain point, that D belongs.

Lets, $\Delta_{X_0} = \Delta_{x_1^0} \otimes \Delta_{x_2^0} \otimes ... \otimes \Delta_{x_n^0}$, where

$\Delta_{x_i} = [a_i, b_i]$; $a_i + b_i = 2x_i^0$ $(i = \overline{1, n})$ a_1 and b_1 – arbitrary real constants; N and K – some positive constants. A symbol $\otimes$ Decart's set product means. In other words $\Delta_{x_0} = \left\{ X = (x_1, x_2, ... x_n) : x_i \in \Delta_{x_i^0} \right\}$,

$i = \overline{1, n}$.

Lets $\Delta_{X_0} \subset D$ (Figure 1).

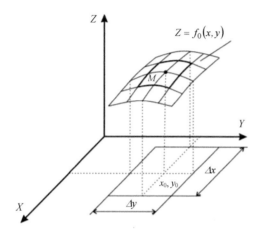

Figure 1. Point vicinity (x_0, y_0) choice.

Function constraints can be written as:
$$\max_{X \in \Delta_{X_0}} |F(X)| \le N \;;$$

$$\max_{X \in \Delta_{X_0}} \left\{ \left| \frac{\partial F(X)}{\partial x_1} \right|, \left| \frac{\partial F(X)}{\partial x_2} \right|, ..., \left| \frac{\partial F(X)}{\partial x_n} \right| \right\} \le K \;, \qquad (1)$$

where N and K – some constants.

Fixing $(n-1)$ variables in turns as show below, we will form the n following functions:

$$f_1(x_1) = F(x_1, x_1^0, ... x_n^0);$$

$$f_2(x_2) = F(x_1^0, x_2, ... x_n^0);$$

....

$$f_n(x_n) = F(x_1^0, x_2^0, ... x_n).$$

We consider that $f_i(x_i) \in C(\Delta_{x_i^0})(i = \overline{1, n})$,

where $C(\Delta_{x_i^0})$ – the set of functions continuous on a segment $\Delta_{x_i^0}$. Obviously, that a function $f_i(i = \overline{1, n})$ is limited on a segment $\Delta_{x_i^0}$, that is

$$\max_{x_i \in \Delta_{x_i^0}} |f_i(x_i)| \le M_i (i = \overline{1, n}),$$

where M – the arbitrary constant.

Lets $G_i \subset C(\Delta_{x_i^0})$, $i = (\overline{1, n})$ – some subspace that belongs $C(\Delta_{x_i^0})$. Do we consider that subspace G_i is counted and everywhere dense in space $C(\Delta_{x_i^0})$, that is for an arbitrary function $f_i(x_i) \in C(\Delta_{x_i^0})$ $i = (\overline{1, n})$, and any number $\delta > 0$ there is some function $g \in G_i$, such, that

$$\max_{x_i \in \Delta_{x_i^0}} |f_i(x_i) - g_i(x_i)| \le \delta_i, i = (\overline{1, n}).$$

Consider some set G_i and subspace $G(n_i) \in G_i$ such that $\dim G(n_i) = n_i$. Thus for an arbitrary function $f \in C(\Delta_{x_i^0})$ have:

$$E_{n_i}(f)_C = \inf_{g \in G(n_i)} \max_{x_i \in \Delta_{x_i^0}} |f(x_i) - g(x_i)| \le C_f \varphi(n_i),$$

where $E_{n_i}(f)_C$ – best approaching of function $f \in C(\Delta_{x_i^0})$ by the elements of subspace $G(n_i)$; C_f – some constant which depend on the function f and does not depend on the number n_i; $\varphi(n_i)$ – some descending function which is determined by approximable subspacious $G(n_i)$ characteristics and such, that $\varphi(n_i) \to 0$ at $n_i \to \infty$.

We will define a next subspace $\Omega(\Delta_{X_0})$, which consists of such kind functions

$$\varphi(X) = \alpha \prod_{i=1}^{n} g_i(x_i), \qquad (2)$$

where α – arbitrary constant and $g_i \in G_i, (i = \overline{1, n})$ – any functions.

We make designation

$$E(F, \Omega(\Delta_{X_0}))_C = \inf_{\varphi \in \Omega(\Delta_{X_0})} \max_{X \in \Delta_{X_0}} |F(X) - \varphi(X)| \text{ is}$$

the best approaching of function $F \in C(\overline{D})$, where $\overline{D}$ – set D closure with subspace $\Omega(\Delta_{X_0})$ elements, and $C(\overline{D})$ – set of all continues function in $\overline{D}$.

Theorem. Lets $\delta = (\delta_1, \delta_2, ..., \delta_n)$ – some of n dimension space point such, that $0 < \delta_i < \varepsilon$, $(i = \overline{1, n})$, $\varepsilon > 0$ – the arbitrary constant and function $F \in C(\overline{D})$ that function F has a first order partial derivative, limited in a domain D and satisfies constraints (1).

Then in set $\Omega(\Delta_{X_0})$ is such function (2)

$$\varphi_*(X) = \alpha_* \prod_{i=1}^{n} g_i^*(x_i),$$ for which the following inequality takes place:

$$E\left(F, \Omega(\Delta_{X_0})\right) \leq \|F - \varphi_*\| \leq C_F \left(\sum_{i=1}^{n} |\Delta_{x_i}|\right), \quad (3)$$

where $\left|\Delta_{x_i}^0\right|$ – length of segment $\Delta_{x_i}^0, (i = \overline{1, n})$; C_F – some constant, which depend on the function F and does not depend on n.

Attention. Location $M = M\left(x_1^0, x_2^0, ..., x_n^0\right)$, $M \in \overline{D}$ point is significant parameter which caused the different kind of function presentations. We propose to make point choice in the way: $x_j = (b_j - a_j)/2$, where a_j, b_i – interval limits, when the additional information about function is absent.

3 SOLVING THE PROBLEM

To testify and estimate the sequence approximation method efficient, solve two tasks. First task concerns the dynamic elastically deformed medium with borehole pressure impulse. The second one concerns the averaging bunker volume obtaining when it works in protective layer regime.

Task one. Dynamic parameters of elastically deformed medium with borehole pressure impulse are obtained with one dimensional wave equation (Krylov 1950):

$$r^2 \frac{\partial^2 u_r}{\partial r^2} + r \frac{\partial u_r}{\partial r} - u_r = \frac{r^2}{v_p^2} \frac{\partial^2 u_r}{\partial t^2}, \quad (4)$$

where $u_r(r, t)$ – radial deflection of elastic medium, m; t – time of deformation process, s; v_p – wave elastic speed, m / s. The internal and external surface boundary conditions are:

$$\sigma_r\big|_{r=r_0} = -\psi(t); \sigma_r\big|_{r=r_N} = 0, \quad (5)$$

where r_0, r_N – internal and external radius.

In plane deformation state radius and circumferential stresses are obtained as (Sapegin & Larionov 2012):

$$\frac{\sigma_r}{q_0} = \frac{2}{\pi} \frac{r_0}{v_p(t_2 - t_1)} \sum_{i=1}^{\infty} \left(\int_0^{a_1 - \delta_1} \overline{\sigma_r}\left(\xi_1, \frac{r}{r_0}\right) sin[\xi_1 \bar{t}]d\xi_1 + ... + \int_{a_{i-1} - \delta_{2i-2}}^{a_1 - \delta_{2i-1}} \overline{\sigma_r}\left(\xi_1, \frac{r}{r_0}\right) sin[\xi_1 \bar{t}]d\xi_1 \right), \quad (6)$$

$$\frac{\sigma_\theta}{q_0} = \frac{2}{\pi} \frac{r_0}{v_p(t_2 - t_1)} \sum_{i=1}^{\infty} \left(\int_0^{a_1 - \delta_1} \overline{\sigma_\theta}\left(\xi_1, \frac{r}{r_0}\right) sin[\xi_1 \bar{t}]d\xi_1 + ... + \int_{a_i - \delta_{i+1}}^{a_{i+1} - \delta_{i+2}} \overline{\sigma_\theta}\left(\xi_1, \frac{r}{r_0}\right) sin[\xi_1 \bar{t}]d\xi_1 \right), \quad (7)$$

where a_1, a_2,..., a_3 – discontinuity coordinates of deflection transforms or stresses; δ_1, δ_2,..., δ_3 – deviation in discontinuity vicinity; $\xi_1 = \frac{\varpi r_0}{v_r}$ – dimensionless parameter of the transformation; $\overline{\sigma}_r$, $\overline{\sigma}_\theta$ – radial and circumferential stress correspondingly; ϖ – transformation parameter, $1/c$; $\bar{t} = \frac{v_p t}{r_0}$ – dimensionless time of the deformation proc-

ess; r – the current variable radius; $t_c = t_2 - t_1$ – real load shut down time, s; q_0 – internal load, Pa; t_1, t_2 – lifting and load shut down time respectively, measured from the origin, s.

As can be seen from (6), (7), the original expression for the determination of stresses and displacements are complicated and inconvenient for practical using. (Sapegin & Larionov 2012).

The main parameters that influence on the process of time-dependent deformation of an elastic medium are: the inner radius of the borehole r_0, the value of the internal applied load q_0, the speed of elastic waves v_p the shut down time of the internal load t_c. Influence degree estimation of each parameter is in a complicated way because they are as in front of the integral and under one (see (6), (7)).

As the function, the presentation of which we are looking for, take the maximum tensile stress value in the first half-wave of change, divided by the value of the maximum amplitude of the internal load q_0:

$$\bar{\sigma}_r = \sigma_r / q_0 = \sigma\left(t_c^{\alpha_1}, V_p^{\alpha_2}, r_0^{\alpha_3}\right), \tag{8}$$

where α_1, α_2, α_3 – the exponents of the tested parameter functions.

The domain of a given parameter is formed with intervals:

The inner cylindrical cavity radius r_0 – 0.05-0.5 m; elastic wave speed – 100-2000 m / s; load shut down time – 0.001-0.1 s.

The base point $M = M\left(x_1^0, x_2^0, ..., x_n^0\right) M \in \overline{D}$ is defined by parameter:

Load shut down time $t_c^0 = 0.01$ c; the elastic wave velocity $v_p^0 = 600$ m / s; the inner radius $r_0^0 = 0.1$ m.

Changing incrementally parameter t_c, we calculate the sequence of function values $\bar{\sigma}_r$. For the sequence $\{t_c, \bar{\sigma}_r\}$ we find the approximation function $g_1(t_c)$ in the form: $g_1(t_c) = a_1 t_c^{\alpha_1}$, where $a_1 = 0.00020186$; $\alpha_1 = -0.944$.

Graphic representation of implementation of these steps is shown in Figure 2.

Thus, the dimensionless tensile stress on the load shut down time will be:

$$\bar{\sigma}_r\left(t_c, v_p^0, r_0^0\right) \approx \varphi_1(t_c) = a_1 t_c^{\alpha_1}. \tag{9}$$

Formula (9) can already be used in the calculations for determining the tensile stress depending on the different shut down time (for a fixed base point and the parameter values v_p^0 and r_0^0).

Perform the steps for the second parameter v_p. Changing incrementally v_p, we compute the sequence of function values $\bar{\sigma}_r$. For the sequence $\{t_c, \bar{\sigma}_r\}$ we find the function approximation $g_2(v_p)$ the form $g_2(v_p) = a_2 v_p^{\alpha_2}$: where: $a_2 = 0.091318$; $a_2 = -0.957$.

Graphic representation of implementation of these steps is shown in Figure 3.

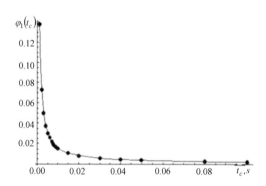 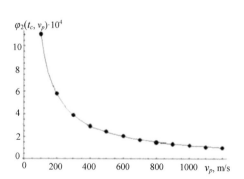

Figure 2. Function $g_1(t_c)$ in tabular form and function of its approximation.

Figure 3. Function $g_2(v_p)$ in tabular form and function of its approximation.

Thus, the dimensionless tensile stress on the load shut down time t_c and the elastic wave velocity v_p will be in the form:

$$\bar{\sigma}_r\left(t_c^0, v_p, r_0^0\right) \approx \varphi_2(t_c, v_p) = a_2 t_c^{\alpha_1} v_p^{\alpha_2}. \tag{10}$$

Equation (10) can be used for special calculations of the stresses at a fixed value of the inner radius r_0.

Perform the steps for the third parameter r_0.

Changing the parameter r_0 with a step, we compute the sequence of function values $\bar{\sigma}_r$. For the sequence $\{r_0, \bar{\sigma}_r\}$ we find the function approximation $g_3(r_0)$ in the form $g_3(r_0) = a_3 r_0^{\alpha_3}$: where: $a_3 = 0.91729$; $\alpha_3 = 1.01$.

Graphic representation of implementation of these steps is shown in Figure 4.

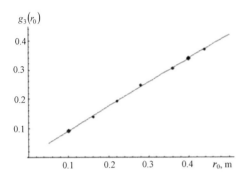

Figure 4. Function $g_3(r_0)$ in tabular form and function of its approximation.

Thus, dimensionless tensile stress of the load shut down time t_c and the elastic wave velocity v_p and the inner radius r_0 will have the form:

$$\bar{\sigma}_r\left(t_c^0, v_p^0, r_0\right) \approx \varphi_3\left(t_c, v_p, r_0\right) = a_3 t_c^{\alpha 1} v_p^{\alpha 2} v_0^{\alpha 3}. \quad (11)$$

At last the final formula for the radial tensile stresses on all tested parameters can be written as:

$$\sigma_r = 0.917 \frac{q_0^{1.0} r_0^{1.01}}{v_p^{0.957} t_c^{0.944}}. \quad (12)$$

Taking into account that all exponents of the study parameters are close to unit (12). Make round off the exponents we obtain a more convenient formula for the calculation of tensile stress on the inner contour of the cylindrical cavity in the first half-wave stress changes in the form:

$$\sigma_r = 0.917 \frac{q_0 r_0}{v_p t_c}. \quad (13)$$

A comparison of the relative errors stresses with formulas (12), (13) and the exact with the Fourier integral method (Sapegin & Larionov 2012) is obtained. Error analysis of the solution (12) has shown that the average error for the shut down time t_c

does not exceed 3.8%, for the elastic wave velocity v_p – 2.0%, for the inner radius r_0 – 4.3%.

Analysis of the results of the comparison of the relative errors stresses formula (13) with the exact showed that the average relative error for the shut down time t_c is not more than 2.7%, for the speed of elastic waves v_p – 2.2 % and for the inner radius r_0 – 4 %.

Task two. Averaging bunker (AB) play an important role in underground coal mine conveyor transport. AB is usually equipped mines in the roadways, particularly in the areas of down hole traffic congestion on prefabricated conveyors, shown in Figure 5.

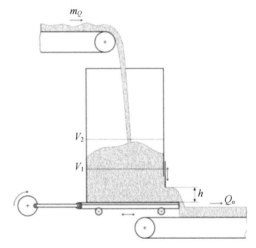

Figure 5. Scheme of the bunker loader (BL).

In order to prevent the destruction of mining equipment due to the fall of large rock pieces in the BL is necessary to maintain a protective layer (PL) of material. To determine the necessary unloading capacity we must know: the feeding parameters, the capacity of the bunker and the PL height.

When bunker works in the PL support mode unload freight flow (FF) from bunker turned off, if the amount of material in the bunker reaches a maximum value V_2 (m^3), and is activated if the amount of material in the bunker became less than the allowable minimum value of V_1 (m^3). During the process loaded freight flow into the bunker does not turn off, even when the amount of material in the bunker reaches the allowable maximum or minimum value.

This work mode simulation results have shown (Kiriya, Braginec & Mischenko 2008) that when the average throughput of incoming FF m_Q (t / min) is

251

greater than or equal unloading productivity Q_n (t / min), ($m_Q \geq Q_n$), the amount of material in the bunker increases indefinitely. If $m_Q < Q_n$, the average amount of material in the bunker has a finite value that depends on the parameters of the incoming FF, and minimum and maximum amount of material in the bunker. Therefore, to maintain the PL of material in the bunker need that inequality $m_Q < Q_n$ was satisfied.

To develop a mathematical model of the AB functioning is in the PL material mode assume that entering the bunker minute FF is a normal random Markov process with mean m_Q, standard deviation σ_Q (t / min) and the correlation function is equal to (Schahmeister & Solod 1976):

$$R_Q(\tau - t) = \sigma_Q^2 e^{-\alpha(\tau - t)},$$

where t, τ – start and end times, respectively, min; α – correlation function parameter, 1 / min.

Unloaded FF from the bunker has a constant value Q_n, that is equal to minute feeder productivity or free discharge capacity of bulk material from the bunker with adjustable damper.

The bunker work can be described by a normal two-dimensional Markov process with the $V(t)$, $Q(t)$ components that write as coupled system of equations (Sveshnikov 1968):

$$\begin{cases} \gamma \dfrac{dV}{dt} = Q(t) - Q_n; \\ \dfrac{dQ}{dt} + \alpha Q(t) = \sigma_Q \sqrt{2\alpha}\, \zeta(t) + \alpha m_Q, \end{cases} \quad (14)$$

where $M[Q(t)] = m_Q$; $D[Q(t)] = \sigma_Q^2$; $V(t)$ – the amount of material in the bunker at time t, m³; $Q(t)$ – minute FF entering to the bunker, t / min; $\zeta(t)$ – white noise, i.e. random function with zero mean and variance equal to the delta function $\delta(t)$; γ – specific volume weight of the material t / m³.

If unloading machine works then in equation (14) $Q_n > 0$, if do not work, then $Q_n = 0$. Equations (14) must satisfy the initial and boundary conditions:

– the initial conditions:

at $t = 0$ $Q(0) = Q_0$; $V(0) = V_{min}$, if $Q_n = 0$;

$Q(0) = Q_0$; $V(0) = V_{max}$, if $Q_n > 0$; $\quad (15)$

where Q_0 – the FF value coming into the bunker at the initial time, t / min.

– the boundary conditions:

at $V = V_{max}$ $\qquad Q_n > 0$;

at $V = V_{min}$ $\qquad Q_n = 0$. $\quad (16)$

It is condition must be satisfied beside that:

$V_{min} \leq V(t) \leq V_{max} (0 \leq t < \infty)$.

Random process described by the system (14) is a continuous Markov process, the distribution function of which is described by Fokker-Planck-Kolmogorov equation of the first kind (Sveshnikov 1968):

$$\frac{\partial f}{\partial t} + (x_2 - Q_n)\frac{\partial f}{\partial x_1} - \alpha(x_2 - m_Q)\frac{\partial f}{\partial x_2} +$$

$$+ \alpha \sigma_Q^2 \frac{\partial^2 f}{\partial x_2^2} = 0, \quad (17)$$

where $f(t, x_1, x_2, \tau, y_1, y_2)$ – conditional probability density function of the two-dimensional distribution of the Markov process of transition from the initial state ($x_1; x_2$) at the initial time t in the state ($y_1; y_2$) at time τ ($\tau > t$).

Here, x_1, y_1 – the values of the random function $V(t)$ at time t, and τ, and x_2, y_2 – value of the random function $Q(t)$ at time t and τ.

Beside that the initial and boundary conditions are to be satisfied:

– the initial conditions:

at $t = \tau$ $\quad f(t, x_1, x_2, \tau, y_1, y_2) =$

$= \delta(x_1 - y_1)\delta(x_2 - y_2)$;

– the boundary conditions:

at $x_1 = V_1$ $f = 0$; if $Q_n = 0$;

$x_1 = V_2$ $f = 0$; if $Q_n > 0$;

at $x_2 \to \infty$ $f = 0$ at any $Q_n \geq 0$.

Furthermore, the normalization conditions are to be carried out, i.e.:

$f(t, x_1, x_2, \tau, y_1, y_2) > 0$;

$\int\limits_{V_{min}}^{V_{max}} \int\limits_0^\infty f(t, x_1, x_2, \tau, y_1, y_2)\, dy_1\, dy_2 = 1$.

To determine the average time filling the bunker with do not working unload machine θ ($Q_n = 0$) and a running discharge ($Q_n > 0$), following (Sveshnikov 1968 & Bolotin 1971), we integrate equation (17) with respect to y_1 from V_{min} to V_{max}, and to y_2 from 0 to ∞ and to time t and from 0 to ∞. As a result, we obtain the equation of Pontryagin:

$$(x_2 - Q_n)\frac{\partial\theta}{\partial x_1} - \alpha(x_2 - m_Q)\frac{\partial\theta}{\partial x_2} + \alpha\sigma_Q^2\frac{\partial^2\theta}{\partial x_2} = -1, \quad (18)$$

where $\theta = \varphi(x_1, x_2)$.

Further, the initial and boundary conditions are to be satisfied:
– the initial conditions:

at $x_1 = V_1$, $x_2 = Q_0$; if $Q_n = 0$;

$$x_1 = V_2, \; x_2 = Q_0; \text{ if } Q_n > 0; \quad (19)$$

– the boundary conditions:

at $x_1 = V_2$, $\theta = 0$, if $Q_n = 0$;

$$x_1 = V_1, \; \theta = 0, \text{ if } Q_n > 0. \quad (20)$$

Equation (18) in the partial derivatives of the two variables can be reduced to Rikati equation. However, the Rikati equation has no analytic solutions in general (Kamker 1971). Taking into account that $-\alpha k_1 \varepsilon m_Q < \alpha(x_2 - m_Q) < \alpha k_1 \varepsilon m_Q$ and that the average FF deviation σ_Q flowing lava into the bunker, far less than the average minute productivity m_Q, i.e. $\sigma_Q \ll m_Q$.

Consequently, the ratio:

$$\varepsilon = \frac{\sigma_Q}{m_Q} \quad (21)$$

is a small parameter. Furthermore, for minute FF entering in the bunker the inequality be fulfill:

$$|x_2 - m_Q| < k_1\sigma_Q, \quad (22)$$

where k_1 – dimensionless coefficient that characterize the relative deviation of the real FF coming into the bunker from the average value.

Instead of a single equation (18) we consider two equations with unknown values θ_1 and θ_2, in which the coefficients in the second term of equation (18) take the maximum and minimum values, respectively:

$$\varepsilon^2\alpha m_Q^2\frac{\partial^2\theta_i}{\partial x_2} - (-1)^{i+1}\varepsilon\alpha k_1 m_Q\frac{\partial\theta_i}{\partial x_2} +$$

$$+ (x_2 - Q_n)\frac{\partial\theta_i}{\partial x_1} = -1, \, (i = 1, 2). \quad (24)$$

Therefore the initial and boundary conditions (19) and (20) where Q_0 in depending on i takes the value

$$Q_{0_i} = m_Q + (-1)^{i+1}k_1\sigma_Q (i = 1, 2).$$

Then the solution of equation (18), because of the small parameter ε, can be approximately represented as:

$$\theta \approx \frac{\theta_1 + \theta_2}{2}. \quad (25)$$

To solve the equations (24), we apply the asymptotic Pade method (Beiker & Greivs-Moris 1986), i.e., represent the solution of equations (24) as the ratio of linear polynomials:

$$\theta_i = \frac{a_0 + a_1\varepsilon}{1 + b_1\varepsilon}. \quad (26)$$

It seems to note that series fraction expansion in small parameter ε (26) is to coincide with one for equation solving (24) up to second power inclusively.

As a result, the approximate solutions of equations (24) take the form:

$$\theta_i = \frac{\gamma(V_2 - V_1)}{Q_{0i} - Q_n}A_i, \, (i = 1, 2), \quad (27)$$

where

$$A_i = \frac{1 + (-1)^{i+1}\left[\alpha\left(1 - \frac{k_1}{2}\right)\frac{\gamma(V_2 - V_1)}{(Q_{0i} - Q_n)^2} + \frac{2k_1}{Q_{0i} - Q_n}\right]\sigma_Q}{1 + (-1)^{i+1}\left[\frac{\alpha\gamma(V_2 - V_1)}{(Q_{0i} - Q_n)^2} - \frac{2k_1}{Q_{0i} - Q_n}\right]\sigma_Q}.$$

Represent whole PL bunker work cycle in two periods. At first period bunker unloading do not work ($Q_n = 0$), material volume in bunker increases from minimum value $V_1 = V_{min}$ to maximal one $V_2 = V_{max}$. At second period bunker unloading work ($Q_n > 0$) and material volume in bunker decreases from maximal material volume $V_1 = V_{max}$ to minimal one $V_2 = V_{min}$.

The average bunker work time in the first period

253

is θ_{load} , and is calculated from (25) and (27) under the value $Q_n = 0$. And the average bunker work time in the second period (the unloading bunker time) θ_{unload} is given by (25) and (27) under the value $Q_n > 0$.

The average bunker work time during one cycle t_c is obtain in form $t_{av} = \theta_{load} + \theta_{unload}$.

The average material volume in the bunker in a steady PL regime according to the assumption of argotic stochastic process (Sveshnikov 1968), is given by

$$V_{av} = \frac{1}{t_{av}} \int_0^{t_{av}} V(t)dt , \qquad (28)$$

where $V(t) = \begin{cases} V_1 + m_Q t, \\ V_2 - (Q_n - m_Q)(t - \theta_{unload})\theta_{load}, \end{cases}$

at $0 \le t < \theta_{load}$;

at $\theta_{load} \le t \le \theta_{unload}$.

Integrating the expression (28) takes the form:

$$V_{av} = \frac{V_1 \theta_{load} + V_2 \theta_{unload}}{t_{av}} +$$

$$+ \frac{m_Q \theta_{load}^2 - (Q_n - m_Q)\theta_{unload}^2}{t_{av}} . \qquad (29)$$

Substituting in (29) the values of θ_{load} and θ_{unload} defined by formulas (25) and (27) we obtain the average material volume in PL bunker work as a function of m_Q , σ_Q , α , Q_n , γ , V_1 and V_2 . In order to investigate the influence of various factors on the average material volume in the bunker represented by approximate relation (29) apply the method described in (Larionov 2011) and as a result we obtain

$$V_{av} = A_{ap} \frac{m_Q^{0.000164} V_p^{0.695} \Delta V^{0.3035}}{Q_p^{0.00263} \alpha^{0.001076} \gamma^{0.001063} \sigma_\theta^{0.90154}} , \quad (30)$$

where A_{ap} – approximation coefficient. Make substitution parameters with negligible small power with average values ones we obtain the formula:

$$V_{av} = A \alpha_0 m_{Q0} \gamma_0 \sigma_{Q0} Q_{n0} V_1^{0.695} \Delta V^{0.3} , \qquad (31)$$

where $\Delta V = V_2 - V_1$; A – coefficient approximation. Here, the subscript 0 denote the average values of the parameters (the basic point M): α , σ_Q , m_Q , γ , Q_n .

From (31) it follows that the average material volume in the bunker is in the PL mode essentially depends on changes in the maximum and minimum V_1 , V_2 volume of the PL material in bunker and slightly affected by changes in the incoming FF parameters, m_Q , σ_Q , α and magnitude of the unloaded material Q_n .

Figure 6 shows a plot of the average material volume in the bunker V_{av} in depending on the discharge parameter Q_n , built in according to formula (29). In this case, the original data values are accepted: $V_2 = 9$ m^3; $V_1 = 4.5$ m^3; $m_Q = 3.7$ t / min; $\sigma_Q = 1.23$ t / min; $\alpha = 0.14$ min^{-1}; $\gamma = 1$ t / m^3; $k_1 = 0.1$.

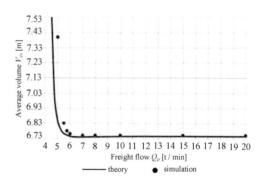

Figure 6. Average bunker volume dependence on unload productivity.

Graph shows that an increase discharge productivity Q_n average material volume in the bunker V_{av} first sharply decreases, reaching a minimum value and then in further increase Q_n the average volume, V_a slightly increase and tends to a limiting value equal to half the sum of the maximum and minimum volumes material in the bunker ($V_c \to (V_1 + V_2)/2 = 6.75$ at $Q_n \to \infty$).

Furthermore, points in Fig. 6 show the results of simulation bunker operation in PL regime, obtained for the same input data. The graph shows that the relative errors in theory values and simulation ones do not exceed 10%.

4 CONCLUSIONS

1. The sequence approximation method proposed to establish the dependences between system parameters confirm the efficiency. Relative boundary errors do not exceed 5-7% what allow to use approximate reconstruction in analytical form function for engineer computation.

2. Thus stress strain state obtaining with formula (13) is essentially simpler then one with formulas (6, 7) in first task.

3. Average bunker material volume obtaining with formulas (30), (31) becomes simpler then with formulas (25)-(29). Furthermore the last formulas are convenient for engineering calculations.

4. It is necessary to notice that the problem of a classes approximation functions choice is one of the major problems not only in applied mathematics, but in technology applications also. The real specialist knows the needed class functions to choice and the boundary allowing errors limits.

REFRENCES

Larionov, G.I. 2011. *The anchor design parameter evaluating.* Dnipropetrovs'k: National metallurgy Academy of Ukraine: 286.

Larionov, G. & Larionov, N. 2012. *Evaluating of metal-resin anchor parameters influence on the support capacity.* Geotechnical Processes During Underground Mining. Tailor & Francis Group, London: 189-194.

Krylov, A.N. 1950. *On some mathematic physic differential equations, applied in technical questions.* Leningrad. SSSR AS: 369.

Sapegin, V. N. & Larionov, G.I. 2012. *On solving task analysis for non steady state elastic medium deformation.* Scientific News. Update problem of metallurgy. Dnipropetrovs'k: National metallurgical academy of Ukraine, 14: 47-59.

Kiriya, R.V., Braginec, D.D. & Mischenko, T.F. 2008. *Average and accumulating bunker work simulation model for coal mine conveyer lines.* Geotechnical mechanic: collection of scientific works. Dnipropetrovs'k: IGTM of Ukraine's NAS, 77: 100-109.

Schahmeister, L.G. & Solod, G.I. 1976. *Underground conveyor machines.* Moscow. Nedra: 432.

Sveshnikov, A. A. 1968. *Applied methods of random functions.* Moscow: Nauka: 464.

Bolotin, V.V. 1971. *Probability and durability methods application in building calculations.* Moscow: Building literature: 255.

Kamker, E. 1971. *Usual difference equation handbook.* Moscow. Nauka: 576.

Baker, G.A. Jr. & P. Graves-Morris. 1986. *Pade approximants. P.1 Basic Theory. P.2 Extension and Applications.* Moscow: Mir: 502.

New technical solutions during mining C_5 coal seam under complex hydro-geological conditions of western Donbass

V. Russkikh, Yu. Demchenko & S. Salli
National Mining University, Dnipropetrovs'k, Ukraine

O. Shevchenko
"Samarskaya" mine, "Ternovskoye mine management", DTEK "Pavlodradugol", Ukraine

ABSTRACT: Mining operations of Western Donbass mines are conducted under complex hydro-geological conditions that leads to worsening the conditions of a deposit mining. The article presents the substantiation of new technical solutions on the development of C_5 coal seam under conditions of increased flooding and weak stability of host rocks. As a result of the works conduction, the measures on the rock massif unloading were substantiated. At Samarskaya" mine, "Ternovskoye mine management", DTEK "Pavlodradugol" the research results were tested for conditions of the 541 longwall.

1 INTRODUCTION

Coal industry of Ukraine in Europe (excluding Russia) takes the second place by production volume. Nevertheless, the production need for high-quality raw material is still a current problem. Minerals consumption volumes increase requires widening of raw material base, provision of necessary quality of marketable ore, maintenance of profitability of coal gaining and coal dressing enterprises.

During a long working time of coal enterprises the mining-geological conditions have become more complex. Saving the production profitability and increasing the efficiency of underground operations is possible basically at the expense of an extraction intensification.

Stoping efficiency and level of mining operations safety will depend on the degree to which parameters of basic construction elements of the technology are optimal.

2 FORMULATING THE PROBLEM

Mastering new technologies by coal companies of Ukraine and introduction of highly productive equipment of high power has allowed to transfer to conduction of mining operations with 2-3 longwalls with production capacity equal to 1.5-2.5 million of tons yearly. However such an organization of production is complicated at mines developing deposits under complex hydro-geological conditions inherent to Western Donbass. These conditions are character-

ized by significant non-uniformity of water inflows into mine workings substantiated by the massif watering and seasonality of atmospheric precipitations.

"Samarskaya" mine, "Ternovskoye mine management", DTEK "Pavlodradugol" in the block #3 that has been mined since the beginning of 2008 has encountered a problem of C_5 seam development under complex mining-geological conditions.

The developed seam has wavy bedding with 0.4-1.01 m thickness, average 0.88 m. Average extracted thickness of the seam makes up 1.05 m and is provided by additional undercut of roof rocks by 20 cm. Coal strength $f = 4$-5 based on Prododyakonov scale. Main roof in the longwall is presented by mudstones and siltstones; immediate roof – mudstone with strength of 2.2-2.5; immediate bottom – mudstone of lumpy texture with strength of 1.3-1.5.

Coal seam sloughing is not occurring in the face. An immediate roof rocks are destroyed with formation of vaults of 0.3-1.8 m-height. There are some cracks in the longwall roof above the support sections. The cracks are formed with step equal to the shearer width of cut and subsidence equal to 0.3 m; apparent signs of the main roof subsidence are noticed.

The longwalls were equipped with mechanized complex 1 MKD-80 with the shearer KA-200.

When developing the block #3 of C_5 seam the mine has encountered the problems of excessive pressure occurrence at lower areas of extraction pillars that led to the mechanized complex subsidence on rigid base (Figure 1).

Changing the parameters of the technology, i.e. transfer from the combined mining method (longwall #535-537) to pillar method, the longwall length reduction, the support sections movement with propping in roof, increase of a face movement rate did not lead to a positive result. Hydro-geological conditions have had a significant influence on it.

Figure 1. Plan of mining operations of C_5 with an indication of places of the complex subsidence on rigid base.

3 MATERIALS UNDER ANALYSIS

During development of longwall #541 that were located between worked-out longwalls the problem only increased. Having conducted quality analysis on the block development it became obvious that one of the basic reasons of the complex subsidence on rigid base was presence of an increased water inflow after the main roof collapse. During development of the longwalls #535, 537, 539, 545 it was established that during the main roof subsidence the load is formed approaching the value of a support section bearing capacity that is confirmed by hydro-props shafts closing down to minimally accepted level of the remained "mirrors" of the support props at height of 50-70 mm at the back row of props and 10-30 mm at the front row of the props.

At the same time separate visual observations point at the development of collapse zone with the height up to 5-7 m, with rocks weight being 120-190 kN / m~ and in 2.6-4.2 times lower than bearing capacity of the 1KD-80 support that, according to the complex technical characteristics makes up 500 kN / m². The following law of the water inflow change was observed:

• between the main roof subsidences (and after the support subsidence on rigid base) the water inflow is relatively little (at about 3-5 m³ / h) as a water "dripping" being quite uniform along the whole longwall length;

• before the main roof subsidence the water inflow decreases down to 2 m³ / h and lower;

• after subsidence of the main roof the water inflow recovers up to the preliminary level of -5 m³ / h;

• after going on the "rigid base" the water inflow rapidly increases up to 40-50 m³ / h.

Based on the works of (Serdyuk 1984) the observation for interaction of the support and wall rocks under conditions of C_5 coal seam of mines at Western Donbass have shown the following picture of rocks caving and their location in the worked-out area (Figure 2).

After a regular cycle of a coal seam extraction and movement of both support models the gradual delamination and subsidence of the entire strata were occurring. After the support movement, the immediate roof collapsed the first with separate layers forming a zone of chaotic collapse with the height equal to the seam thickness. Then the above-situated strata were collapsing as separate blocks (zone of ordered collapse). The dimensions of these blocks made up 2-6 m.

Figure 2. Rocks dislocation model during C_5 coal seam mining.

With the face farther advance the main roof were collapsing. The first subsidence of the roof has occurred after retreating from the face entry by 30 m. As the face moved the rocks dislocation of the main roof occurred along the cracks with the worked-out area equal to 10 m.

Cracks dip angle made up 60-70⁰ to the plane of bedding. Zones of the crumpled rocks of 0.5-2 m were situated along the contacts between the blocks.

During development of the 541 longwall the following behavior of the massif during the main roof subsidence was noticed (Figure 3). Primary subsidence of the main roof took place at a distance of 28 m from the face entry and did not lead to the complex subsidence of rigid base. Besides, the water inflow up to 5 m³ / hour was noticed. It could indicate about the water-bearing horizon not being undercut. The following four subsidences of the main roof have led to the complex subsidence on rigid base and was followed by the water inflow of more than 30 m³ / hour.

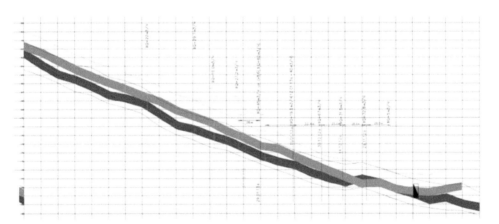

Figure 3. Matched profile of the 541 longwall with indication of the main roof subsidence step and water inflow.

4 TECHNOLOGICAL PART

The closest drainage horizon (at the height of 30 m from C_5 seam) was the C_6 seam that has an exit to Buchagsk water-bearing horizon. It is apparently that when the main roof of the C_5 seam subsides the formed cavity (Figure 2) was getting filled with water. At the secondary roof subsidences the water was under pressure and were transferring to the longwall working along the cracks transferring an additional load on the mechanized complex from the above-located rock layers that led to complex subsidence on rigid base.

To decrease water factor influence on the mechanized complex extra load it was decided to drill unloading holes in the formed cavity (Figure 4). The holes were drilled in the haulage drift from the face side to the backfilled one using the following parameters: length – 40 m, diameter – 92 mm, decline angle on the backfilled area is 45°, decline angle on the longwall is 45°. Installation spacing is 9 meters.

With further subsidences of the main roof the water inflow from relief holes made up about 30 m³ / hour. This has allowed to reduce the load on the mechanized support and, furthermore, to develop the longwall without emergency subsidences of the complex.

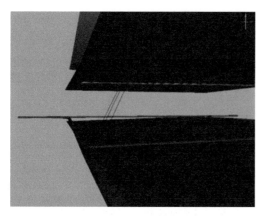

Figure 4. Model of technological solutions realization.

5 CONCLUSIONS

Considering an accumulated experience on the Western Donbass' deposit exploitation and the results of carried out researches it can be concluded the following:

– one of the basic reasons of the complex subsidence on rigid base is a large water inflow into the worked-out longwall area in connection with undercut of water-bearing horizon;

– relief-exploratory holes are a local means of the mechanized complex relief from an increased rock pressure;

– for further effective exploitation of the block it is necessary to conduct water drainage along the above-situated C_6 coal seam by corresponding mine workings.

Based on the gained results, the 541 longwall has been successfully developed under the conditions of host rocks weak stability and increased watering.

REFERENCES

Serdyuk, V.P. 1984. *Candidate of technical sciences dissertation "Substantiation of console-free longwall mechanized support.* Dnipropetrovs'k.

Effect of degasification efficiency of gas-emission sources under complex degassing on maximum load on a stope as for gas factor

O. Mukha, I. Pugach & L. Tokar
National Mining University, Dnipropetrovs'k, Ukraine

ABSTRACT: An approach to identify maximum load on a stope as for gas factor taking into account degasification of certain gas-emission sources is represented.

1 INTRODUCTION

Mining deepening as well as increase in gas concentration of coal beds and rocks mined and contiguous results in the problem that it is impossible to keep non-hazardous methane concentrations in a mine air with the help of ventilation only. Under such conditions, degasification of gassy coal beds and mined-out space is important procedure favouring increase in mining safety, and load increment on a stope.

The topic is relevant as the Program of Mining Enterprises Development involves intensification of mining, and load increment on stopes. The problem can not be solved without complex degasification under gas concentration within working areas and complicated mining and geological conditions.

2 FORMULATING THE PROBLEM

The paper objective is to improve mining safety within working areas owing to rational parameters of the system of complex degasification.

To succeed, it is required to solve following problems:

1. To analyze available ways and means of different methane-emission sources degassing.

2. To validate a choice of methane-emission sources degassing on the basis of the analysis.

3. To determine load dependencies on a stope according to gas factor from degassing efficiency.

4. To produce a procedure identifying maximum load on a stope on a gas factor taking into account degassing effect of certain gas-emission sources.

5. To determine load behaviour on a stope according to gas factor depending upon operating mode of degassing system.

6. To boost shearer production on a gas factor, and to cut own costs of mining.

7. To minify probable gaseousness of working area, and to improve labour safety.

The paper concept is to improve degasification efficiency owing to rational parameters while providing planned load for a stope.

The paper subject of research is gas balance of working area, and degassing system of "Mine Belozerskaia" MD.

The subject at hand is to specify efficiency gas emission sources on maximum load of a stope as for gas factor.

3 RESEARCH

Gas safety of "Mine Belozerskaia" MD is provided by means of mined-out space degassing with the help of gas-pipe spurs (so-called "candles") left in unsupported share of windway (cave-in).

Vacuum-pump station (VPS) equipped with six BBH2-150 pipes is applied for degasification.

Besides independent methane removing out of worked-out space is provided. The process is performed by means of rigid pipeline with the help of gas-suction plant with ВМЦГ-7M ventilator.

The approach is: gas-air mixture coming from worked-out space in mine workings is sucked out by means of ventilator; specific pipeline is used for its transportation to a common return. In a mixing chamber it is reduced to safe concentration.

The approach is efficient when reverse mining with entry and airway leaving takes place. The matter is that some share of methane will get to the worked-out space.

Roof is degassed by means of holes drilled out from airway towards a stope. The holes are bored in groups – two holes from a group with a turnback to a stope and without it. The parameters are in Table 1.

Table 1. Degassing hole parameters.

Parameters	Hole #1	Hole #2
Entry off-axis turnback, degrees	32	90
Inclination to a level, degrees	46	63
Length, m	51	41
Bottom diameter, mm	76	76
Tubing depth, m	10	10
Drilling interval, m	20	20

Degasification of coal beds brings in coincidence with the preparation and mining processes. Besides sizeable drainage of massif takes much time. Not always the conditions give provide desirable gas-emission drop into entries when advance is accelerated.

If advance is $V_{ost} = 2.8$ m / day, and distance between holes is 20 m, one hole life is seven days while expenditures connected with drilling and a hole hookup is about 8.000-10.000 UAH. To extend holes life, they are not shut-in but stay operating in a cave-in. to prevent destroying hole mouths, chocks are left under them. However, even such measures can not solve a problem of control of methane concentration being caught from holes left in worked-out space. In concentration in a pipeline drops lower than permissible limits, then it is required to shut-in the whole group of holes left in cave-in.

One extraction pillar which length is 1.800 m and roof degasification with the help of holes drilled out from airway towards a stope involve 90 hole groups which total length is 90×(51+41) = 8.280 m; with it, working from which the holes are drilled is left. However, drilling costs of a hole running meter is 200-300 UAH.

According to Standard (Degassing...2004), such degasification efficiency is 40% (it is hardly achievable in practice). Still, when a stope nears a hole mouth the latter often ruins. That results in considerable air inleakage as well as in a premature need to shut the hole in due to law methane concentration.

To solve the problems it is required first to identify parameters of working area ventilation and to study degasification system.

Air consumption and methane concentration within ingoing and outgoing ventilation streams are identified with the help of standardized telecommunications system of centralized control and automated control of mining equipment and process systems (STS). The system sensor installation is performed in accordance with requirements (Safety rules... 2010 & Instruction book... 2003). Readings are recorded centralized in a Control Centre.

Degasification system parameters are identified experimentally.

The experiment approach provides measurement parameters of gas-air stream within degasification pipeline, and identification of degasification flight geometry.

The key task of field studies is to determine methane concentration in gas-air mixture moving in the pipeline, temperature of mine air within experiment, vacuum within degasification pipeline, and pressure difference on orifice plates, gas pipeline diameter and length, and atmospheric pressure in the mine.

4 JUSTIFICATION OF RATIONAL PARAMETERS OF COMPLEX DEGASIFICATION

Gas balance of working area according to the results of methane concentration estimation without degassing is in Table 2.

Table 2. Gas balance of working area before degasification.

Methane-emission sources	Expected average methane emission, m³ per minute	Methane-emission shares, %
Working seams	11.18	18.94
Tapping seams	33.72	57.13
Enclosing rocks	10.31	17.47
Overworking seams	3.81	6.46
Total	59.02	100
Worked-out space	38.98	66.05

According to instruction manual (Guidance... 1994), expected methane emission from a stope (q_{st}) and working area (q_{lw}) when ventilation plan is 1-M are identified by:

$$q_{ost} = \left(q_{r.s} + q'_{r.c} + q''_{r.c}\right) \cdot \left(1 - k_{d.s}\right) + k_{w.s} \cdot q'_{w.s};$$

$$q_w = \left(q_{r.s} + q'_{r.c}\right) \cdot \left(1 - k_{d.s}\right) + q'_{w.s}$$

where $q'_{r.c}$ – relative methane emission of broken-down coal in a face, m³ per ton; $q''_{r.c}$ – relative methane emission of broken-down coal in a belt entry, m³ per ton; $q_{r.s}$ – relative methane emission from a stope, m³ per ton; $q'_{r.c}$ and $q''_{r.c}$ – identified according to item 3.3.1.4 (Guidance... 1994); $k_{d.s}$ – efficiency factor of working seam degasification, fraction units; and $k_{w.s}$ – factor taking into consideration methane emission from worked-out space to face space, fraction units.

Expected methane emission from worked-out area

($q'_{w.s}$) within working area is determined by:

$$q'_{w.s} = \left[k_{e.s} \cdot (x - x_0) \cdot (1 - k_{d.s}) + \left(\Sigma q_{cn.ni} + q_{por}\right) \times \right.$$
$$\times (1 - k_{d.s.s}) + \Sigma q_{cn.ni} \cdot (1 - k_{d.s.n})\right] \times$$
$$\times (1 - k'_{i.m.e}) \cdot (1 - k_{d.w.a})$$

where $k_{d.s.s}$ – factor taking into account efficiency of contiguous tapping seams, fraction units; $k_{d.s.n}$ – factor taking into account degasification efficiency of contiguous overworking seams, fraction units; $k_{d.w.a}$ – factor taking into account degasification efficiency of worked-out space, fraction units; and $k'_{i.m.e}$ – factor taking into account efficiency of independent methane extraction, fraction units.

Maximum allowable load on a face (tons per day) as for gas factor is determined by

$$A_{max} = A_p \cdot I_p^{-1.67} \cdot \left[\frac{Q_p \cdot (c - c_0)}{194}\right]^{1.93},$$

where I_p – average absolute methane concentration of a stope (I_{st}) or working area (I_{lw}), m³ per minute; it is accepted on Table 7.1 (Guidance... 1994); according to Table 7.1 (Guidance... 1994) it is for mine ventilation diagram of working area with successive methane diluting on sources $I_p = I_{lw} = 7.03$ m³ per minute; Q_p is maximum air consumption within a stope (Q_{st}) or working area (Q_{lw}) which can be used to dilute methane up to permissible rates by Safety Rules, m³ per minute; it is accepted on Table 7.1; according to Table 7.1 (Guidance... 1994) it is for mine ventilation diagram of working area with successive methane diluting on sources $Q_p = Q_{st.max} \cdot k_{e.w}$, m³ per minute.

Taking into account above-mentioned expressions, one can demonstrate average expected methane emission within working area in the form of Table 3.

Table 3. Average expected methane-emission within working area on intake sources.

Methane-emission sources	Average expected methane emission	
	Relative, m³ per ton	Absolute, m³ per ton
Broken-down coal, $q_{r.c}$	3.86	4.83
Coal broken-down in a face, $q'_{r.c}$	1.99	2.49
Working face, $q_{r.s}$	4.37	5.47
Enclosing rocks, q_{por}	8.25	10.31
Tapping seams, $q_{t.s}$	26.98	33.72
Working seams, q_s	8.94	11.18
Overworking seams, $q_{o.s}$	3.05	3.81
Worked-out space, $q'_{w.s}$	38.98	48.73
A stope, q_{st}	9.55	11.94
Working area, q_w	45.35	56.68

Analyzing the expression of expected methane emission from worked-out area ($q'_{w.s}$) determining it is possible to conclude that methane comes to worked-out space from:
1. Broken-down coal left in worked-out space.
2. Overworking seams.
3. Tapping seams.
4. Enclosing rocks.
Taking into account the fact that only some share of the methane source is degassed, and pure methane inevitably gets into worked-out space it is quite necessary to apply independent removal of methane, and to degas the worked-out space with the help of gas-pipe spurs (so-called "candles").

Hence, application of above-mentioned degasification approaches gives ability to degas each source with the exception of overworking complemental rock and working seam. Share of getting methane from overworking stratum is negligible (6.56% of total gas balance of working area according to Table 2). For this reason, the source can not influence gas concentration of the area greatly.

Methane-emission share of working area is 18.54%. In this context, the methane gets into:
1. Worked-out space from broken-down coal left in cave-in.
2. Stope from an exposed surface of a seam.
3. Belt roadway while transporting broken-down coal.

Absolute methane-emission, m³ per min.

- Broken-down coal
- Coal broken-down in a face
- Working face
- Enclosing rocks
- Tapping seams
- Working seams
- Overworking seams
- Worked-out space

4.83
2.49
5.47
10.31
48.73
33.72
11.18
3.81

Figure 1. Diagram of working area gas balance.

Thus, to decrease volume of methane getting from working seam to working area, it should be degassed.

Assume a plan of working seam degassing with the help of parallel and single holes turned about towards a stope.

Parameters of complex degasification on methane-emission sources are:

1. Working seam: holes are turned about towards a stope $k_{d.s}^{max} = 0.4$.

2. Tapping seams: holes are drilled out towards a stope from an airway to be left $k_{d.s.s}^{max} = 0.4$.

3. Worked-out space: degasification of worked-out space is performed with the help of gas-pipe spurs (so-called "candles") left in unsupported zone of airway (cave-in) $k_{d.w.a}^{max} = 0.6$.

4. Independent methane removal from worked-out space outside working area by means of rigid pipeline with the help of gas-suction plant with ВМЦГ-7М ventilator $k_{i.m.e}^{max} = 0.8$.

Reproducibility will be estimated on fidelity value of R^2 approximation.

Initially, application of each degasification approach is considered separately. Then, load on a stope is calculated. Following curves are the calculations result.

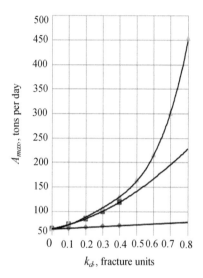

Figure 2. Dependence of load on a stope as for gas factor upon efficiency of certain methane-emission source.

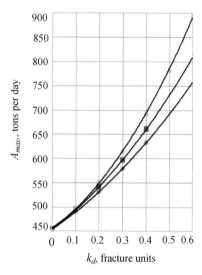

Figure 3. Dependence of a stope load on efficiency of degasification of certain methane-emission source if $k_{i.m.e} = max$.

On the basis of the results, degasification of worked-out space by means of independent gas removal is assumed as basic approach.

The approach can not provide specified load on a stope which results in necessity to apply complex degasification. Consider application of the basic approach in combination with other approaches particularly.

Having analyzed dependences obtained, assume complex degasification of worked-out space (independent removal of methane and degasification of worked-out space with the help of pipeline spurs) for subsequent calculations. In this context, degasification of both working seam and tapping seam is considered as optional part.

According to Figure 1, degasification of worked-out space is followed by degasification of tapping seams as their share in gas balance ranks next being 33.7 m³ per minute.

However, investigation of Figure 4 helps to come to following conclusion. First it is required to degas working seam which gas emission is 11.2 m³ per minute as will help to achieve greater stope load as for gas factor.

Figure 5 demonstrates dependences of stope loads on degasification factor when degasification efficiency of tapping seams ($k_{d.s} = const$) and degasification efficiency of working seam ($k_{d.s.s} = const$) vary.

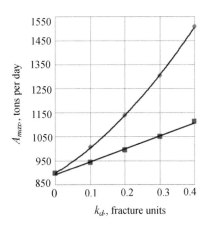

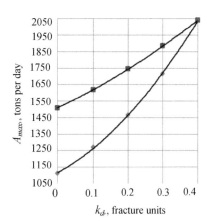

Figure 4. Dependence of a stope load on degasification efficiency of certain methane-emission source if $k_{i.m.e} = max$ and $k_{d.w.a} = max$.

Figure 5. Dependence of stope load on degasification efficiency of tapping seams and working seam degasification.

5 CONCLUSIONS

The dependences of maximum stope load as for gas factor upon degasification efficiency of gas-emission sources give ability to identify rational parameters of degasification system. In such a case, to provide specified face load it is required to know degasification approaches available for the mine-technical and mine-geological conditions, and information concerning gas balance of the working area.

The results are of practical value for:

1. Identification of nature of a stope load variation on gas factor depending upon operation mode of degasification system.

2. Increase in efficiency of a shearer on gas factor, and cost reduction of mining.

3. Reduction of possible gas content of mine workings, and improvement of labour safety.

REFERENCES

Degassing of coal mines. Requirements for methods and schemes of degassing. SOU 10.1.00174088.001-2004. Kyiv: Mintopenergo of Ukraine: 162.
Safety rules in coal mines. NPAOP 10.0-1.01-10. 2010. Kyiv. Derzhgirpromnaglyad: 432.
Instruction book for Safety rules in coal mines. Vol. 1. 2003. Kyiv: Osnova: 480.
Guidance for designing of coal mines ventilation. 1994. Kyiv: Osnova: 311.

Modification of cement-loess mixtures in jet technology during mastering underground area

S. Vlasov, N. Maksymova-Gulyaeva & E. Maksymova
National Mining University, Dnipropetrovs'k, Ukraine

ABSTRACT: The results of experimental researches of cement-loess mixtures are shown to develop me-thods of bearing capacity of strengthening engineering structures increase.

Loessial depositions are disseminated all around Ukraine's territory except for flood-lands, first river terraces, utmost western, northern-western regions and cover land forms of various genesis with continuous covering. Its thickness, in average, varies from 10 to 20 m reaching 50-60 m at some areas. Peculiar feature of loessial soils is strength loss when soaked, i.e. their degradation.

Intensive urbanization peculiar for many regions of Ukraine has substantiated increase of anthropogenic load on territories that led to negative consequences of people business activities (Vlasov & Maksymova-Gulyaeva 2010). Problem of territories flooding has gained specific currency. Deformation increase and loessial massifs strength reduction are in the direct dependence on flooding and, hence, activation of landslide processes.

Building of shallow underground structures-subways, transport intercharges, collectors of various purpose and so on can lead to emergency damages of structures located on undermined territories. This is caused by deformation of fundaments due to additional uneven subsidences of ground base and loads on structures.

Together with it, underground building often is connected with operations in weak soils and on underflooded territories. In connection with this, ensurance of normal conditions of building and structures exploitation is fundamentally impossible without strengthening and hydroisolation of structures outlines and their surroung soil massif.

Implementation of jet solidification of soils opens new possibilities in solving these problems. Such a new technology allows to solve tasks of mine workings stability increase driven in loessial strata during ventilation shafts sinking, escalating tunnels, shallow subway stations, during strengthening tunnels on potentially dangerous by landslides slopes.

With its help it is possible to build such a structure as "wall in ground" based on implementation of splitting cement-loess poles. This is current for preventing deformations of fundaments bases of existing engineering structures during building of new underground objects in open foundation pits under conditions of compact urban building (Baykov & Strogin 1980).

Use of jet technology can provide creation of reliable antifiltering protection on under flooded area around shallow objects such as underground parking lots, stowage areas and other buildings.

It is worth of special attention that loess soils are both host environment for underground structures and a structural material. This substantiates ecological cleanness and efficiency of cement-loess structures building with help of jet technology of soil solidification.

To substantiate jet methods of protective structures elements building the necessity to carry out complex of works has emerged that contains development of solidifying mixtures formulation and other means of cement-loess mixtures modifications. During implementation of these works, the dependences of solidifying mixtures on loess soils strength properties were established. Thus, unconventional technology of structures building was created for rocks of Prydniprovs'k region loess complex.

Improvement of jet technology consists in regulating the properties of cement-loess mixtures due to resource-saving technological techniques. It defines scientific direction of the work. Complex of technological means for mixtures modification is quite wide: use of additives, vibrational activation, armoring, etc. One of the most effective and universal methods is an introduction of additives. In one concrete case to strengthen bearing capacity the loess ground properties are used along.

That is why the purpose of the present article is to analyze the study's results for cement-loess mixtures modifications due to additives use in cement mixture when implementing jet technology of solidification.

Taking into account specific properties of loess grounds when choosing additives the process of their interaction with cement used as a binding matter was researched. As large ground fractions do not influence it, it was decided to use ability of fine-dispersed component for active chemical and physical-chemical interaction with binding materials. These phenomena play deciding role and substantiate strength of solidified soils.

There are three stages of cement-loess structures formation. The first stage causes destruction of unbalanced contacts between particles. The second stage – cement interaction with water in solvate shells edging cement granules. Hence, the most favorable conditions for crystal hydrates formations occur. As their concentration rises around unhydrated cement kernels the helium coats are formed. Farther, in hardening stage that is characterized by contraction of helium volume around cement kernels the reaction edges emerge that bind separate cement kernels between each other.

Together with this, as a result of loess soil capability to interaction (adsorbing, chemical, adhesive) with products of cement hydration – volumetric structure, basically coagulation is developed significantly faster than in ordinary concretes and possesses little strength. Such coagulation structures possess thixotropic properties that have the ability to reverse recovery after mechanical destruction.

Basic type of the particles interaction with each other is forces of Van-der-Waals. The third stage involves increase of strong contacts number substantiated by crystals coalescence into random structure – hardening crystalline structure that then is compacted and strengthened because of crystals accumulation. It can be assumed that there are continuous hydration processes going on in the system and crystalline structure formation begins gradually. It is not possible to allocate clear time bound between the stages. It can only be about prevalence of one process over the other (Figure1).

(a)

(b)

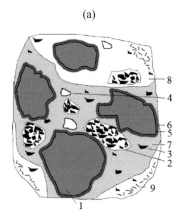 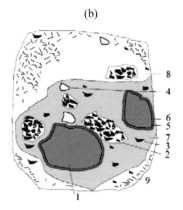

Figure 1. Fragment-scheme of structure formation of cement-loess mixture: (a) early stage; (б) late stage; 1 – unhydrated cement particles; 2 – soil micro-aggregates; 3 – soil particles; 4 – structural macro- and micro- pores; 5 – solvate shells;6 – helium covers; 7 – reactional edge; 8 – coagulation structure; 9 – crystallyzed structure.

As a result a quite complex and branched grid-like carcass is formed in the mixture strength of which is much higher than strength of separate soil macro-aggregates included in this carcass. Cohesion between the particles in soils should be considered as a weak link in the system (Maksymova-Gulyaeva, Nalbandyan & Solopov 2005). Strength of cement-loess mixtures, basically, depends on cohesion of cement stone with loess soils as a filler. It is substantiated by physical-chemical interaction due to adhesion properties of the filler and by chemical – its chemical activity relatively to cement hydration products.

Basic strength parameter of cement-loess muxture, as for concrete – compresson strength was defined by uniaxial compression in various time spans (Figure 2).

Strength decrease occurs after 42 days, its maximal duration makes up 28 days for the ratio of cement/lime 12/2% and minimal – 14 days for the ratio of 50/5%. Then strength gradually rises, i.e. chemical processes in the system occur all the time. The largest value makes 3.0 MPa for the ration of cement/lime 12/2% and 10 MPa – for 50/5% were recorded after 500 days.

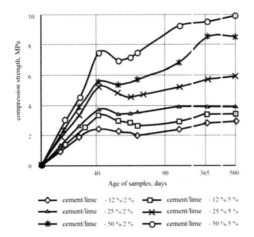

cement/lime   - 12 %/2 %   cement/lime   - 12 %/5 %
cement/lime   - 25 %/2 %   cement/lime   - 25 %/5 %
cement/lime   - 50 %/2 %   cement/lime   - 50 %/5 %

Figure 2. Dependence of compression strength on mixture formulation and hardening time.

The strength decrease phenomena is explained by the fact that at the early stages of hydration process when crystalline structure only begin to form, brittle and quite large crystals of hydroaluminates and cal-

cium oxide hydrates form. At this period separate crystals move coagulation structure elements preventing their growth. Stresses occurring during it are insignificant and do not show practically any influence on the structure strength. At latter stages when small-crystal structure of calcium hydrocilicates forms, rigidity of the structure rises. Due to increase of crystalline contacts based on Rebinder A.P., high internal stresses are developed that trigger widening of micro-cracks and, hence, strength decrease. Researches of various quantity of cement and lime influence on the mixture strength shows the following. Regardless percentage content of cement, introduction of 2% of lime (the quantity needed to saturate absorbing capacity of soils) increases strength of samples on average by 11.9% compared to the ones having no lime. Addition of 3% more of lime increases their strength on average by 36.5%. At permanent lime content and cement content increase in two times (from 12 to 25% and from 25 to 50%), strength increase makes up 48.7-50.9% (Table 1).

Table 1. Compression strength increase depending on mixture formulation.

Cement conten, %	Lime conten, %	Rated compression strength, MPa	Strength increase when increasing cement content and lime content being permanent, %	Strength increase when increasing lime content and cement content being permanent, %		
				от 0 до 2	от 0 до 5	от 2 до 5
12	0	2.2		-	-	36.0
			59.1			
25	0	3.5		-	-	39.5
			42.8			
50	0	5.0		-	-	33.9
			50.9*			
12	2	2.5		13.6	-	-
			54.0			
25	2	3.8		10.0	-	-
			47.4			
50	2	5.6		12.0	-	-
			50.7*			
12	5	3.4		-	54.5	-
			55.9			
25	5	5.3		-	51.4	-
			41.5			
50	5	7.5		-	50.0	-
			48.7*			

* rated values.

Analysis of received results shows that lime introduction significantly influence binding strength increase of cement-loess mixture. Based on our data, addition of 3% of lime is more effective than increase of cement content in two times. Consider-

ing insignificant difference in price of 1 ton of cement and lime one can count for economic efficiency and resources saving when building cement-loess constructions with lime addition.

Analysis of cement-loess mixtures strength definition has shown that strengths dependences character on mixtures compjsition and hardening time is similar to compression strengths on the same parameters. With hardening time increase the samples binding strength rises but by lower values (Figure 3).

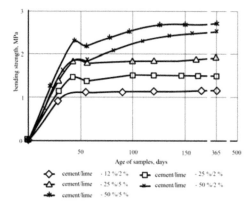

Figure 3. Dependence of strength on mixture composition and hardening time.

It should be mentioned that with introduction of 2% of lime the binding strength changes by relatively larger values than compression strength. Value increase makes up 71-75% against 10-13.6% (Table 2). With lime content increase up to 5% difference in increase gets leveled.

Shearing strength is one more important mechanical characteristic. Studies of cement-loess mixtures were carried out by slant cut method with compression.

Table 2. Bending strength increase depending on the mixture formulation.

Cement content, %	Lime content, %	Rated bending strengt h, MPa	Bending strength increase when lime content rises, %	
			From 0 to 2	from 2 to 5
12	0	0.63	-	-
25	0	0.86	-	-
50	0	1.09	-	-
12	2	1.10	75.0	-
25	2	1.47	71.0	-
50	2	1.89	73.4	-
25	5	1.83	-	24.0
50	5	2.30	-	22.0

Scheme of one-plane shearing with compression in inclined matrixes excludes occurrence of extension stresses reducing material resistance to shearing. There are no supports for the samples that lead to stresses concentration in fixing places influencing overall stress state.

At uniaxial compression of the material samples the largest tangential stresses emerge on areas inclined to compression loads axis under 45-degree angle and equal to $\tau = \dfrac{\sigma}{2}$ (Belyaev 1969).

Since until now cement-loess mixtures had not been used for structures building and there are no data on their shearing strength we have conducted experimental researches on shearing based on methodology for rocks testing (Khatsaurov 1966, Maksymov A. 1973, Baklashov & Kartoziya 1975). The results of experiments are shown on Figure 4.

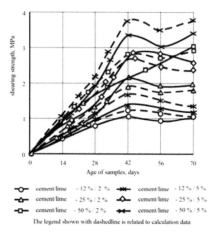

Figure 4. Dependence of shearing strength on the mixture formulation and hardening time.

Comparative analysis of the researches results and calculation values testify that the difference makes up 20%. This value is within admissible range of technical calculations errors and calculations during tests. Insignificant errors of the samples linear dimensions during their preparation and angles between the edges can lead to deviations of results.

CONCLUSIONS

1. Laws of mechanical properties change have been gained for the first time (limits of compression, bending and shearing strengths) of cement-loess mixtures at various quantities of lime and cement depending on hardening time.

2. Lime addition to cement-loess mixtures within 5% with cement content from 5 to 50% allows to increase compression strength of these mixtures by 42 days and hardening by 1.5 time compared to mixtures without lime. Maximal value makes up 7.5 MPa.

At the same values of cement-loess mixtures components and lime the bending strength rises in

2 times, maximal value make up 2.3 MPa.

Shearing strength limit rises in 3 times for 42 days of hardening and cement content increase from 12 to 50% and lime – from 2 to 5%. Its maximum value – 3.3 MPa.

3. During strength indices research of cement-loess mixtures it was noticed that strength decreases in 1.05-1.3 time depending on cement and lime percentage content. This phenomena is substantiated by development of high internal stresses causing expansion, formation of micro-cracks and, as a sequence, strength decrease at the beginning of intensive formation of metal crystalline rigid structure.

The authors' job represents the results of scientific researches allowing to substantiate methods of jet technology improvement. It is a logical consequence of scientific trend that is actively developed in special methods laboratory of dispersed rocks strengthening at underground mining department in the National Mining University.

REFERENCES

Vlasov, S. & Maksymova-Gulyaeva, N. 2010. *Stability increase of slopes inclined to landslides with help of jet technology for soils strengthening*. Monography. Dnipropertrovs'k: National mining university.

Baykov, V. & Strogin, S. 1980. *Structural constructions*. Moscow: Building issue.

Maksymova-Gulyaeva, N., Nalbandyan, L. & Solopov, N. 2005. *Basics of theory of cement-loess mixtures structure formation*. Col. Sc. Papers NMU.

Belyaev, N. 1969. *Materials resistance*. Moscow: Higher school.

Khatsaurov, I. 1966. *Rocks mechanics*. Moscow: MNU issue.

Maksymov, A. 1973. *Rock pressure and roof support*. Moscow: Nedra.

Baklashov, I. & Kartoziya, B. 1975. *Rock mechanics*. Moscow: Nedra.

Mining of Mineral Deposits – Pivnyak, Bondarenko, Kovalevs'ka & Illiashov (eds)
© 2013 Taylor & Francis Group, London, ISBN: 978-1-138-00108-4

The modernization of ways of treatment of coal stratums for rise of safety of underground mine work

V. Pavlysh, O. Grebyonkina & S. Grebyonkin
Donetsk National Technical University, Donetsk, Ukraine

V. Ryabichev
Volodymyr Dahl East Ukrainian National University, Anthracite, Ukraine

ABSTRACT: The problem of development of theory and technology of treatment of coal stratums in order to raise of security of underground coal extraction is considered. The using of two-phases pneumatic and hydraulic treatment as a way of reduction of gas and dust extraction is proposed.

On the classification of academician A.A. Skochinskyi (Skochinskyi 1956), the main hazards in underground coal mining are outgassing, dust, sudden coal and gas emissions, endogenous fires. The complex of methods of solving tasks of control of major hazards in underground coal mining processes occupy a very important place of exposure to coal seams that allow you to change their status and thereby reduce the intensity of dangerous and harmful properties.

In the suggested article describes the approach to reduce dust and gas emissions through integrated hydropneumatic exposures than is determined by the relevance of the work.

The aim of this work is the development of theoretical bases of processes and support integrated mode hydropneumatic effects on anisotropic coal seams.

Currently, there are a number of ways of hydraulic impact on coal seams, which differ both in purpose and in running operations, they are compulsory and regulated by normative documents (DNAOP 1.1.30-XX-1.04. 2004). For the relevant process the methods of calculation of parameters are designed (Pavlysh & Grebyonkin 2006).

All parameters of hydraulic impact can be divided into two groups: the layout parameters and parameters of the pumping. Parameters, belong to the first group: length, diameter and depth of the hermetic part of drill-holes, well spacing or the effective radius, for short is the irreducible wells, lead to long-distance from the treatment plant to scarification first well. The second group includes: liquid consumption on the well, pressure, temp and time.

Look at the schema and settings on a local, regional, and opening the way to discharge coal seams.

Location scheme short (up to 25 m) drill-holes, perpendicular bisector of the line of control, the slaughter is shown in Figure 1a. The length of the boreholes is usually selected a multiple week moving slaughter. This allows for testing of, at first, position the filter part of the limit of drill-holes reference zone pressure and increase the uniformity of treatment, and secondly, significantly reduced compared with the howling of the short-holes schema dependency injection works from sewage treatment works.

Hydraulic impact on coal seam in the reservoir through a preparatory drill-holes, located on one of the schemas shown in Figure 1b, c. The length when using both charging boreholes as anti-coal-out recommended within 8-11 m (DNAOP 1.1.30-XX-1.04. 2004), with the suppression of dust – 30-80 m (DNAOP 1.1.30-XX-1.04. 2004). Feature diagram (Figure 1b) is imposing zones of influence of wells due to the limited size of the preparatory slaughter mean.

Basic layout of long wells, parallel shall slaughter, are shown in the Figure 2. The scheme shown in Figure 2b applies if it is not possible to drill wells in one working-out as the full length of the lava. When hollow-steep reservoir excavating pillars on the dip long wells are drilled on the nozzle at right angles to the uprising of the slaughter equipment.

The main disadvantage of long wells, limiting their diffusion used nowadays, is the difficulty of drilling and hermetic sealing due to lack of reliable equipment. In addition, the use of long wells is possible only when sufficient lead preparator of workings, i.e. almost in-pillar development systems.

If you cannot provide simultaneous operation of three wells are recommended to use the technology of continuous cascade process, which consists in the following:

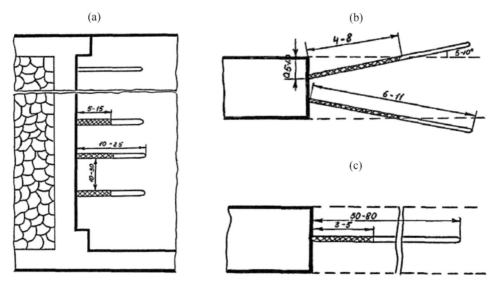

Figure 1. Borehole locations (a) short in the sewage hole; (b) short in preparatory design; (c) longest is preparatory design.

Pressure is one of the three wells temporarily, one of them located on the part of the treated area is supplementary and is used to create a flowing counter drain fluid in this area from the remaining two wells (injection). After working for you at discharge wells for half the estimated time delivery helper hole is minor, and is adjacent to fuel the next hole plugs. Thus resulting in a group of three wells again one is supporting two-pressure, and the process is re-peated. Rate of fluid in injection wells must correspond to the natural conditions of injection of reservoir in pursuit of achieving the required flow. A secondary wellbore pressure enough to maintain an approximately equal pressure on the next delivery.

The use of such technology is largely preventing the outflow of fluid turns the predominant part of the extreme wells with water in between larger areas or under intense interaction flows.

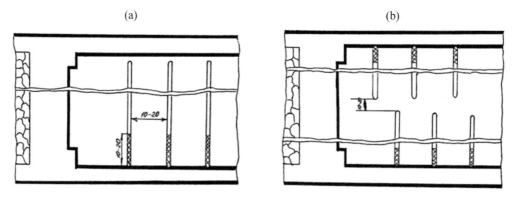

Figure 2. Long boreholes: (a) from a preparation; (b) from entries.

Application of cascade processing (Figure 3) allows, according to a simulation of the modeling and natural experiments, reduce the area of rough areas on the 50-80% depending on the technological scheme -2 in 1.5 times and reduce the saturation intensity variation of liquid array on the effects of the injection with the same rate of liquid reduces the values of these indicators, respectively, at 35-60% and 25-30% compared to the outflow of the unregulated temp (Pavlysh & Grebyonkin 2006).

The local way. When a discharge in the preparatory slaughter geometric parameters to show in Figure 1b, c, the effective radius is chosen from the processing conditions for zone 4-metre contour.

The length of the wells drilled from a treatment plant slaughter, usually made of lava moving is a multiple of the week, but not more than 25 m, diameter 60 mm 45-wells. The minimum lead time for short wells was adopted the length of filtration compartment.

Calculation of process parameters is performed according to (Pavlysh & Grebyonkin 2006).

A regional way. Wells are drilled with a diameter of 75-100 mm depending on the use of the equipment used. The length of the wells located on the scheme of Figure 1a: $l_s = L_l - 20$ m; on the scheme of Figure 1b: $l_s = \dfrac{L_l}{2} - 20$ m, where L_l – length of lava, m.

Seal depth is usually 10-20 m.

Calculation of rate and time of discharge is made assuming the angular motion of the one-dimensional nature of the liquid from the wells.

Discharge time:

$$T_N = \frac{53Qm\mu}{l_F k_x (P_H - P_z)}(0.13Q + 1) \times$$

$$\times \left(\frac{6.6}{m^2} + 1 \right)\left(\frac{4.5 \cdot 10^{-3}}{n_z} + 1 \right)\left(1.7\sqrt{A} + 1 \right) \text{, h}$$

Calculation of parameters in cascade hydraulic action.

Discharge time of liquid in every blow injection well (group wells):

$$T_N = \frac{70Qm\mu}{l_F k_x (P_H - P_g)}(0.13Q + 1) \times$$

$$\times \left(\frac{6.6}{m^2} + 1 \right)\left(\frac{4.5 \cdot 10^{-3}}{n_e} + 1 \right)\left(1.7\sqrt{A} + 1 \right) \text{, h}$$

Thus, the hydraulic impact is a normative activity with sufficiently advanced theory and technology.

Studies on the development of ways of dealing with methane in Moscow Mine University mountain is the idea of forcing air into the coal seam in filter mode through wells drilled from the mountain (pneumatical action) with the aim of increasing coal degasification of the array. Mechanism of natural gas the reservoir reducing the discharge air is to output of free methane in air flow ending well, resulting in displacement of the sorption equilibrium in the system of "free-sorption gas" and methane desorption and release it. Technological scheme pneumatic action is shown in Figure 4.

The simulation results show that the maximum efficiency to reduce gas content is achieved by cyclic pneumatic action with the lowest possible discharge pressures.

When cyclic pneumatic action and $P_H = 2.0$ MPa methane volume 35-40% higher than for degassing. In this case the duration of discharge cycles was 1-2 days, spontaneous end – 3-4 days.

Air rate determined in accordance with the law of Darcy taking gas compressibility.

(a)   (b)

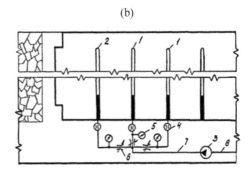

Figure 3. Technological schemes of the Cascade with injection subsidiary wells: (a) at a known location field; (b) by continuous technology; 1 – injection wells; 2 – auxiliary wells; 3 – pump installation. 4 – counter-flowmeter; 5 – pressure gauge; 6 – adjustable choke; 7 – high-pressure hose; 8 – district water pipe.

The initial duration of the discharge cycle is determined by the conditions of near-total removal of free methane from coal volume filtration.

It is clear that the right choice of discharge pressure and cycle times can be reduced gas containing.

Over time, the duration of discharge cycles is reduced because of diminished concentration of free methane formation.

Pneumatic action should check out when the air discharge of methane increases little takeaway. As shown by the results of modeling, this moment corresponds to the equilibrium concentration reduction, on average, by an order of magnitude.

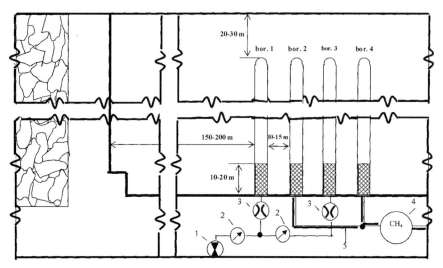

Figure 4. The base variant of technological scheme of pneumatical action on coal stratum: 1 – compressor; 2 – manometer; 3 – counter-flowmeter of air; 4 – measurer of concentration of gas; 5 – mine gas-way.

On this point thanks to the results there are grounds to challenge the development of theoretical bases and technology integrated hydropneumatic impact on coal seams, which includes two stages.

1. Pneumatic processing of dry stratum. At this point is take away free and desorption methane. In addition, this method has a perspective regarding the change of physical and chemical state of the reservoir and, perhaps, would reduce the ability of the seam to spontaneous combustion.

2. Hydraulic impact. This kind of exposure through the use of technology allows the seam saturation liquid that reduces dust, reducing gas emissions and thus has a positive impact on working conditions in underground coal mining.

The wells used for circulating the air, can be used for subsequent injection of fluids.

CONCLUSIONS

Complex hydro-pneumatic impact provides targeted state change in the coal stratums to increase the load on the purge of culling, the pace of mine workings and labour protection.

Proposed integrated hydropneumatic impact on coal seams, which includes two phases:
– pneumatic processing of dry seam;
– hydro-treating using wells of pneumatic pumping.

REFERENCES

Skochinskyi, A.A. 1956. *Whipping up the water into coal seam to be an effective tool in reducing dust undercutting of coal.* Coal, 8: 31-34.
DNAOP 1.1.30-XX-1.04. 2004. *Safe mining works on the seam prone to dynamic phenomena (1-st Edition).* Kyiv: Ministry: 268.
Pavlysh, V.N. & Grebyonkin, S.S. 2006. *Physics-technical basis of hydraulic impact on coal seams.* Monograph. Donets'k: "VIC": 269.

Mining of Mineral Deposits – Pivnyak, Bondarenko, Kovalevs'ka & Illiashov (eds)
© 2013 Taylor & Francis Group, London, ISBN: 978-1-138-00108-4

About the influence of stability of workings on the parameters of their ventilation in terms of anthracitic Donbass mines

P. Dolzhikov, A. Kipko, N. Paleychuk & Yu. Dolzhikov
Donbass State Technical University, Alchevs'k, Ukraine

ABSTRACT: The article presents the results of mine researches of stability of the intake airways of sheth and parameters of their ventilation are driven. Conformity to regularity of change of air expense and depression is set from the workings sustainability indicators. Work out the practical recommendations for stabilization of ventilation parameters in the intake airways of sheth.

1 INTRODUCTION

A modern mine vent network is a difficult, constantly time-space-varying structure. The state of every workings, as an element of mine vent network structure, in a great deal determines stability of ventilation and management complication a vent mode. As, since many domestic mining enterprises mastered of the actual mining depth in a 1000 m and more, the new class of tasks was identified in area of geomechanics and ventilation, related to the dynamic worsening of the state of the development and main workings. If the state of workings determines from the point of geomechanics view, mainly, an economic and financial aspect: worsening of the state of workings entails material expenses on maintenance of such workings and corresponding losses in speed of transporting of useful minerals that, from the point of ventilation of mines view is stability of ventilation and safety of miners. An existent normative document (DNAOP 1.1.30-6.09.93 1994; SOU 10.1-00185790-002-2005 2006; NPAOP 10.0-1.01-10 2010) though envisage the revision of project of mine ventilation or his part at stated intervals, but does not take into account the change of ventilation parameters at the change of the workings state. The researches (Baymukhamyetov 1984) sent to the ground of rational aerodynamic parameters of the sheth workings are presently conducted, influence of roof supports parameters is studied on aerodynamic descriptions of workings (Makshankin 2012), however for maintenance of project ventilation parameters and providing of the protracted stable state of every separately taken working complex approach is needed.

In this regard, actual is realization of research of influence of workings stability on the parameters of their ventilation in the terms of deep anthracitic Donbass mines.

2 FORMULATING THE PROBLEM

To the basic parameters of ventilation, as is generally known, belong: depression of workings and amount of air, passing through unit of cross-sectional area of working in time unit. A calculation of prognosis and determination of actual values of these parameters are the basic task of mines ventilation as sciences. Exact determination of project values of depression and air expense sizes is a very intricate problem as, since the extent of the mine workings of modern mines arrives at a 30 000 m and more, and to take into account all losses of air amount and pressure is not possible. Sheths behave to the most difficult objects of ventilation, therefore the serve of air to them must be provided in full, which not always maybe by reason of the unsatisfactory state of the intake airways.

The aim of work is research of influence of sheth intake airways stability of Donbas deep coal mine on the actual values of air expense and depression in them.

Under reaching the put aim next tasks decided:
– research of sheth intake airways stability during a year;
– research of air expense in working of sheth (intake airways) and joining to them for a year;
– study of influence of sheth workings stability on the air expense and depression in them;
– development of recommendations on the improvement of terms of providing of sheth the necessary amount of air by the increase of stability of the intake airways workings.

3 RESEARCH OF INFLUENCE OF STABILITY OF WORKINGS ON THE PARAMETERS OF THEIR VENTILATIONS

The object of investigation were selected main and development workings of the mine "Partizanskaya" of the State Enterprise "Antratcite" directly being air intakes for the sheth of 204-th western longwall of seam h_{10} : 18-th east haulage drift, auxiliary slope, step slope #2, east air connection and 204-th sub-drift (Figure 1).

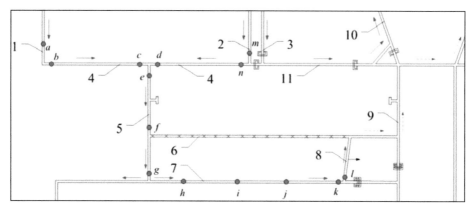

Figure 1. Chart of the sheth ventilation: 1 – east air connection; 2 – step slope #2; 3 – crossheading #8; 4 – 18-th east haulage drift; 5 – auxiliary slope; 6 – 203-th sub-drift; 7 – 204-th sub-drift; 8 – 204-th western longwall; 9 – a conveyer slope; 10 – a sloping crossheading; 11 – 18-th western haulage drift; a-n are points of measuring of air expense.

A mine is not dangerous on methane, dust, sudden outburst of coal and gas. Coal is not predisposed to spontaneous combustion. Angles to the dip of rocks in the studied workings is 2-19° in the depth range 810-950 m. The taken out power of coal seam h_{10} is a 1.3 m. The coal-face mechanized complex of 3KD-90 is set in longwall. The calculation amount of entering air mine is equal 3813 m^3 / min, and actual is 4202 m^3 / min. Performance reserve of the main ventilation VC-31.5 is fixed at the level of 10-15%. The project value of amount of air for ventilation of sheth makes 674 m^3 / min at the coefficient of ventilation loss through mined-out area equal 1.77, and for ventilation of 204-th western longwall of h_{10} seam is 415 m^3 / min. Temperature of air on the entrance of a longwall is 25 °C and on an exit from longwall is 32.5 °C. The sheth workings is envisaged by arch frame support from rolled metal profile SVP-22 with the area of cross-sectional in clear of 10.2 square meters.

The subject of investigation is the workings stability and ventilation parameters: air expense and depression. Stability of workings was estimated by the index ω_S, defined as the ratio of the actual minimum to project of working area cross-sectional, and also through the index ω_N, calculated as the ratio of the number of able-bodied frame of metal roof supports to their total number in working. The results of research of workings stability in time are shown in the Figure 2.

As follows from the presented charts, in all researched workings there is reduction of stability indexes in time, that, in principle, expectantly and comports with the results of many researches (Baymukhamyetov 1984; Makshankin 2012; Dolzhikov, Kipko & Paleychuk 2012). Index ω_N in the sheth intake airways of 204-th western longwall of h_{10} seam during a year changed from 0.65-0.95 to 0.56-0.67, and index ω_S – from 0.775-0.925 to 0.52-0.67, resulting complications in the transport, ventilation settings and, consequently, reducing the actual loading on a cleansing coalface.

From the brought regularities it ensues that the least the steady working is a 204-th sub-drift, as indexes of stability in him have minimum values, that it is related to being of sub-drift in a zone influence of the coal-face works.

Measuring of air movement speed in points a - n (Figure 1) was produced through the portable mine electronic anemometer APR-2 by the technical staff of The Ventilation and Safety Mine Department. The air expense Q was determined through expression

$$Q = v \cdot S, \tag{1}$$

where v – middle air speed in working, m / min; S – actual area of working area cross-section in clear, sq. m.

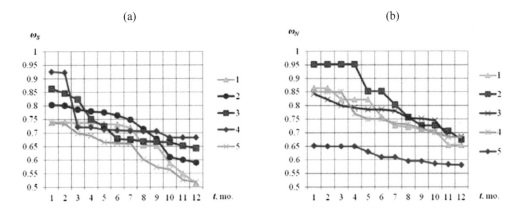

Figure 2. Regularities of time-history of working stability indexes: (a) index ω_S; (b) index ω_N; 1 – 18-th east haulage drift; 2 – auxiliary slope; 3 – step slope #2; 4 – east air connection; 5 – 204-th sub-drift.

For ease of analysis of the results of the mine measuring we will take advantage of relative size of air expense, i.e. the ratio of the actual amount of Q_f toward project Q_p. As a research period makes one month, and measuring was executed every day, the relative air expense settled accounts as an arithmetical mean for a month. In workings with a few points of measuring amount of air Q was determined the as a geometric mean from air

expenses at the working beginning and end for the account of the dispersed ventilation loss.

The results of the mine measuring of relative air expense for a year are presented in Figure 3; it ensues from that during a year the deviation of air expense in the sheth intake airways arrive at 14%, that, taking into account the aerodynamic losses resulted in a reduction of air expense in the longwall at 43%.

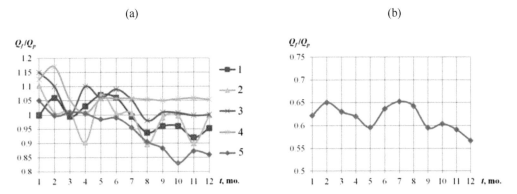

Figure 3. Time-history relative of air expense: (a) in the main and development workings: 1 – 18-th east haulage drift; 2 – auxiliary slope; 3 – step slope #2; 4 – east air connection; 5 – 204-th sub-drift; (b) in longwall (l-point; Figure 1).

Initially, the relative air expense in all investigated development workings exceeds unit (Figure 3a), i.e. the actual amount of air exceeds a project. This phenomenon is due to the fact that the main ventilation fan has the performance reserve, and all investigated workings are air intakes. The air expense deviations during the year due to, firstly, deviations in atmospheric pressure, and secondly – by the change of airflow resistance of working, because

$$Q = \sqrt{\frac{h}{R}}, \qquad (2)$$

where h – depression of working, $(N / m^2) \cdot 10$, which depends on the atmospheric pressure; R – airflow resistance of working, $kg \cdot s^2 / m^8$, depending on the state of working.

Also air expense deviation depends on external

and internal ventilation losses and error of measuring. Longwall value of Q_f / Q_p less than one due to the air loss through which the mined-out area.

One of tasks of research is determination of regularity of change of relative air expense from the indexes of working stability. For determination of presence and type of this regularity the methods of

mathematical statistics were used, in particular is a cross-correlation and nonlinear regressive analysis. Application of the foregoing methods realized in the program Microsoft® Excel allowed setting polynomial dependence of relative air expense from the indexes of stability of sheth air intake workings, which is presented in Figure 4.

(a)

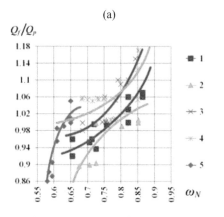

(b)

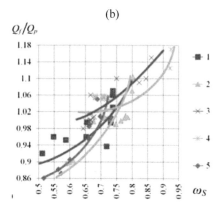

Figure 4. The Dependence of relative air expense from: (a) index of stability ω_N ; (b) index ω_S in working: 1 – 18-th east haulage drift; 2 – auxiliary slope; 3 – step slope #2; 4 – east air connection; 5 – 204-th sub-drift.

With most authenticity the got dependences are approximated by the polynomials of the second degree of kind

$$Q_f / Q_p = a + b \cdot \omega_i + c \cdot \omega_i^2 , \qquad (3)$$

where a, b, c – coefficients at i-member of polynomial; ω_i – index of stability: for regularities of Figure 4a – ω_N, and for dependences in Figure 4b – ω_S.

Data of regressive statistics for dependences in Figure 4 are shown to the Table 1.

As follows from the Table 1, the most values of index of authenticity of R^2 during approximation are observed the polynomials of the second degree at dependences of $Q_f / Q_p = f(\omega_S)$. This fact evidently demonstrates influence of local airflow resistance in the least cross-section of working, as an air expense can be expressed through depression and all types of airflow resistances through expression

$$Q = \sqrt{\frac{h}{\left(\dfrac{\alpha Pl}{S^3} + \dfrac{\xi}{2gS^2} + k_s \dfrac{\gamma S_{mid}}{S(S - S_{mid})^2} \right)}} \cdot k_1 ,$$

where α – the coefficient of the airflow wall-friction resistance of the mine working; P – perimeter of working, m; l – length of making, m; ξ – coefficient of local airflow resistance; g – acceleration of the free falling, m / sq. s; k_s – coefficient of head-resistance; γ – volume weight of air, kg / m³; S_{mid} – area of midship section, sq. m; k_1 – coefficient, taking into account reserve of vent equipment and network.

Consequently, stability of the sheth air intake workings, which is estimated by the corresponding indexes ω_N and ω_S, renders direct influence on the air expense. Coming from physical nature, on the wall-friction resistance that is described to the first elements of denominator in expression (3), most influence renders the index of stability ω_N, and on the local airflow resistances, described by the second element of denominator in (3) is an index ω_S. On the air expense has influence also encumbered of working by machines and mechanisms, which is described to the third elements of denominator in equalization (3).

Table 1. Data of regressive statistics for dependences of relative air expense from the working stability indexes.

Number of workings	$Q_f/Q_p = f(\omega_N)$				$Q_f/Q_p = f(\omega_S)$			
	Coefficients			R^2	Coefficients			R^2
	a	b	c		a	b	c	
1	1.33	-1.51	1.39	0.81	1.50	-2.15	2.04	0.68
2	-0.35	2.79	-1.39	0.66	0.62	0.28	0.31	0.82
3	4.47	-10.04	7.23	0.75	-0.35	3.19	-1.71	0.85
4	4.35	-8.75	6.01	0.51	3.31	-6.00	3.97	0.85
5	-13.21	43.61	-33.43	0.89	-0.34	3.25	-1.90	0.89

In addition, analysis of materials of the depressed surveys executed on a mine "Partizanskaya" allowed setting influence of workings stability on their depression. For ease of analysis of the depressed surveys results we will avail, as in the case of air expense, depression relative size, i.e. ratio of the actual overfall of pressures h_f toward project h_p. The research results of influence of workings stability on their depression are presented in Figure 5.

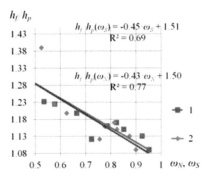

Figure 5. Dependence of relative depression from the stability of working indexes: 1 – index ω_S ; 2 – index ω_N .

As can be seen from the Figure 5, at the values of workings stability indexes more than 0.8, deviation of actual value of depression size from a calculation makes 9-17%. At the values of indexes ω_N and ω_S less than 0.8 depressions in the researches workings increase by 12-39%. It is necessary to take into account, what the researches workings are intake airways and increase of depression in them less than, than in other mine workings. In the mine, general, actual depression exceeds planned at 43-46%, i.e. to ensure that all mine workings required amount of air the main ventilator must overcome the overfall of pressures to 46 percent anymore than project.

Thus, for minimization of workings stability influence on the ventilation parameters of every separately taken working it is necessary to provide in all her length the project value of cross-sectional area during all term of exploitation. Then the deviations of amount of air in the main and district working not subject to direct influence of coal-face works will be determined by encumbered of cross-sectional, and the depression changes will depend from the overfalls of atmospheric pressure.

4 RECOMMENDATIONS FOR STABILIZATION OF THE VENTILATION PARAMETERS OF THE SHETH INTAKE AIRWAYS

As can be seen from the preceding paragraph, to one of the basic directions of providing of sheth the required amount of air to belong supporting of the air intake workings in the proper state, i.e. with the values of stability indexes not below 0.8, that corresponds to the project size of cross-sectional area in clear at the maximal sinking of roof support due to the movement of the rocks and possible amount of defective frames of metal roof support. However, this task behaves in a greater degree to the geomechanics and mine building, than to ventilation of mines; therefore this task can be decided complex, i.e. with the use of positions of such sciences of mine profile as geomechanics and aerology.

At the ground of events on the increase of workings stability it is necessary to take into account the tensely-deformed state of the rocks massif, volume and direction of maximal component of rocks displacements. To this end, the Donbas State Technical University was the development of an appropriate methodology, which is implementation of "UKRNDIPROEKT" and "LUGANSKGIPRO-SHAKHT" institutes (Dolzhikov, Kipko & Paleychuk 2012).

According to this methodology, the calculation of convergence and size of roof supporting pressure can be carried out in accordance with paragraphs 4 and 6 of SOU10.1.00185790.011:2007 industry standard. Deviation from the vertical of the roof support prevailing pressure's defined by the formula

$$\delta = \begin{cases} \varphi + \alpha, & if \ \left| r_L(-x) \right| > r_L(x), \ \varphi > \alpha, \\ \alpha - \varphi, & if \ \left| r_L(-x) \right| < r_L(x), \ \varphi < \alpha, \\ \varphi - \alpha, & if \ \left| r_L(-x) \right| < r_L(x), \ \varphi > \alpha. \end{cases} \quad (4)$$

where α – the angle to the dip of rocks, deg.; $\left| r_L(-x) \right|$ and $r_L(x)$ – are relative radiuses of the failure zone accordingly on the left and on the right of normal to the plane of bedding rocks; φ – the angle of deviation from the normal to the plane of bedding rocks maximum components of their displacements, deg.

$$\varphi = arccos\left(\frac{r_{L\,min}}{r_{L\,max}}\right), \quad (5)$$

where $r_{L\,min}$ and $r_{L\,max}$ – therefore the smallest and largest values radiuses of the failure zone. These values may be defined in mine terms or are based or expected on the general formula of prof. Shashenko A.N.

$$r_L = \left(exp\left[\sqrt{\frac{\gamma H}{2 \cdot R_c}} - 0.5 \right] \right) \cdot a + b, \quad (6)$$

where γ – weighted average specific gravity of rocks, MN / m³; H – depth of working buildings, m; R_c – weighted average compressive strength of rocks, MPa; a and b – are empiric coefficients, taking into account the lithological type of rocks that in the terms of "Partizanskaya" mine are 0.79, 1.55 for rocks of h_{10} seam and 0.36, 0.70 – for seam h_8 rocks.

The definition of these parameters will allow choosing the shape of working cross-section and type of roof support that will be better for supporting power, size of pliability and degree of perception of the asymmetric compressive from the side of roof. In this regard, it is appropriate to use the new technical level supports the KMP-A3 (A4, A5)-R2-type (Figure 6), which mass-produced of West-Donbass Research and Production Center "The Geomechanics".

Efficiency of application of this metal roof support consists in that it provides the protracted maintenance-free supporting of main and development working being in the affected of coal-face works zone, that in turn creates terms for stabilizing of ventilation parameters due to minimization of the local airflow resistance created at reduction of cross-sectional area of working and reduction of airflow wall-friction resistance, as in case of applica-

tion of this type roof support the specific metal content of working's reduced by 20-50%, which is justified in following the work of (Dolzhikov, Kipko & Paleychuk 2012).

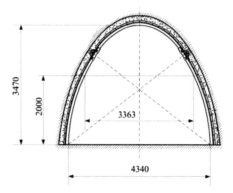

Figure 6. Original appearance and geometrical sizes of metal roof supports the KMP-A3-R2-11.4.

5 CONCLUSIONS

1. Development of fracturing in rock massif around the workings, as a result of flowing of geomechanical processes and influence of coal-face works defined the change of workings stability indexes for a year. So value ω_N in the sheth intake airways of 204-th western longwall changed accordingly from 0.65-0.95 to 0.56-0.67, and index ω_S – from 0.775-0.925 to 0.52-0.67, that entailed complications in-process transport, change of ventilation parameters and, as a result, decline of the actual loading on a cleansing coalface.

2. According to mine research established that deviations of actual air expense from a project in the sheth air intake workings made 14%, taking into account the aerodynamic losses and leaks caused a reduction of the air expense in the longwall at 43%.

3. It is set at the analysis of materials of the depressed surveys found that depression of the sheth air intake workings is linearly dependent from the indexes of their stability.

4. The results of scientific researches in area of increase of workings stability in the intensely fractured rocks of deep mines have made it possible to develop practical recommendations for stabilization of ventilation parameters of the sheth intake airways.

REFERENCES

DNAOP 1.1.30-6.09.93. 1994. *Guidance on planning of ventilation of coal mines*. State normative act about a labour protection, 312. Kyiv.

SOU 10.1-00185790-002-2005. 2006. *Rules of technical exploitation of coal mines.* Standard of Ministry of coal industry of Ukraine, 354. Kyiv.

NPAOP 10.0-1.01-10. 2010. *Safety rules in the coal mines.* A Normatively-legal act is from a labour protection, 432. Kyiv.

Baymukhamyetov, S. 1984. *Management by an offgassing and choice of rational aerodynamic parameters of sheths workings at development of powerful and middle power of coal seams.* Candidate of technical sciences' dissertation on specialty 05.26.01 – Safety and fire-prevention technique, 188. Karaganda.

Makshankin, D.N. 2012. *Substantiation of fastening of mine workings by metallic support from mine profile.* Abstract of Thesis of the candidate of technical sciences' dissertation on specialty 25.00.22 – Geotechnology, 19. Kyemyerovo.

Dolzhikov, P.N, Kipko, A.E. & Paleychuk, N.N. 2012. *Stability of workings in intensity fracturing rocks in deep mines.* Monography, 220. Donets'k.

SOU10.1.00185790.011:2007. 2007. *The Development workings on declivous seams. Choice of fastening, methods and facilities of support.* Standard of Ministry of coal industry of Ukraine, 116. Kyiv.

Mining of Mineral Deposits – Pivnyak, Bondarenko, Kovalevs'ka & Illiashov (eds)
© *2013 Taylor & Francis Group, London, ISBN: 978-1-138-00108-4*

In-stream settling tank for effective mine water clarification

V. Kolesnyk, D. Kulikova & S. Kovrov
National Mining University, Dnipropetrovs'k, Ukraine

ABSTRACT: An original design of in-stream settling tank for effective treatment of mine waters from suspended solids of polydisperse composition in the low-rate stream by gravitational settling technique. The shape and the structure of partitions of the settling tank that provide water flow mode close to laminar are analyzed. Main geometric parameters of the proposed settling tank with consideration of real coal mine drainage conditions are selected. The required quantity of perforated partitions is substantiated. The efficiency indices for mine water clarification are presented.

1 INTRODUCTION

Horizontal settling tanks are widely used in the coal industry of Ukraine. They allow extract coarsely dispersed solids and partially organic pollutants without pretreatment techniques. In such ponds are mainly settled coarse suspended solids (particles of coal or rock). The effect of water clarification after sedimentation is relatively low and reaches about 30%. Such efficiency does not meet the up-to-date requirements of water protection legislation in Ukraine, which have become more strictly over the last decade.

2 FORMULATING THE PROBLEM

The technology of mine waters treatment by its retention at settling tanks (clarification) is the simplest, least time-consuming and relatively inexpen-

sive. It provides a selection of coarse suspended impurities from wastewater which density exceeds the density of the liquid.

Commonly used traditional settling tank constructions are adequately studied. It is needless to find the way of significant increasing their effectiveness. Therefore, one of the ways of settling intensification and increasing efficiency at sedimentation tanks is improvement of their design.

3 GENERAL PART

The authors proposed the construction of in-stream horizontal settling tank for treatment of industrial wastewaters polluted with suspended solids of polydisperse composition by their gravitational settling (Kolesnyk & Kulikova 2012). General view of the settling tank is presented in Figure 1.

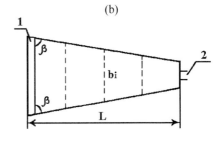

Figure 1. The advanced design of the horizontal settling tank: (a) side view, (b) top view.

The main distinctive feature of the proposed settling tank design is that its body is made in the form of a flume, which getting narrow towards the drain

hole while simultaneous increasing the depth – *H*. The cross-perforated vertical partitions are consistently placed inside the flume with the height h_i and the width – b_i, which correspond to the area of the cross section of the settling tank body. Partitions divide the settler into several sections and have holes that break the wastewater flow into numerous separate streams. This furthers flattening of flow velocity in sections of the settler body, and provides intense sedimentation of suspended particles in the flow which mode is close to laminar. The coarse particles are settled to the bottom already in the first section. Smaller particles penetrate into the next section, where they continue to move towards the bottom.

Wastewater enters the tray 1. Clarified (cleaned) water is run off through the drain hole 2. The collected sludge sediment slips down on a sloping bottom. There is clearance between the bottom and the wall provided for this purpose. Compacted sediment in the form of sludge is output through the opening 3 without the shutdown of the settler.

The proposed settling tank design ensures the formation of a uniform settling velocity and the laminar flow mode for the treated water. At the same time, the cross section for the flow is changed – from not deep but wide – at the intake of polluted mine water, to a narrow but deep – at the point of the discharge of treated water. As far as the water moves forward, suspended particles settle along the increasing depth. Therefore, a sufficiently high layer of clarified water at the drain area is formed, that allows run it off with a minimum capturing the sludge that is collected closer to the bottom of the settling tank. Settler has a high value of the index of volume use and high intensity of sedimentation of suspended solids. It is caused by the geometrical form of the settler and the application of the perforated partitions. In the end, the efficiency of clarification (treatment) process of mine waters is generally increased.

Practice shows that the index of volume use for settling tanks with perforated partitions is higher than those without partitions. Perforation in the partitions can be made in the form of rounded holes, and also in the form of horizontal or vertical slits.

Also, perforated metallic partitions with rounded openings, which got together in the form of hexagons, are recommended to install in the proposed settling tank (Figure 2).

It is known, the mode of the water is determined by the Reynolds number (Re). Therefore, the condition of the laminar flow of water through the partitions of the settler is presented as: $Re \leq Re_k$ .The values of Re in the openings of the partitions is determined by the formula:

$$Re = \frac{Q \cdot d}{S \cdot k \cdot v}, \tag{1}$$

where Q – the intensity of the mine water pumping, m³ / s; S – space partitioning settler m²; v – kinematic viscosity coefficient, which depends on the water temperature, cm² / s; d – diameter of the holes in the partition, cm; k – coefficient of water transmittance through the holes in partitions per 1 m² of their area.

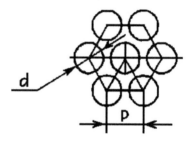

Figure 2. Design of rounded perforations with misaligned rows as hexagon: d – hole diameter, cm; p – distance between centers of the holes, cm.

The authors carried out a simulation of the hydraulic mode at the settling tank with the hole diameter (d) in the partitions varying in the range 2.5-9 cm. The coefficient k is defined as the ratio of the area of holes to 1 m² of the partition. The dependence of the coefficient k values on the selected diameter d of the holes in partitions as a second order polynomial is presented below:

$$y = -0.0014 \cdot x^2 + 0.0282 \cdot x + 0.5684 ,$$

$$R^2 = 0.9982 . \tag{2}$$

where y – value of the coefficient k , and x – hole diameter d in partitions, cm.

Equation (1) allows choose such diameter of the holes in partitions, which gives the Reynolds number that does not exceed the critical value ($Re \leq Re_k$), and the flow mode remains laminar with a steady stream. Re_k values for real tanks are recommended to take up to 2300 (Rabinovych 1980).

Cross-sectional areas of each subsequent partition (from the wastewater inlet to the point of release of clarified water in the area of the settler end wall) is gradually increasing. This leads to appropriate decreasing in the flow rate in each subsequent section, and the Reynolds number for the holes in partitions. This also furthers maintaining the laminar flow.

To ensure effective operation of the proposed settling tank it is important to choose its main geometrical parameters. The authors established the interconnection between the basic design parameters and the settler technological parameters: the coefficient of volume use for the settler in-stream part (K_{set}), retention time of the wastewater cleaned (t) and the efficiency of water clarification (P).

The obtained dependences allowed justify the basic geometric parameters of the settler with consideration of actual conditions of the existing coal mine drainage. Thus, the inclination of the bottom is accepted as $\alpha \approx 30°$, and the convergence angles in plane are adopted as $\beta \approx 84°$. With an initial width of the settler $B_0 = 10$ m, as the basic dimension for project design, its length will be $L = 20$ m, the terminal width $B_k = 6$ m and the maximum depth $H = 11.5$ m (Kolesnyk 2012).

Recommended angle $\alpha \approx 30°$ allows minimize the length of the settling tank and reduce retention time, and provide sliding sludge collected at the bottom in the best way towards the drain hole. Angle $\beta \approx 84°$ provides the best conditions for deposition of suspended particles at a high value of the coeffi-

cient of use of the settling tank in-stream part $K_{set} = 0.944$, which is 1.9 times higher than the same value for traditional horizontal tanks.

The number of partitions is proposed to choose so that the distances between them were no more than their width. At the same time the number of mounted partitions may be changed because the proposed settling tank narrows in plane.

The authors carried out a comparative analysis of the impact of partitions number on the hydraulic flow mode for wastewater treatment and settling suspended particles. The obtained results suggest the feasibility of installation of five partitions in the settling tank which ensure the highest effect of water treatment to remove suspended particles. The first partition (at the water input point) and the last one (at the water output point) should be placed at a distance of 4 m from the settler end walls. Three intermediate partitions should be placed on the same distance from each other.

Efficiency of mine water purification (clarification) from suspended solids was determined on the basis of the data of sedimentation analysis for particles typical for mine waters.

The timing for sedimentation of suspended particles was carried out experimentally by settling samples of mine waters in static conditions. For this purpose the mine water that contains suspended solids at the concentration of 200 mg / L was poured into a laboratory measuring cylinders-sedimentators. At regular time intervals, samples of water were taken from the cylinders, and the value of residual concentration of suspended solids was determined. The obtained test results are presented in Table 1.

Table 1. Dependence of the effect of mine water clarification on the duration of the settling process in a cylinder-sedimentator.

Efficiency of purification (clarification) of mine waters P, %	10	24	35	50	60	73
Retention time of mine water in the cylinder-sedimentator t, min	5	15	30	60	120	300

According to obtained results, the retention time of water in any cross-section of the settling tank for achievement of desired clarification effect P can be determined.

Recalculation of the duration of mine waters settling process in cross-sections of real settling tank at the height h_i is carried out as follows:

$$T_i = t \cdot \left(\frac{h_i}{h_s} \right)^n, \qquad (3)$$

where h_s and h_i – heights of the cylinder and variable depth of the settling tank, respectively, m; t and T_i –retention time in the cylinder-sedimentator and settling tank at a selected depth respectively, min; n – index which describes the ability of the particles to aggregate in quiescent state and for mine waters is accepted as $n = 0.35$ (Koghanovs'kyy 1974).

According to the obtained values T_i the average rate of deposition of suspended particles (average values of particles hydraulic size) as $U_0 = h_i / T_i$, mm / s, were calculated.

In Figure 3 a family of values $U_0 = f(h_i)$ that correspond to the efficiency-driven effect of water purification (clarification) P, %.

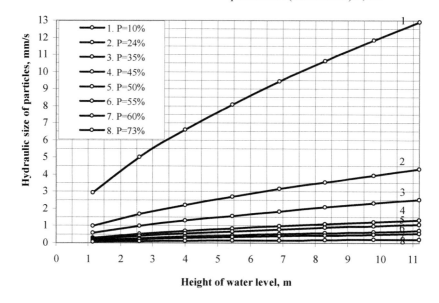

Figure 3. Dependencies $U_0 = f(h_i)$ as values of hydraulic size of suspended particles (U_0) on the depth (h_i) under efficiency-driven effect of water treatment (P).

Through this relationship, the hydraulic size U_0 of suspended particles for achievement of the desired clarification effect (P) of mine waters (see Table 1) at any settler depth h_i can be determined.

Knowing the suspended particles sedimentation velocity in the cross-sections of a real settler, we can calculate the distance l, at which the particles settle onto the bottom of the tank, while ensuring the specified treatment efficiency of mine waters. The calculation of this distance is carried out by the following formula:

$$l = \frac{b_i}{4 \cdot ctg\beta} - \sqrt{\left(\frac{b_i}{4 \cdot ctg\beta}\right)^2 - \frac{Q}{2 \cdot k \cdot U_0 \cdot ctg\beta}} \quad , \quad (4)$$

The calculated l values often exceed the commonly accepted total length L of real designed settling tank. Therefore it is important to evaluate the actual effect of purification (clarification) of mine waters P, for example, under given length $L = 20$ m, and angles $\alpha \approx 30°$, $\beta \approx 84°$ and the typical flow rate of mine waters $Q = 0.0225$ m^3 / s. In this case, not all the particles reach the bottom in the deepest part of the settler (at the end wall). The depth of their

sedimentation under given clarification efficiency (P) can be estimated from the dependency shown in Figure 4.

At the point of clarified water discharge from the settling tank, the value P defined for particles of different sizes U_0 is achieved at different depths h_{os}.

Dependence shown in Figure 4 allows determine the expected effect of clarification (purification) of water P and then hydraulic size of suspended particles U_0, which sediment with specified efficiency. For example, the particles with the hydraulic size $U_0 = 12.918$ mm / s at the water output point, under $P = 10\%$ are almost completely settle onto the bottom in the deepest part of the tank, i.e. at the depth of $h_{os} = 10.9$ m. At the same time the particles with $U_0 = 0.538$ mm / s will settle down only at the depth $h_{os} = 6$ m which corresponds to the treatment efficiency $P = 60\%$.

Finally, the expected effect of mine waters treatment in the proposed settling tank reaches $P = 80\%$. Hydraulic size of suspended particles that settle in the deepest part of the settling tank is about $U_0 = 0.15$ mm / s. These particles will be deposited at the depth of $h_{os} = 0.225$ m. This will allow drain easily the most clarified upper layer of mine waters from the settling tank.

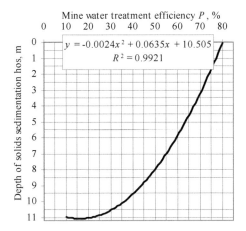

Figure 4. Dependence of the depth of suspended solids sedimentation with various hydraulic size – h_{os} on the effect of mine water treatment efficiency – P for selected geometric parameters of the settling tank.

4 CONCLUSIONS

1. The proposed design of the in-stream settling tank for effective mine water clarification ensures even distribution of water flow over its cross section, as well as better use of its volume. Partitions being installed in various sections of the tank provide laminar flow mode of purified water. This increases the intensity of the gravitational settling suspended par-

ticles.

2. The recommended angle of inclination of the bottom is $\alpha \approx 30°$, and the convergence angle in plane – $\beta \approx 84°$. These angles determine other geometric dimensions of the proposed settling tank (length, width and height).

3. To create a stable hydrodynamic flow, five perforated partitions are recommended to install along the cross-section of the settling tank.

4. Optimization of geometrical parameters of the settling tank and the number of partitions along its length will improve the efficiency of mine water purification from suspended solids, which reaches up to 80%.

REFERENCES

Kolesnyk, V.Ye. & Kulikova, D.V. 2012. *The equipment for treatment dumping from suspended solids.* The patent UA 98382 Ukraine, MPK (2006) B01D 21/02; declared by October 8, 2010; published on May 10, 2012, Bulletin, 9: 6.
Rabinovych, Ye.Z. 1980. *Gidravlika* [Hydraulics]. Moscow: Nedra.
Kolesnyk, V.Ye. 2012. *Justification of geometric parameters of the improved horizontal sedimentation tank for conditions of dewatering the operating mine.* Collection of Scientific Papers of the National Mining University, 39: 229-239.
Koghanovs'kyy, A.M., Kul'skyy, L.A. & Sotnikova, Ye.V. 1974. *Ochistka promyshlennych stochnych vod.* Treatment of Industrial Wastewaters. Kyiv: Tekhnika.

Mining of Mineral Deposits – Pivnyak, Bondarenko, Kovalevs'ka & Illiashov (eds)
© 2013 Taylor & Francis Group, London, ISBN: 978-1-138-00108-4

Rationale of method of unloading area rocks around of developments workings for her repeated use

O. Remizov
Stakhanov a training and scientific institute of mining and educational technologies Ukrainian engineering and pedagogics academy, Stakhanov, Ukraine

ABSTRACT: The problems of protection and support of openings within unloading zones by the example of Almazno-Marievski mines of Donbass geological and industrial district are considered. Simulation under the mining and geological conditions has been carried out; calculations characterizing effect of unloading zone on extraction working have been performed; diagrams of stresses and deformations with their time progress while extraction pillars mining under longwall advancing gently sloping coal seams have been described.

1 INTRODUCTION

The majority of total production costs are mine openings maintenance costs. The costs share as well as their amount depends on a number of interconnected natural and geological, and mine technical factors. The latter creates general difficulties, mining science and practice face while solving problems of mine workings maintenance.

As in actual coal mining practice, freedom in choosing and potential effect on natural geological factors are restricted, then most of all solving the problems of mine workings maintenance come to rational barring selection, and to the method of mine workings protection.

Together with barring, such different methods of drop of stress within rock mass are used: overworking and underworking (Zborshchik 1991), development of relieve slots (Cherniaev 1978), massif unloading with the help of compensation bags (Poturaiev 2000; Bondarenko 2000; Ageiev 2000; Kyrychenko 2000 & Baisarov 2008), and drilling and relief holes camouflet blasting. Control methods for mine pressure based on workings section (Kyrychenko 2000), and mechanical methods of massif strengthening with the help of strained ankers (Bulat 2002) are widely used.

However, it is rather difficult to solve the problems; not always that results in the intended effect wherein mine working maintenance needs both labour costs and material ones.

Thus, under the conditions of Eastern Donbass, total value of 1 m of working construction is UAH 7.000 to 16.000 depending upon cross area and mining and geological environment. As a rule, costs of 1 m of working retimbering with arch support replacement are 20-50% than those of constructing. As it is impossible today to reach maintenance-free workings, it becomes obvious that it will seriously affect cost value of 1 ton of coal.

In addition to labour and material costs for maintenance, poor state of workings being arteries of modern mine also adversely affects a number of coal mining processes: transport, ventilation, and organization of production as a whole damaging operations. The damages can hardly be estimated in terms of money.

Taking into account difficulties of problems of barring and workings protection, the work problems solution was carried out on the basis of experimental and industrial research of a complex of basic factors having an affect on a state of workings.

2 FORMULATING THE PROBLEM

In the process of wall advance, zones of high stresses are formed. The zones follow rock shift on a working contour. Sizeable deformations of mine working support as well as soil heave are seen in a zone of intensive effect of mining.

In the process of further mining, after immediate roof have been caved, hard rocks of upper roof flex towards mined-out space in the form of consoles giving secondary stresses of side rocks. That results in degrading stability of mine workings.

Methods of stresses decrease are based on formation of specific unloading zones for redistribution of maximum supporting force from walls of a working into massif or into worked-out space.

We propose innovative approach of reusable mine workings protection within mining affected zone in which application of binding material on the basis of cement and mineral mixture with yielding supports is foreseen. Declaration Patent of Ukraine confirms it (Shtanko 2010).

The closest design adopted as a technical prototype was an approach on certificate of authorship (Shtanko 2010); it involves hard independent supports erecting. Unloading pockets are prepared on coal seam in advance of a face. The distance should go beyond the bearing pressure zone. Within the pockets, cement-block supports are lined, or cast supports are erected with the help of solidifying materials. The approach disadvantage is the large number of pockets, enormous costs and labour intensity, and formation of stresses within walls of a working when coal is mined.

Such tasks have been set to improve protection of mine working: to avoid labour-intensive protective structures, to prevent serious deformations of steel arch against wall rocks shift, and increase stability of the working during its service.

The approach provides working construction under coal seam with nor less than 30-m advance of working face; coal mining above the working is performed synchronous with a face. The artificial space is supported with the help of individual barring, and is packed with foaming material. A line of wood chocks is made on the face brow from worked-out space. Over the row, cut timber runs. Central hole of the chock is also packed with rapid-hardening material.

Hence, the design provides unloading zone over the working. It enables the working to be free of rock mass supporting force. Stress behavior around mine working is in Figure 1.

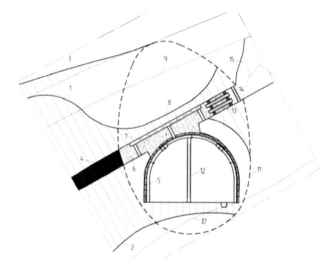

Figure 1. Behaviour of stress distribution around a mine working under the conditions of its protection within unloading zone: 1– a curve of supporting force over virgin coal; 2 – a curve of increased stresses under virgin coal; 3 – nature of supporting force curve for toprock; 4 – coal seam; 5 – steel arch; 6 – saponific material; 7 – individual support; 8 – a curve of lower stresses over a mine working; 9,10 – smooth deflection of roof rocks and soil with zero aperture of discontinuity; 11 – unloaded rock mass around a mine working; 12 – wood prop (false stull); 13 – wood chock; 14 – timber; 15 – worked-out space.

Besides, protective structure size reduction to the dip enables roof rocks fall. That downsizes overhanging consoles providing negligible rock pressure perceived by steel arch without its structure braking. In this case, the mine working will be stable during the whole operating life.

3 MATERIALS UNDER ANALYSIS

Use of any technique aimed at maintenance and protection of mine workings involves a number of research to identify features of compatibility of developed support system and actual mining and geological characteristics of rock mass (Kuznetsov 1987). Not only common bearing ability of barring structure should be estimated but also variations in stress-strain state of "barring-rock mass" system depending upon a working contour deformation, and, as a result, internal forces redistribution. Such an analysis may be made with the help of field studies, laboratory modeling, and computational experiment. From the viewpoint of preparation and implementation period, and with due account for making cost and abilities to vary parameters of barring and rock mass interaction, computational experiment is the most advantageous (Kuznetsov 1990 & Vynogradov 1989).

Application of finite-element method (FEM) is widely used in the process of carrying out computational experiments to solve geomechanical problems (Amusin 1975). Its fundamental idea is to divide design object into finite quantity of elements limited by "simple" geometric surfaces. In this context, each element may have its own mechanical data including an alternative of their variation describing.

Hence, FEM helps to solve any geomechanical problem within selected idea of physical medium with regard to limitations set by a researcher (Gospodarikov 2007).

To make analysis concerning behaviour of mine working supports parameters and mining and geological characteristics of rocks there were calculated conditions of mine workings in such mines of "Pervomaiskugol" SE as "Zolotoie" (coal seam m_3), "Karbonit" (coal seam $k_8^в$), and "Pervomaiskaia" (coal seam k_8). Both structure and mechanical data of rock mass correspond to real mining and geological conditions of Almazno-Marievski geological and industrial district of Donbass.

On the basis of the problem formulation, the computations were performed for elasticoplastic medium taking into account rock rheology. Computational experiment was implemented volumetrically with simulation of two meters of mine working along its axis. Under computations, the pattern midsection was sixteen meters apart, behind the face. The only schema of a working protection ap-

plied under field conditions was simulated in each variation of performing computations.

As a result, there were developed curves of stresses and deformations characterizing the effect of unloading zone on a mine working stability under different mining and geological conditions as well as their time progress.

While making primary analysis of the results, consider stress intensity curves obtained for different mining and geological conditions of Almazno-Marievski district. The curves are in Figure 2.

As the stress intensity helps to identify zones of maximum difference between basic stresses $\sigma_1 - \sigma_3$ without selection of stress projection vector then their maximums indicates places of the most probable rock mass destruction if stress is loading is complex (Tkachev 2008). Absolute intensity values σ_{int} are used to identify stress concentration value within specific element of simulation model.

In this respect, stress distribution behaviour within "barring-roc mass" system is common for all computation versions. Values of stress concentration R_σ are not more than 12, and absolute values of σ_{int} maximums within rock mass are: 37 MPa for "Zolotoie" mine, 28 MPa for "Karbonit" mine, and 36 MPa for "Pervomaiskaia" mine. Schematically, each of the curves in Figure 2 is divided into the three analysis sections: 1 – unloading zone, and rocks adjoining it from above; 2 – rock volume under chock, and right side of a working; 3 – left side of a working, and adjoining it rock.

(a) (b) (c)

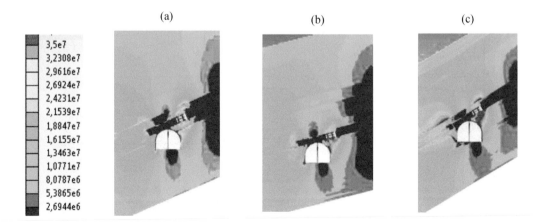

Figure. 2. Stress intensity curves obtained for mining and geological conditions of such mines as: (a) "Zolotoie"; (b) "Karbonit" and (c) "Pervomaiskaia".

Section one consideration is the analysis of features of the barring elements within unloading zone, and rock console pressing it from above. It is obvious that in all the three computation versions, stress distribution in rocks gives rise to supporting force formation on the peripheries of console. Low pressures are formed in the central part. Throughout the height, the area stays approximately similar to unloading zone height for different mining and geological conditions; however, it has different dimensions. Both width and geometry of the area depends on the parameters of rock seams within border zone of rock mass. Successive study of Figures 2c, a, b shows that increase in slope angle results in the area dimensions rise. In this context, stress concentration on the right edge of the console increases 3.

In all computation versions breaker props and wood chock are at the distance close to critical. However, it is true for Figure 2a where adjoining from above rocks retain original level of strength to rock pressure. Then, slope angle is not a determinant when rocks transit to marginal state and overmarginal one. Effect of geometric data as well as mechanical data of rock seams on stress distribution in a roof of a mine working is understandable. As Figure 2c shows, effect of a mine working on stresses within roof rocks propagates up (to 16 m) resulting in formation of large bearing area with weak stress gradients. In this case similar situation is in Figure 2b; however, stress average for the version is only 11 MPa to compare with 20 MPa for version in Figure 2c.

Besides, when slope angle increases clear dependence is seen as for the stress decline in a left leg of individual barring: from 21 to 14 MPa. Probably, this effect is directly connected with redistribution of internal forces towards formation lines. As a result, motion of σ_{int} in a support of individual barring means deficiency of primary structure of individual barring as well as transition from internal forces concentration to increase in deformations on the surface of unloading zone (Figure 3). Qualitatively, dependence data can not be a function of rock mass structural features except slope angle; they are of quite universal nature for the system of mine working maintenance. Values within the whole range of slope angles shown in the diagram have been developed by means of square extrapolation of dot computation data.

Simultaneous analysis of curves in Figure 3 helps to deduce that considered schematic of the mine working maintenance is the most efficient within 17-32°. Ideal from the viewpoint of strength factors, stresses-deformations balance shows that. With increase in slope angle, "working barring-rock mass design system progressively transforms to marginal state and overmarginal one. That increases significantly possibility of unloading zone roof fall. Accordingly, it becomes critical to consider conditions of main cracks formation in walls of the working in the context of maintaining axisymmetric deformation of elastic and isotropic multilayer semispace with cylindric hollow having rigid support.

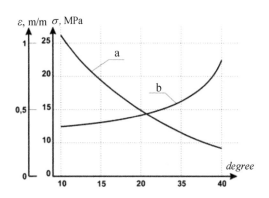

Figure 3. Dependence diagram of variations of stress intensity maximums σ_{int} in a left support of individual barring (a) and deformation value in a roof of unloading zone (b).

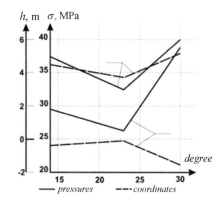

Figure 4. Variation in stress intensity maximum σ_{int} and vertical coordinate of its location as for the bottom for left (a) and right (b) walls of mine working.

Consider integrally conditions of both left and right walls of mine working with the help of comparative analysis. The first and principal our conclusion which makes it possible to analyze stress curves in Figure 2 is essential effect of the rock mass geological structure on stress-strain state of the areas. Moreover, the result comparison with the stress pattern enables to predict that if a border of rock seams being higher than geometric center of a working cross-section it results in supporting force loss having no effect on stress value in the left-side rocks. If the border is lower than the centre, it causes stress jump in the left-side rocks without change of stress concentration in a right side of the mine working.

To make detailed analysis consider diagrams in Figures 3 and 4. The diagrams show that rock pressure value as well as its location as for the mine working is perfectly balanced if slope angle is 23° (Figure 2a). The matter is that within zones under consideration, dilute stress concentrations take place, and their effect on the barring is concentrated within contact zones of both left and right barring supports. In fact, availability of thin rock seam in an arch under unloading zone provides rock pressure balancing in its left and right walls. Hence, stress redistribution into rock mass takes place. As a result, stability of the working arch increases (Vynogradov 1989).

Availability of rock seams border in a lower part of the working (Figure 2c) factors in formation of increased rock pressure in a lower part of the mine working left wall. Taking into account absolute stress values within the zone as well as its geometry, one should conclude that in this variation of stress-strain state system, conditions are created to form rock outburst of the mine working left wall. Computations of the research predict increase in unloading zone width by means of coal mining towards dip by 0.3-0.7 m. That should result in stress concentration factor lowering to 29-32 MPa.

Another negative feature of computation variant on "Pervomaiskaia" mine is availability of increase rock pressure in a right wall of the mine working. Just in that case, combination of the two zones of stress concentration takes place: a zone neighbouring the mine working contour near frame support flexibility centre, and a zone of supporting force under the base of a wood chock. Under such conditions when stress intensity σ_{int} is within 37-40 MPa, issues of energy theory (Tkachev 2008) originate a situation for accumulated energy of deformations to transform into kinetic one with destruction of rock mass area integrity. To minimize such a scenario probability, it is required to decrease a level of stress concentration. As extra computations show, that can be achieved owing to availability of another wood chock in worked-out space, or erecting two or three lines of breaker props below available one.

Generally, diagrams in Figures 3 and 4 show that effect of rock seams border passing through a contour of a mine working can not result in drastic changes in development trend of weakening zones growth around a mine working. That is while developing a trend of variations in "support-massif" stress-strain state system, average relative deviations are 9% for stresses, and 6% for deformations. Nevertheless, as it is demonstrated above such effect minimization is possible owing to increase in unloading zone dimensions, or if extra support elements are erected in the mine working wall to the rise.

To select any measures to maintain and protect a mine working under considered conditions, one should clearly understand general geomechanical idea used to form in "support-massif" stress-strain state system (Vynogradov 1989). As the stress curves in Figures 2 and 5 show, development of unloading zone results in rock console formation. On one side the console is rigidly fixed in undisturbed rock mass, and on the other side it is freely supported with wood chock and breaker props. In this context, application of individual support in combination with "Karbofil" foam helps to cut deformation gradient growth on the rock console bottom face rather that to increase formed arch rigidity (Zborshchik 1991). Therefore, stress concentration in filler is impossible under the conditions of the problem, and stresses in supports and individual prop point at geometry of lowered deformations zone within the rock console. Hence, the element of a mine working support is shock reducer in the process of force transfer to frame support, and that also corresponds to operational conditions.

To consider behaviour of elements supporting a mine working in unloading zone, and to estimate their efficiency, examine a stress pattern in the form of curves of vertical stresses σ_y (Figure 5). First of all, curves σ_y give ability to make a cold evaluation of performed computations quality by means of comparing them with conventional theories of rock mass state (Poturaiev 2000).

Distribution of vertical stresses in unloading zone support elements shows availability of both compressive internal forces and extending ones. Maximums of compressive stresses are 32-37 MPa, and extending ones are not more than 5 MPa. Hydrostatic pressure for Figures 5a, b is 9.6 MPa, and for Figure 5c it is 7.4 MPa to be equivalent to the mine workings depth. In this context, extending stresses concentration in the latter computation variation is 1.7-2.2 times higher to compare with previous ones.

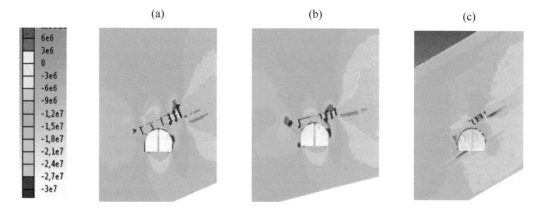

Figure. 5. Curves of vertical stresses obtained for mine and geological conditions of such mines as: (a) "Zolotoie"; (b) "Karbonit"; (c) "Pervomaiskaia".

However, performed computations as well as analysis of "support-massif" stress-strain state system variation helps to deduce patterns of displacement of points of roof and bottom, taking into account slope angle of rock seams during considered time interval. To identify variations of value H of a mine working, use value U_H being a sum of maximum displacements in a bottom and in a roof. The value is described in the form of created function of slope angle at the certain time moment (Fester 1983) for specific mining and geological parameters of rock mass. It is described by

$$U_H = \frac{t_d^{1.2}}{\left(\alpha + \dfrac{\pi}{3}\right)^{0.08}} \left(\frac{1}{\dfrac{sin(\alpha)}{K_1} + K_2 \cdot cos(2\alpha)} - log\left(\frac{\alpha}{4}\right)\right); \qquad (1)$$

where K_1 – coefficient of structural unfastening of marginal rock mass which value range is $1.6 \leq K_1 \leq 18.2$; and K_2 – coefficient of rock mass rigidity on Protodiakonov scale.

In turn, horizontal displacements of walls of a mine working are used to identify value U_B variation. The value shows how values B change for different factors of slope angle of rock seams during examined period of the mine working operations. As a result, we obtain following empirical dependence

$$U_B = 0.8 \cdot t_d \left(\frac{3.61 \cdot K_2}{t_d^{cos(2.127\pi/\alpha)}} + 4.62^{lg(1.3\alpha - 0.675\pi/t_d)} \right), \qquad (2)$$

where K_2 – coefficient of rock mass rigidity on Protodiakonov scale.

4 CONCLUSIONS

We propose innovative approach of protection of reusable mine workings in a zone of effect of mining. The approach means application of binding material on the basis of cement and mineral mixture with yielding supports.

Effective algorithm for stress-strain state computations to solve geomechanical problems with the help of modern computer systems has been developed on the basis of boundary-element method and Solid Works software.

Hence, that provides to predict changes in a contour of a mine working protected with the help of unloading zone under mining and geological conditions of Almazno-Marievski geological and industrial district of Donbass.

REFERENCES

Zborshchik, M.P. & Nazimko, V.V. 1991. *Protection of workings of deep mines within unloading zones.* Kyiv: Tekhnika: 248.

Cherniaev, V.I. 1978. *Basic parameters of relieve slots.*

IHL Proceedings. Mining Magazine, 7.

Poturaiev, V.N., Zorin, A.N., Vynogradov, V.V. & Bulat, A.F. 2000. *A rule of critical and stressed rocks under small disturbances.* Diploma #1 Scientific discoveries, hypotheses, and ideas. Moscow: SBR: 20.

Bondarenko, V.I, Pivnyak, G.G. & Zorin, A.N. 2000. *A rule of loose water-saturated rocks concreting under the effect of electric current.* Diploma #12 Scientific discoveries, hypotheses, and ideas. Moscow: SBR: 27-88.

Ageiev, V.G, Kuzhel, S.V, Sdvizhkova, E.A, Tulub, S.B. & Shashenko, A.N. 2000. *A rule of outcrop stability variations in workings.* Diploma #131. Scientific discoveries, hypotheses, and ideas. Moscow: SBR: 114.

Kyrychenko, V.Ya., Zvagilski, E.L., Lishin, A.V., Usachenko, V.B. & Khalimendik, Yu.M. 2000. *Phenomenon of traversing shear zones formation within strained rocks.* Diploma #188. Scientific discoveries, hypotheses, and ideas. Moscow: SBR: 62-63.

Baisarov, L.V, Iliashov, M.A., Levit, V.V., Palamarchuk, T.A., Sergienko, V.E., Usachenko, V.B. Yalanski A.A., & Pototski, V.V. 2008. *A rule of soil and rock masses self-organization in the neighbourhood of elongated underground workings.* Diploma #318. Scientific discoveries, hypotheses, and ideas (1992-2007). Research and information review. Moscow: MMAANOI: 298-299.

Bulat, A.F. & Vynogradov, V.V. 2002. *Supporting and rock anchor of mine workings of coal mines.* Dnipropetrovs'k: IGTM of NASU: 372.

Shtanko, L.A. & Remizov, O.V. 2010. *Protection method for mine working.* Patent of Ukraine for useful model #52896 (UA) of 07.04.2010. #201004071. Published 10.09.2010. Bulletin #17.

Kuznetsov, G.N., Ardashev, K.A., Filatov, N.A. et al. 1987. *Methods to solve problems in rock mechanics.* Moscow: Nedra: 248.

Vylegzhanin, V.N., Egorov, P.V. & Murashev, V.I. 1990. *Structural models of rock massif in a mechanism of geomechanical processes.* Novosybirs'k: Nauka: 295.

Vynogradov, V.V. 1989. *Rock mechanics of control of massif state in the neighbourhood of mine workings.* Kyiv: Naukova Dumka: 192.

Amusin, B.Z. & Fadeiev, A.B. 1975. *Finite element method in the process of solving geomechanical problems.* Moscow: Nedra: 143.

Gospodarikov, A.P., Syrenko, Yu.G. & Zatsepin, M.A. 2007. *Computation algorithm for stress-strain state of a roof taking into account optimum choice of parameters of process diagrams under the conditions of Startobinskoie deposit.* Proceedings of Mining Institute, SPB, volume 170, part 1: 106-110.

Tkachev, V.A. & Kompaneitsev, A.Yu. 2008. *Providing of openings stability under the complicated mining and geological conditions.* Growth prospects of western Donbass: collection of scientific papers. Part 1. Shakhtinsk Institute (a branch of YRGTU) NPI. Novocherkas'k: PTC "Nabla" YRGTU (NPI): 145-151.

Fester, E. & Renz, B. 1983. Methods of correlation and regressive analysis. Moscow: Finance and statistics: 302.

Technological parameters of cutoff curtains, created with the help of inkjet technology

O. Vladyko
National Mining University, Dnipropetrovs'k, Ukraine

ABSTRACT: The article is devoted to the creation of impervious curtains around excavations shallow burial. The patterns of the depth of penetration of high-pressure jets of fixing solutions on the technological parameters is based on the analyze. How to create elements of impervious curtains, allowing determining the volume of fixative solution for dispersed species, which depend on the filtration and mechanical properties of the veil. On the basis of experimental studies of the influence of established quality and quantity of fixing solution and the injection pressure on the filtration coefficient and the strength of the rock-cement elements is determined . The regularities of changes in the filtration coefficient to changes in the density of reinforcement solutions, which allows predicted the physical and mechanical properties of the new screens are settled.

1 STATEMENT OF THE PROBLEM

The task of providing the necessary conditions of underground workings in weak rocks watered is important for the development of industry in Ukraine. The presence of water inflows complicates the operation of existing mines and underground structures. The main factors that have an adverse operating condition are the height of head of groundwater, the permeability of water-bearing rocks and power.

Under pressure water filtration for the construction of impervious curtains the most effective way is to use high-pressure jets of fixing solutions. However, until now researchers have not been established laws governing the formation of veils in weak rocks watered.

2 ISOLATION OF UNSOLVED PROBLEMS

Since inkjet technology is based on securing the hydraulic fracture of rocks, it was considered modern state of research in this area. Significant contribution to the study of the processes of destruction of rocks have high-pressure jets of V.K. Kooley, D. Artingstol, K. Modi, S.E. Shavlovs'kyi, N.V. Dmytriev, M.V. Hasin, B.S. Fedorov, L.V. Petrosyan, G.I. Abramovych.

To solve the problems of formation of impervious curtains secure way jet was necessary to perform complex investigations, including the establishment of:

– changing patterns of penetration of high-pressure jet fixing solution, depending on its density, toughness rock experimental function C, the nozzle diameter and primary jet velocity;

– raising the speed of the influence of the monitor to the depth of penetration of high-pressure jets of fixing solution to disperse rock, establishing new relationships change in thickness cutoff curtains on the density of the solution and fixing species, the discharge pressure fixing solutions, the diameter of the nozzle and the jet-speed lift the monitor;

– of the influence the quality and quantity of fixative solution, the pressure of his discharge on the strength and permeability coefficient rock-cement elements, allowing defining the rational parameters of formation thickness cutoff curtains.

3 ANALYSIS OF RECENT RESEARCH

The essence of inkjet technology fixing species is at a hydraulic fracture and fixing stirring solution of dispersed rocks fixing or replacing their composition. This technology enables to form a fixed area rock strictly defined shapes and sizes. Construction impervious curtain using inkjet technology may fasten in two ways: without rotating jet monitor ($\omega = 0$), and its rotation ($\omega > 0$). In the first method as a result of lifting the monitor there is formed one or two-piece curtain. Sections pitch angle of 1300-1500 to each other for better clamping. In the second method produces a variety of piling, which depends on the design parameters of the monitor and the technology used.

At the Institute of Foundations (G. Nosov) studies have been conducted on the application of inkjet technology for cutting frozen rocks (Fedorov 1983).

Rational parameters were defined and cutting of frozen ground, the minimum size of the diameter of the nozzle jet monitor, where possible destruction of rocks. Great contribution to the theoretical development inkjet technology has made the study of the laws fixing the flow of water jets conducted G.I. Abramovych (Abramovych 1948). He solved the problem of movement in the wake of a turbulent jet flow. Developing theoretical and applied bases consolidate dispersed rocks in mining operations, S.F. Vlasov, applying the theory of turbulent jets, calculated process parameters fixing c given the dependence of C functions included in the theory of turbulent jets, the ratio of the densities of the two mixing media (Vlasov 1999). It is possible to establish new patterns of penetration of high-pressure jets of the lifting and rotation of the tool.

4 SORTING OUT THE UNSOLVED PART OF THE OVERALL PROBLEM

Under pressure water filtration for the construction of impervious curtains the most effective way is to use high-pressure jets of fixing solutions. However, until now researchers have not been established laws governing the formation of veils in weak rocks watered.

In this regard, it was necessary to justify the creation of rational technological parameters cutoff curtains with high-pressure jets of fixing solutions. To do this it was necessary to perform a series of studies, including the assessment of the effect of various process parameters on the thickness of grout curtain and the depth of penetration of the fixing solution to disperse rock establishment of filtration and strength properties of impervious curtains depending on the amount of reinforcement solutions in unit volume, as well as to develop the technology of formation of impervious screens, providing service objects dispersed in weak rocks.

5 FORMULATING OF GOALS

The main purpose of the study and development of process parameters for the formation of impervious curtains in weak dispersed rocks with high-pressure jets of fixing solutions that enhance the effectiveness of the protection of underground facilities and waterworks. Goals are also working to determine a rational rate of uplift jet monitor at which high-pressure jet penetrates to the maximum distance and does not cause cost overruns fixing solution. Set the patterns of change in the depth of penetration of the high-pressure jet of density fixing solutions, density

and toughness of the treated rock with the sound speed lift the monitor. Determine the necessary amount of fixative solution mixing with the rock, allowing providing the required filtration coefficient constructed curtains.

6 THE MAIN MATERIAL

At present there are a large number of specific ways to create a cutoff curtains around underground facilities. One of the key is the use of specific methods, altering the physicochemical properties of the rocks. To protect underground facilities from water inflows in the last decade, most often used are: chemical and electrochemical methods, "slurry wall" smolizatsiya and jet fixing species A large number of ways caused by that the efficiency of formation of protective curtains depends on the nature and technological factors.

The main technological parameters of inkjet technology include: the diameter of the nozzle jet monitor the depth of penetration, flow and supply pressure fixing solution, the density of the solution, and the rate of uplift of the monitor, the toughness and density of the rock. Research on how to destruction by high-pressure jets were carried out for different rocks and a variety of process parameters (Vladyko 2001).

The presence of aquifers and water-bearing rocks disperse significantly complicate the operation of underground and hydraulic structures, and as mentioned above, the most effective way to improve the operation of such facilities is to provide a cutoff curtains around them. Rate filtration rate cutoff curtains, constructed using inkjet technology is proposed on the basis of the carried out laboratory and theoretical studies.

To confirm the relationship of volumes fixing solution and process parameters in a fixed jet consolidation scope were conducted with the use of fine-grained sand watered with toughness $\eta = 90$ J / m^2 and loess with toughness $\eta = 447$ J / m^2 with humidity $w = 14 \%$. When using the monitor with a jet nozzle diameter $d = 0.003$ m fixing solution was pressure $P = 0.4$ MPa and speed of a jet lifting monitor $- v = 0.01$.

Results (Figure 1) show that the sand with increasing density fixing solution 1400 to 1800 kg / m^3 content increases in the fixing solution attached rock volume from 47 to 80%. For sands with 30% moisture content of the fixing solution was increased from 20 to 60%. Increasing the sand moisture from 14 to 30% volume reduction occurs in the fixing solution of the sample is 1.5-2 times. The correlation coefficient for the sample size for different moisture $n = 15$ sand was $r = 0.6$-0.65.

Interesting results were obtained when fixing elements curtains with rocks having different particle size distribution. The studies were conducted at the facility to secure the jet loess, fine-and coarse-grained sands with humidity %. For loess with a grain size of

0.01-0.1 mm pinning volume of the solution was 70% for fine-grained sand with a grain size of 0.1-0.25 mm – more than 60%, and medium-grained sand with a grain size of 0.25-0.5 mm – more than 50% (Figure 2). The correlation coefficient with the volume of the sample $n = 15$ was $r = -0.6$.

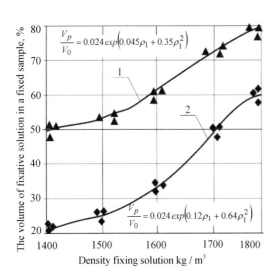

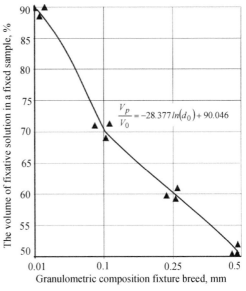

Figure 1. The dependence of the fixing solution in a fixed sample of fine-grained sand on the density of the solution: 1 – 15% – moisture rocks, 2 – 30% – moisture rocks.

Figure 2. The dependence of the stirring of the soil granulometric composition of the rocks.

The content of cement per cubic meter of air curtains received as accepted technology of concrete structures is determined by the volume of one meter built by the veil. Next, define the initial water content of the veil, and soil as a placeholder method of absolute volumes.

To determine the flow properties need to know:
– the content of cement per cubic meter of soil gets hooked;
– to determine the initial water content of the fixing solution and humidity of the breed;
– the number of species in cutoff curtains.

Cone fixed breed is shown on Figure 3.

By adjusting modes, inkjet technology and calculating process parameters using the following formulas:
– Raising the speed jet monitor v_n, where the penetration depth is maximized and will not cause cost overruns fixing solution:

$$v_n = \frac{32 c \eta}{2.73 \pi \rho_1 d_0 u_0 \left(1 + \dfrac{\rho_2}{\rho_1}\right)}, \qquad (1)$$

where η – toughness of the fixture rock J/m²; ρ_1 – density of the fixture breed, kg / m³; ρ_2 – density of the fixture breed, kg / m³; d_0 – diameter of the nozzle jet monitor, m; u_0 – initial jet velocity, m / s; c – experimental feature.

– The depth of penetration h of the high-pressure jets of matter in securing a dispersed breed:

$$h = \sqrt{\frac{7.45 \pi \rho_1 d_0^3 u_0^2}{32 c^2 \eta}}. \qquad (2)$$

– Thickness of the grout curtain t_c on the densities of the fixing solution and rocks, the discharge pressure of the solution, the diameter of the nozzle and the lifting speed monitor. It allows you to theoretically calculate the thickness of the grout curtain in the penetration of high-pressure jets of fixing solutions in the dispersed species:

$$t_c = \frac{2 \cdot d_0}{d_1} \sqrt{\frac{7.45 \pi \rho_1 d_0^3 u_0^2}{32 c^2 \eta}} . \qquad (3)$$

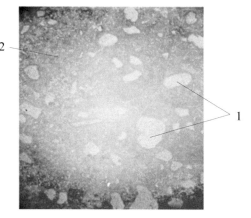

Figure 3. The picture of the section of the fixed species (Lens 3.6x (-)): 1 – sand rocks, 2 – fixing solution.

Using the formulas, we can calculate the theoretical process parameters for the cutoff curtains. Further, the design parameters curtain is to make an adjustment on soil moisture, which is different for coherent and non-cohesive soil. Feature inkjet technology is shown here that the soil moisture is not compensated by a separate dosing of mixing water, in concrete technology, and changing the content of all components of the grout curtain.

In contrast to the non-cohesive (granular) soils, cohesive soils for water in the pores of the particle aggregates is not free. Cohesive soil moisture exerts its influence on the properties of the veil is not a change in the W / C soil-concrete (that's been in the existing models), and through the influence of porosity, density and strength of the aggregates. The principal difficulty is the model proposed that virtually unknown as there are particle aggregates in the original cohesive soil in the natural state and how much it remains after the erosion jet water cement solution and subsequent mixing in a particular construction of curtain, which in size and structure, they have a porosity how this will affect the porosity of all the properties of grout curtain. It is possible, in the framework of the proposed model, based on the known dependence of concrete, to assess the filtering properties of the veil.

7 CONCLUSIONS AND PROSPECTS FOR THE DEVELOPMENT WORK

The study results allows to:
 – calculate laboratory preparations for creation of curtains;
 – technological parameters of formation of impervious curtains in weak dispersed rocks with high-pressure jets of fixing solutions to help you manage strength and filtration properties produced cutoff curtains are substantiated;
 – its actual composition and, if necessary, make the appropriate adjustments by changing the water-cement mortar flow by varying the speed of raising the water jet;
 – the dependence of the maximum depth of penetration of the high-pressure jet of raising the speed monitor that provides an effective solution flow is settled;
 – depend on the influence of the hydraulic resistance and the amount of fixing solution in rock-cement elements that identify the rational parameters of formation thickness cutoff curtains in solving various mining tasks is determined.

REFERENCES

Fedorov, B.S. 1983. *The use of high-pressure jets at the device bases in the North*. And Foundation Soil Mechanics, 2: 19-21.
Abramovych, G.N. 1948. *The turbulent free jets of liquids and gases*. Moscow: Gosenergoizdat: 288.
Vlasov, S.F. 1999. *Theoretical and practical bases of the jet fixing weak disperse rock during mining operations*. Dis ... Doctor. Tehn. Sciences: 05.15.09. Dnipropetrovs'k: 375.
Vladyko, O.B. 2001. *Justification technology of cutoff curtains fixing jets of high-pressure fluids*. Dis ... candidate Tehn. Sciences: 05.15.09. Dnipropetrovs'k: 161.

Mining of Mineral Deposits – Pivnyak, Bondarenko, Kovalevs'ka & Illiashov (eds)
© 2013 Taylor & Francis Group, London, ISBN: 978-1-138-00108-4

The investigation of coal mines influence on ecological state of surface water bodies

A. Gorova, A. Pavlychenko & S. Kulyna
National Mining University, Dnipropetrovs'k, Ukraine

O. Shkremetko
St. Petersburg Energy Institute, St. Petersburg, Russia

ABSTRACT: The features of coal mines impact on water quality in groundwater and surface water bodies are analyzed. Bioindication assessment of mine water quality in storage ponds is conducted.

The problem of water bodies' protection becomes increasingly important in Ukraine every year due to the intensive development of mining industry. Coal production is accompanied by significant negative impact on the quality of both surface and groundwater. Groundwaters got into mining, actively interact with rocks and coal that are crushed at the coal production and leach macro-and microelements from them. Mine waters that are saturated with mineral salts, suspended solids, sulfates, petrochemicals and other pollutants are pumped from the mine to the surface and dumped in ponds. Today the production of 1 ton of Ukrainian coal is accompanied by loss of approximately 3 m^3 of groundwater which enter the mine workings (Grebenkin 2008; 2009 & Riznyk 2006).

Long-term coal mining has led to pollution of ground and surface waters. The situation that has developed with the state of surface water bodies causes concern because exceeding of maximum permissible concentrations of pollutants is observing in them. This results to a deterioration of water quality in surface water bodies and increasing of diseases frequency in the mining areas population (Gorova 2009, 2012).

The coal mining is conducted in the Chervonograd mining region (CHMR) over 50 years. Today, the majority of mines has exhausted the main part of reserves and are in the process of damping. The hydro system of CHMR is formed by Western Bug River and its left affluents – Bolotnya, Spasivka, Zheldets, Merezhanka, Bilyj Stik. The largest of them – the rivers Rata and Solokiya are situated in the area of influence of region mining companies and the Central Concentrating Plant (CCP). Coal mines activity causes water pollution in rivers Rata and Western Bug.

During the development of coal mines in CHMR mine waters are pumped by pipeline to shared storage ponds, which are located in the Gorodyshche village and Chervonograd town. The part of the water from storage ponds after using is the technological needs of CCP (for coal flotation) is directed by the pipeline to sludge pit, which is located between the rivers Western Bug and Rata.

The storage pond in Chervonograd is located on the territory of liquidated mine #1 "Chervonograds'ka" and has four sections which are separated by a dam (two section with size 185 by 920 m and two sections with size 50 by 150 m).

In the operation rate of the storage pond the dams has been repeatedly mounded, expanded and now their spine constitutes 10 m. Today the "Chervonograds'ka" mine dumps 27.8 m^3 of mine water per year into the storage pond in Chervonograd town.

Seven operating mines and one mine that is in the closure process discharge mine waters in the storage pond of Gorodyshche village.

The pond has two sections with size 300 by 800 m. The total inflow of water is 327.9 m^3 per year. An average 4000 m^3/year of mine water are pumped from storage pond of CCP. Waterproofing of pond bottom and sides with plastic wrap and clay was supposed to prevent water filtration. Unfortunately, only 30% of the surface is damp-proofed at the moment.

The sludge pit of CCP with size 500 by 1250 m is located in the area between the rivers Western Bug and Rata in the fields of mines "Mezhyrichans'ka" and "Velykomostivs'ka". The distance to the Western Bug river accounts for 50 m and to the Rata rives – 200 m. Waterproofing of bottom with plastic wrap and clay was performed to prevent filtration losses. At the present moment waterproofing screen is constructed only at 25% of surface of pit.

It should be noted that the issue of mine sewage

treatment has been neglected for decades. Construction, reconstruction and modernization of treatment facilities almost not funded, both by the companies and the state. This situation causes concern because the Western Bug River -is the cross border river that flows not only on the territory of Ukraine but also on the Belarus and Poland territories. Ukraine, according to Border Cooperation Programme "Poland-Belarus-Ukraine" must monitor the state of water in the Western Bug River and its affluents. It should be noted that the Western Bug River – the only river in the state which flows into the Baltic Sea, and a third part of the population of Polish capital uses this water for drinking consumption.

Therefore, the aim of the work is the study of water quality in surface water bodies of Chervonograd mining region to improve water protection activities in coal enterprises.

Water samples from ponds of Chervonograd town and Gorodyshche village as well as from sludge pit of CCP were s selected for studies.

Water sampling was performed twice a year in spring and autumn during 2007-2011 years. The pooled tap water (not less than 7 days) was used as a control.

Toxicity of water in storage ponds and sludge pit of CCP was assessed using a growth test with phyto indicator *Allium sulphur L.* The advantage of growth test is that it can be used to assess the impact of water-soluble forms of pollutants. This method is easy to conduct and sensitive for the detection of minimal concentrations of toxic substances in the water. The inhibition of root growth of *Allium sulphur L.* is the indicator of toxicity because it is established that this process is inhibited at lower concentrations of toxicant than sprouting plants (Rudenko 2003).

For each water sample 12 test-tubes were prepared that are further were filled with 25 ml of water samples. Bulbs previously cleared of onion-skin were placed into the each test-tube, so that the bottom touched the liquid in a test tube. Water in the test-tubes was changed to a new one every day. After two days of experiment 2 bulbs with very short roots were removed from each variant. The experiment lasted 72 hours. The length of root and stem system of 10 the most typical bulbs was measured at the end of the experiment (the longest and shortest roots were not taken into account). Thus, 12 bulbs were germinated for each of the water samples and from 4 to 30 roots of them were measured. The reliability of each experiment was confirmed by three times reproduction for 48 test-tubes in each study period. Obtained data were worked on using mathematical and statistical analysis as well as arithmetic error and coefficient of Student were calculated at the results of studies.

Phytotoxic effect (PE), namely degree of oppressing of growth processes was determined in percents in relation to control (by mass and by sprouts or roots length of test-culture) according to the formula (Bilyavs'kyi 2006):

$$PE = \frac{m_o - m_x}{m_o} \cdot 100\%, \qquad (1)$$

where m_o – mass or length of sprouts (root or above-ground part) in control; m_x – mass or length of sprouts in variants of research.

Rating scale was used to determine the toxicity of water using growth test using, Table 1 (Rudenko 2003).

Table 1. Scale of assessment of water toxicity level.

Phytotoxic effect, %	The toxicity level
0 – 20	No or weak toxicity
20.1 – 40	Average toxicity
40.1 – 60	Above average
60.1 – 80	High toxicity
80.1 – 100	Maximum toxicity

The results of assessment of mine water toxicity from storage ponds and sludge pit of Chervonograd mining region for the period 2007-2011 years are presented in Table. 2.

Table 2. The levels of toxicity in the water of storage ponds and sludge pit, 2007-2011 years.

Place of sampling	Phytotoxic effect, %		
	min	max	average
Storage pond of Chervonograd town	36	56	44
Storage pond of Gorodyshche village	46	68	56
Sludge pit of CCP	56	63	61

Analysis of the data of Table 2 showed that the lowest toxicity of water in the samples was observed in the pond located in Chervonograd town and it was rated as "average".

The highest toxicity was detected in samples of water from the sludge pit of CCP and storage pond of Gorodyshche village. The level of toxicity of water was rated as "high".

Analysis of the average values of phytotoxic effect revealed that it was in the same numeric range for storage ponds of Chervonograd town and Gorodyshche village and the level of toxicity of water was rated as "above average". The water in the sludge pit of CCP has "high" level of toxicity.

High levels of water toxicity can be explained by the fact that the constant excess of maximum permissible concentrations is observed in investigated water samples: for iron in 1.1 times, for chlorides in 3.1-4.18 times, for manganese in 1.1-1.8 times.

The following conclusions can be drawn at the results of the study: water from storage ponds and sludge pit of CCP is characterized as "above average" toxicity level and it is one of the pollution sources not only for natural water bodies, but also for soil and groundwater because isolation was not performed in full. In addition, highly mineralized mine water pipes which are laid in the mine fields, are affected by physical and chemical effects due to deformation of the earth's surface, the aggressive action of water and mining waste, leading to their destruction and consequent leakage of pipelines.

Due to the fact that the water in storage ponds and sludge pit has high levels of toxicity its surface impoundment possible only after increasing the efficiency of sewage treatment plants.

We consider that it is necessary to improve the technology of cleaning and disinfection of mine water in order to improve the quality of mine water in Chervonograd mining region. It is also recommended to assess the effectiveness of mine water treatment using highly sensitive bioindicaton methods.

Taking into account the results of this research will reduce the negative impact of discharge of mine water from storage ponds, which are located in Chervonograd town and Gorodyshche village as well as sludge pit of CCP on the ecological status of waters of the Western Bug River and its tributaries.

REFERENCES

Grebenkin, S.S., Kostenko, V.K. & Matlak, E.S. 2009. *Conservation of the natural environment in the mining operations:* Monograph. Donetsk. VIK: 505.

Grebenkin, S.S., Kostenk,o V.K. & Matlak, E.S. 2008. *Physic and chemical fundamentals of demineralization of mine waters:* Monograph. Donetsk. VIK: 287.

Riznyk, T.O., Luta, N.G. & Sanina, I.V. 2006. *Assessment coal enterprises impact to changes of stream flow parameters in Donbas territory.* Bulletin of Kyiv National University named after Taras Shevchenko. Geology: 38-39: 73-76.

Gorova, A. & Pavlychenko, A. 2009. *Integral assessment of the socio-ecological state of mining regions of Ukraine.* Mining Journal, 5: 49-52.

Gorova, A., Pavlychenko, A. & Kulyna, S. 2012. *Ecological problems of post-industrial mining areas.* Geomechanical processes during underground mining. Leiden, The Netherlands : CRC Press/Balkema: 35-40.

Rudenko, S.S., Kostyshyn, S.S. & Morozova, T.V. 2003. *General ecology: a practical course.* Chernivtsi: Ruta, 1: 320.

Bilyavs'kyi, G.O. & Butchenko, L.I. 2006. *Elements of ecology: the theory and the practice.* School-book. Kyiv: Libra: 368.

Mining of Mineral Deposits – Pivnyak, Bondarenko, Kovalevs'ka & Illiashov (eds)
© 2013 Taylor & Francis Group, London, ISBN: 978-1-138-00108-4

Experimental researches of geomechanical characteristics of railway and point switches of underground transport

V. Govorukha

M.S. Polyakov Institute of Geotechnical Mechanics, Dnipropetrovs'k, Ukraine

ABSTRACT: The results of experimental geomechanical characteristics of railway and point switches of underground transport are received during static load with help of special devices and during dynamic load caused by rolling stock.

1 INTRODUCTION

Bearing capacity of railway and point switches defines realization of rolling stock movement and technical characteristics of locomotives, wagons and section trains. Change of technical characteristics of rolling stock, movement speed increase, weight increase of moving units leads to necessity of technical level improvement and structure if integral parts of railway including intermediate couplings, crossties, bars, point switches and access tracks.

Definite attention is paid to rolling stock characteristics improvement including locomotives, wagons and section trains (Melnikov 2009 & Berezhyns'kyi 2012). It contributed to trains' speed increase up to 25 km / h, load capacity increase, and mass of transport means in 2-3 times.

Bearing capacity of railway structure means should become such a technical achievement since discrepancy formation between rolling stock influence and railway construction strength has led to exploitation expenses increase, trauma occurrence rise and safety measures violation when exploiting mine transport (Berezhyns'kyi 2012 & Ivanov 1999) and others.

Given data define currency of the works directed to gaining geomechanical characteristics of railway and point switches of mine transport that are used as initial data for complex studies conduction.

2 BASIC PART OF STUDIES

According to basic tasks of the work, the author together with specialists of IGTM NAS of Ukraine conducted experimental researches on railway construction elements, point switches and access tracks narrow track of 900, 750 and 600 mm in underground conditions of the following mines: "Ukraine", "Russia", Donets'k coal basin; #3 "Velykomostivs'ka" of Lvovs'ko-Volyns'kyi coal basin; named after "Artem" of Kryvorizhs'kyi ore basin (Govorukha 1992).

Experimental researches were conducted under conditions of static and dynamic load on railway structure. Special loading device shown on Figure 1 was used for static influence research. Special calibrated rolling stock was used for dynamic influence determination including locomotive, wagons of various construction and stationary tensometric gauge installed on railway structure elements (Figure 2).

Loading structure was used to create variable vertical and transverse loads on researched elements of railway structure. The device contains two vertical (1; 2) and one horizontal (3) power jacks (4). In upper part vertical jacks hit the canopy of mine support (5). There are manometers installed in power jacks together with force-measuring rods (7) with tensoresistors.

To define rails displacement, crosses, points или fasteners (8) to cross-ties (9) the indicators of an hour type and deflectometer (10) are used and installed in horizontal and vertical directions of power influence.

Measurement devices location scheme is shown on Figure 2. At the rail neck there are neck sensors (2) and (4) are located for defining transversal and (3) for vertical loads transferred from undercarriage of rolling stock to rails. Vibration velocity gauge of IS-318 type (5) is installed on rail bottom to register vibrations and accelerations. In lower part of the rail bottom there are plate deflectometers (6) with tension sensors (7). Support of deflectometer plates was implemented by bolts (8) having independent nipping in rock massif (9).

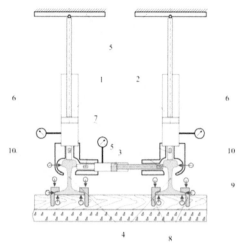

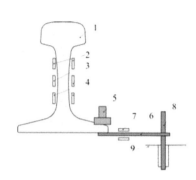

Figure 1. Device for rigidity measurement of railway elements, point operations и access tracks.

Figure 2. Scheme of measurement devices location.

Elastic deformations of railway, point operations and и access tracks are defined as rail dislocation function and another element from load with help of special device shown on Figure 1. Measurement are conducted in vertical and horizontal planes including sections above cross-ties and in the aisle between them, jointing areas and link middle area, pointer, point operation and cross parts of point switches and access tracks. Basic characteristics of elasticity in vertical and transversal directions is accepted rigidity and corresponding to them elasticity modulus (Govorukha 1992).

Graphs of static loads of railway elasticity in vertical plane are shown on Figure 3, and in transversal plane – on Figure 4.

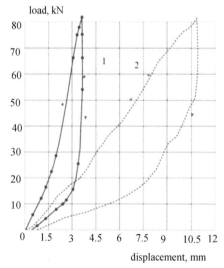

Figure 3. Dependence of vertical dislocations of rails on loads.

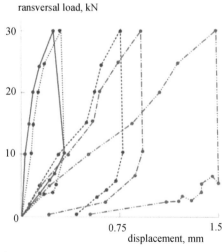

Figure 4. Dependence of transversal dislocations change on load.

Figure 5 shows dependence of dislocations on loads for railways with armored concrete (1) and wooden (2) cross-ties. Figure 5 shows character of vertical way rigidity change along rail length having wooden (Figure 5a) and reinforced concrete (Figure 5b) cross-ties.

Results of static processing of railway rigidity values and elasticity modulus of under-rail basis are shown in Table 1. From Figure 5 and Table 1 it is seen that rigidity and elastic modulus values have a significant difference in cross-sections above the support and aperture between them. Along the reinforced concrete cross-ties and weak under-the-cross-tie basis the difference makes 25-63% and at areas

with rigid under-the-cross-tie basis 48-87%. Along the wooden cross-ties, the difference between elasticity values on the cross-tie and in aperture makes up 12-20%.

Distance increase between cross-ties axes from 0.7 to 1.1 m leads to reduction of rigidity in aperture by 25-50% and above the supports (cross-ties) by 7-20%.

Relative rigidity on supports and in aperture makes up 4-35%.

(a)

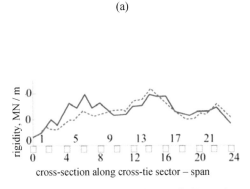

cross-section along cross-tie sector – span

(b)

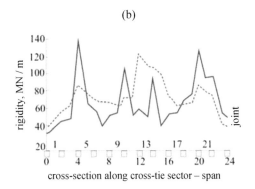

cross-section along cross-tie sector – span

Figure 5. Dependence of vertical rigidity of rail way change along the rail length.

Table 1. Results of static processing of railway rigidity values and elasticity modulus of under-rail basis.

Characteristics of a test area	Application place	Interval of active forces, kN	Rigidity of rail joints, MN / m		Elasticity modulus of underrail basis, MPa	
			Average values	Mean square deviation	Average value	Mean square deviation
Mine railway, rails of type P33, wooden cross-ties 1440 pieces / km, flooded basis	support	0-10	12.0	2.8	6.0	1.6
	passing	0-10	10.0	2.4	5.0	1.4
	support	10-35	25.7	2.7	15.7	2.3
	passing	10-35	24.0	4.8	14.0	3.6
Mine railway, rails of type P33, wooden cross-ties 1440 pieces / km, flooded basis	support	0-25	48.5	9.2	36.0	9.0
	passing	0-25	33.0	5.0	22.0	4.0
	support	25-45	58.0	8.6	46.0	8.8
	passing	25-45	46.5	6.8	34.0	6.4
Mine railway, rails of type P43, wooden cross-ties 1440 pieces / km, dry basis	support	10-15	81.0	17.5	71.0	22.0
	passing	10-15	50.5	6.6	38.0	7.2
	support	15-30	98.0	7.5	92.0	10.0
	passing	15-30	73.0	10.0	63.0	11.6

Dynamic characteristics of elasticity are defined with help of force-measuring gauges and deflecto-meters (see Figure 2) at interaction of railway and rolling stock.

Figure 6 and Table 2 show characteristic dependences between railway joints dislocations and value of static and dynamic load from rolling stock relating to railway areas with wooden (Figure 6a) and reinforced concrete (Figure 6b) cross-ties. Dependences of deformations on static load are shown by

solid line (1), and on dynamic – dotted (2).

At various intervals of forces influence the values between static and dynamic rigidity change in 1.5-2 times. Dynamic rigidity values is larger compared to static by 8-37% for wooden and by 10-26% for ferroconcrete cross-ties.

Rigidity of pointers elements for point and switch rails, basically, corresponds to railway rigidity, and rigidity in cross area is larger by 1.3-1.5 times.

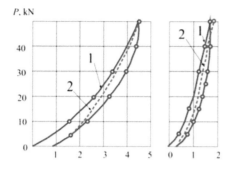

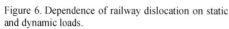

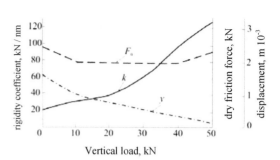

Figure 6. Dependence of railway dislocation on static and dynamic loads.

Figure 7. Dependence of railway static characteristics in transversal plane on vertical load.

Table 2. Characteristic dependences between railway joints dislocations and value of static and dynamic load from rolling stock.

Loading interval, kN	Rigidity, MN / m			
	static		dynamic	
	Wooden cross-ties	Ferroconcrete cross-ties	Wooden cross-ties	Ferroconcrete cross-ties
5-10	9.2	23.7	10.0	30.0
10-30	11.1	31.7	15.2	40.0
30-50	19.6	47.6	20.0	52.6

Dependence of railway static characteristics (elasticity k, dislocation y and dry friction forces F_0) in transversal plane are partially shown on Figure 7 and their values are shown in Table 3. These data show that vertical forces value significantly influences characteristics of railway in transversal plane. At that vertical forces increase leads to reduction of transversal dislocations value and transversal rigidity.

Inelastic forces of railway in vertical plane consist, basically, of constant friction force and variable components proportional to dislocation and movement speed.

Mentioned friction forces are defined by the following methods: hysteretic loop recorded during loading and unloading processes of railway joints; by lagging value of maximal deflection of railway joints relatively to center of moving load; by graphs of railway joints asymmetry during single load movement (wheel pair); by ratio of amplitudes and damping halfcylces of railways vibrations.

Table 3. Dependence of railway static characteristics in transversal plane.

Loading type	Elasticity coefficient, MN / m				Constant force of dry friction, kN	Coefficient of dry friction forces proportional to dislocation, kN / cm	Coefficient of viscous resistance forces, kN·s / m
	0-10	0-20	0-30	0-70			
Loading	8.33	10.00	10.42	14.58	1.50	22.50	11.0
Unloading	4.42	5.81	7.54	–	–	–	–
Loading	7.41	9.30	9.52	14.14	0	21.25	–
Unloading	4.65	5.81	7.46	–	–	–	–
Loading	7.14	8.69	9.04	11.99	5.38	6.25	6.25
Unloading	3.59	5.09	6.29	–	–	–	–

Diagrams and corresponding dependences of dislocations on forces at full cycle "loading-unloading" were gained with help of force gauge and clock type indicators, force-measuring gauges and deflectometers with independent systems of bolts installation in gallery bottom.

Parameters characterizing friction are gained during measurement of forces, dislocations and railway elements acceleration during rolling stock movement along the railway. Friction forces determination proportional to dislocation is conducted based on lagging value of railways maximal deflection relatively to moving loads center (Figure 8a). At this, rail deflection line under moving load influence is defined with deflectometers and forces influencing the railway – with help of neck rail tensoresistors.

Linear coefficient of inelastic resistance of railway elements in vertical direction n correspondence with Figure 8b is defined from the following:

$$\varphi_z = \frac{81\pi^4 EI_y}{128}\left(\frac{1}{x_1^4} - \frac{1}{x_2^4}\right),\qquad(1)$$

where x_1 and x_2 – distances from railway maximal deflection point to points of dislocation line transfer through zero point (without loading) correspondingly in front and behind the moving load; EI_y – rail rigidity at deflection in vertical plane.

Linear coefficient of inelastic resistance in transversal direction is defined analogically by expression (1).

(a)

(b)

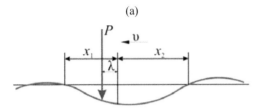
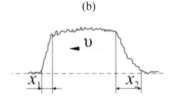

Figure 8. Scheme of rails elastic deflection line under moving load influence (a), copy of oscillogram of railway dislocation record.

To define viscous resistance and railway mass the forces transferred from wheel to rail, dependence of railway joints dislocation and acceleration in studied cross-sections of the railway on rolling stock influence at various movement speed (see Figure 2) (Ivanov 1999 & Govorukha 1992). At this, railway mass taking part in striking process of interaction is defined according to the following:

$$m'_p = \frac{P_{dyn}}{a},\qquad(2)$$

where P_{dyn} – dynamic force caused by wheel influence on the rail; a – acceleration of railway elements in researched direction.

Coefficient of the system viscous resistance "railway – undercarriage" is defines as:

$$\beta_{gen} = \frac{2m\delta}{T_\kappa},\qquad(3)$$

where δ – logarithmic decrement of the system vibrations dumping, defined as $\delta = ln\,A_i / A_{i+1}$; A_i and A_{i+1} – values of successive amplitudes.

The results of studies of friction forces and viscous resistance coefficients, and railway mass are given in Table 4.

Table 4. Results of studies of friction forces and viscous resistance coefficients.

Index	Characteristics of railway area			
	Rails of type P33, ferroconcrete cross-ties, 1440 pieces / km		Rails of type P33, ferroconcrete cross-ties, 1440 pieces / km	
	Connection joint	Middle of joint	Connection joint	Middle of joint
Railway resistance coefficient, kN·s:				
Average value	10.7	15.8	24.4	26.7
mean squared deviation	7.0	5.2	6.2	9.8
Railway mass, kg:				
Average value	18.5	30.7	27.4	45
mean squared deviation	2.7	3.2	5.7	12.2
Friction coefficient, MN / m:				
Train movement speed 1.5 m / s	–	1.6-1.7	–	3.1
Maximal value	–	4.7-5.6	–	9.6

3 CONCLUSIONS

1. Railway and point switches geometrical characteristics are established by experimental way under conditions of coal and ore mines. Values of elasticity, viscous resistance and dry friction of track structure are presented considering static and dynamic load influence from loading devices and rolling stock.

2. Values of track structure static elasticity in vertical and transversal plane have non-uniform character on supports (cross-ties, bars) and in gap between them.

3. Elasticity in vertical plane in pointer area of point switches is, basically, identical with railway indices and in crosses area – larger in 1.3-1.5 times.

4. During dynamic researches, the friction forces and viscous resistance coefficients were received together with railway structure mass.

REFERENCES

Melnikov, S.A. Budishevs'kyi, V.A. & Berezhyns'kyi, V.I. 2009. *Improvement of existing and creation of new mine locomotives.* Coal of Ukraine, 5: 2-15.

Berezhyns'kyi, V.I. & Babakov, S.V. 2012. *New developments for working safety increase on mine transport and hoisting.* Coal of Ukraine, 6: 17-20.

Ivanov, O.I. 1999. *Evaluation of miners' professional risk.* Coal of Ukraine, 11: 46-47.

Govorukha, V.V. 1992. *Physical-technical basics of railway transport elements creation of mine and open-pits:* monography. Kyiv: Sc. Thought: 200.

On the question of implementation prospects of selective mining for exploitation unconditional coal seams

D. Astafiev & Y. Shapovalov
National Mining University, Dnipropetrovs'k, Ukraine

ABSTRACT: The analysis of current situation in coal mining branch in all over the world and separately in Ukraine is realized. Application of selective mining for development low-thickness unconditional coal seams is substantiated. The main conclusions about advantages and disadvantages of separate mining of coal and rock are stated.

1 INTRODUCTION

Coal is the most widespread energy source in all over the world that far exceeds oil and gas reserves. As of 01.01.2013 volume of coal reserves is equal to 840 billion tons. At current consumption level availability in reserves is equal to 116 years than in two times more than availability in gas and in three times more than oil reserves. Coal provides 29.6% of world power demand and produce 42% of electrical energy (Arczukevych 2011).

2 MAIN PART

For the last few tens of year's extraction of coal increased on 80%. The highest growth fell on some

European Union countries, China, Republic of South Africa, but, for example, in India, Spain and Poland in recent years volumes of extraction decreased. In Table 1 structure of coal production and consumption in the world is given.

Table 1. Structure of coal production and consumption in world, million tons (according to the data of BP statistical review of world energy: http://www.bp.com/subsection.do?categoryId=9037151&contentId=7068607).

Region	Production	Consumption	Share,%
Middle East	1.0	8.8	0
South and Central America	53.8	23.8	1.4
Africa	144.9	95.3	3.9
Europe and Eurasia	430.9	486.8	11.5
North America	591.0	556.3	15.8
Asia – Pacific	2509.4	2384.7	67.3
Total in the world	3731.0	3555.7	100

Coal reserves are distributed almost evenly on three coal mining regions, such as Asia-Pacific, North America, Europe and Eurasia (Figure 1).

At present day for majority of world countries the question about using nuclear power is stay dispute. For example, in Germany 23 new coal power plants with a total capacity more than 24 megawatt are constructed. Works of eight from seventeen nuclear power plants at present moment are already stopped. From all types of the fuel using for electricity production the share of coal is more than 40%. At present time using coal as energy material is considered

like transitional stage from nuclear power plants to alternative energy sources. In 2012 the volume of produced electric energy reached the highest rate for the last quarter of century.

Management of Association of Germany Coal Mining Enterprises reports that efficiency of coal power plants reached nearly 50% and they are ready to ensure the country by electricity for all transition time. However, in view of insufficient development and high cost of technology (expenses exceed 100 billion €) further transition to renewable sources most likely fail to materialize. Therefore the choice of

Germany inclines to the cheapest energy resource – coal.

Coal industry is the most important raw material resources base for power industry and metallurgy of Ukraine. Further development of state economy and its energy safety depends upon reliability and effi-ciency of its functioning. As of 01.01.2013 Ukraine overproduced 22% of coal, brought to grass 85 million 745 thousands of "black gold". Excess in comparison with previous year is 3 million 754 thousand tons.

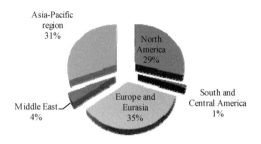

Figure 1. Actual coal reserves in the world (according to the data of BP statistical review of world energy: http://www.bp.com/subsection.do?categoryId=9037151 &contentId=7068607).

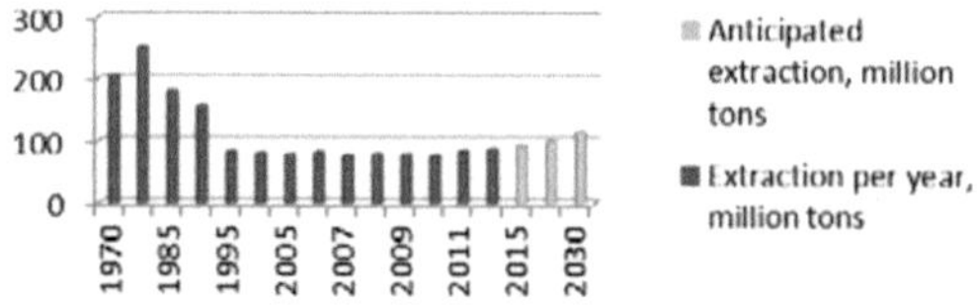

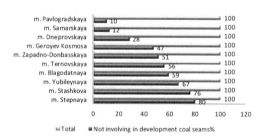

Figure 2. Level of coal extraction in Ukraine in the 1970-2030 timeframe.

According to the project of the up-to-date power strategy till 2030 plans call for increase electricity production by 45.5% to 282 billion kW. At the end of 2010 the program "Ukrainian coal" ended and due to this fact the project of future growth of coal branch till 2030 was created. According to this program the following main objectives were laid: volume of extraction has to be 92 million tons per year, including 63.5 million tons of thermal coal. Further from 2015 to 2020 plans call for increasing level of extraction to 100 million tons / year and already to 2030 to 115 million tons / year from which 75 will fall on high quality thermal coal (Ukraine miner) (Figure 2).

However the question arises as to whether how we will be able to increase coal mining if the most part of reserves are concentrated in unconditional low-thickness seams? For example, from 671 million tons of industrial reserves of "PJSC (Power Joint-Stock Company) DTEK Pavlogradugol" 341 million tons (51%) belong to inexpedient for development (Figure 3).

Percentage of low-thickness and not involving in development coal seams is shown on Figure 4.

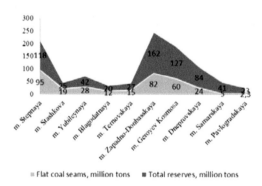

Figure 3. Total quantity of coal reserves in "PJSC DTEK Pavlogradugol" mines.

Figure 4. Percentage of low-thickness and not involving in development coal seams.

Certain attention should be devoting to the questions of coal seams conditions. Conditions are the set of requirements of industry to quality of coal and mine-geological parameters of field for a delinea-

tion and calculation of reserves in subsoil riches and right distribution on balance and non-commercial. In fact conditions are temporary category and it depends upon demands of state economy from coal, reserves conditions and perspectives of coal extraction. At underground coal field extraction the main parameters are:
– minimal extracting seam thickness;
– maximal industrial coal ash content;
– maximum depth of the seam.

Today's volumes of standards bedding within the limits of 0.45-0.7 m, the major part (80%) accrue to 0.55-0.6 m. Average dynamic thickness of developing levels is presented in Table 2. For many mines, despite essential distinctions in mine-technical and mine-geological conditions of developing seams standards on thickness stayed the same as well as 45 years ago (Petenko 2009).

Table 2. Comparative analysis of average dynamic thickness of standards of developing seams of some European countries and USA.

Country	Thickness, m
Ukraine	0.9
England	1.16
Germany	1.37
France	1.4
USA	1.65

As for standards of ash content, current standards completely reflect possibilities of coal industry. Inherent ash and operational ash contents very differ between each other, and this difference fluctuates in the range from one to ten percent. It can explain only that coal blinded with rock in extraction process because of necessity of coal cutting with stone and transportation on mine workings. 95% of Donbass coal has inherent ash within the limits of 8-15%. At bulk technology of coal extraction on some longwalls in Ukrainian mines the operational ash content can reach 55% (Petenko 2009).

Analysis shows that for the purpose of improving quality of products and decreasing constantly growing ash content it is effectually to apply selective mining.

Selective mining method was created by the end of the 1980s, however technology didn't gain a wide expansion, because priorities of coal mining enterprises are directed on increasing volumes of extraction instead the high quality of initial products. Technology of selective mining of thin and very thin flat coal seams is devoted for improvement quality of extracted product in coal cutting with stone longwalls (Buzylo 2012).

Essence of selective mining consist that first of all

shearer takes out coal and then on return movement takes out rock. Sections of mechanized support retruded after the first pass of shearer and conveyor after the second one. Rock separates from coal immediately in mine and backfill with the aid of pneumo-stowage complex is made. Given technology allows to extract valuable coking coals from seams with thickness 0.4-0.7 m with minimal coal cutting with stone (Denisov 2010).

On example of Western Donbass we can see how will increase mines lifetime with involving unconditional reserves (Figure 5).

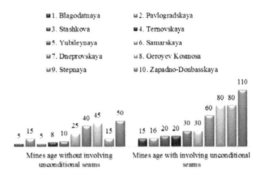

Figure 5. Comparative characteristic of mines age of Western Donbass with and without involving of unconditional seams.

In addition selective mining has a lot of advantages, such as: (Bondarenko 2013):
– decreasing of prime cost of coal extraction for 10-15% in comparison with bulk technology;
– coal extraction with ash content not higher than 15-18%;
– abandonment in goaf dozen millions of rock;
– decreasing emissions of harmful substances in atmosphere during burning of coal;
– increasing rates of preparation in longwall operations, ensuring considerable economy during carrying out and maintenance of mine workings by means of reuse;
– liquidation of essential deformations of surface from mining operations.

Using selective mining would not lead to essential increasing of extraction, however from the energy point of view it allows to extract better coal and leave rock in mine that will allow to save considerable resources on land allocation.

Inexpedient on a factor "geological thickness" seams can lead to change for the worse technical-economic indices of mines work because of increasing rate of amortization of general mine funds (Sokolov 2011).

3 CONCLUSIONS

1. At present the question about implementation of selective mining is very actual and demands further development. Analysis of question of coal role in world energy resources proved that "black gold" was and remains the most important energy material. There will come time when leading coal mining countries of the world will be needed to involve low-thickness seams in development and not remained them. Transition to alternative energy sources still remains very expensive for today.

2. Analysis of coal mining branch of Ukraine has shown necessity of involving low-thickness seams in development for the purpose of increasing level of valuable coking coal extraction.

3. Present working mines of Ukraine are provided with balance reserves (thickness > 0.8 m) at average on 35 years. But at current ratio of coal mining volumes from standard and unconditional seams (95 to 5%) in 15-20 years reserves of productive coal seams will be completely exhausted.

4. Proposed technology of selective mining means separated in time and space extraction of coal seam and coal cutting with stone including separate transportation of empty rocks and mineral deposit. Using of current mine equipment makes technology rather flexible, i.e. gives a chance in required torque cross from separate extraction to bulk and vice versa, without any additional expenses. This technology will allow to receive coal with inherent ash content, to use backfill in worked-out area, take into account ecological aspect and many others.

REFERENCES

Arczukevych, E. 2011. *The current state and prospects of Ukrainian coal industry development and problems of energy dependence.* Economic analysis, 9. Part 3: 35 – 39.

Bondarenko, V., Sulaev, V., Shapovalov, Y. & Astafiev, D. 2013. *Substantiation of selective technique of coal extraction in flat seams with thickness of 0.5-0.8.* Cracow: Materialy Konferencyine Szkola Eksploatacji Podziemnej: 10.

Buzylo, V. & Koshka, O. 2012. *Technology of selective mining of thin coal seams.* Dnipropetrovs'k: National Mining University: 23.

Denisov, S. & Mamaykin, O. 2010. *Specialties of development low thickness seams in conditions of Western Donbass.* Scientific messenger of NMU, 7-8: 18-21.

Petenko, I. & Belyavtsev Y. 2009. *To the question of exhaustible resources.* Donetsk: DonNTU. Collections of scientific papers: 1060-1063.

Sokolov, A. 2011. *Reserves life, extraction and consumption of carbonic minerals in World and Russia.* Electronic scientific magazine "Petroleum Engineering": 400-414.

Ukrainian Miner. Newspaper of mining union: http://shu.prupu.org/category/7/, http://shu.prupu.org/category/21/ BP Statistical review of world energy: *http://www.bp.com/subsection.do?categoryId=9037151 &contentId=7068607*

Study of mechanical half-mask pressure along obturation bar

S. Cheberyachko, O. Yavors'ka & T. Morozova
National Mining University, Dnipropetrovs'k, Ukraine

ABSTRACT: Mathematical model of force distribution along respirator RPA obturation bar is developed. It was determined that the largest forces are in place of chin while in place of nose bridge they are not too large. It leads to untighted half-mask fixing to the face and leak-in of unfiltered air increase.

1 FORMULATING THE PROBLEM

Mechanical pressure of half-mask on the head and face skin plays an important role and is among hygiene factors characterizing negative impact of self-contained breathing apparatus (SCBA) on people.

Because of the local pain and pressure sore arising on the face connected with local blood flow disturbance workers often refuse to use respirators. This is due to the fact that pressing forces are irregular distributed along respirator obturation bar. Besides, on the places of weak pressing during operational process especially due to filter blockage by dust aerosol there is increased unfiltered air blow-off along obturation bar. It reduces SCBA efficiency (Mironov 2002). It makes workers to keep respirator tight. Thus, the pain is increasing. Dependence of protective efficiency on local mechanical pressure on soft parts of the face shows that for leakage removal along obturation bar pressing force should be within the range of 4-10 H, mechanical pressure to the skin will be 2.5-5.2 kPa (Ennan & Schneider

1994). However, laboratory study showed that its distribution along obturation bar of respirator RPA is not regular: maximum pressure is fixed in the places of the nose bridge and chin.

It's very hard to evaluate this index due to absence of available information concerning character of distribution of local loads along obturator depending on separate elements of SCBA as well as recommendations concerning pressure value on soft parts of the face. Therefore, the task is to study dependence of distributing mechanical pressure along obturation bar on various constructive parameters of SCBA with the aim of pain soothing and minimizing leakage of polluted air along obturation bar.

Such problem was solved by authors for light respirators Snezhock with a set of assumptions (Ennan & Baidenko 2002). However, given results cannot be concerned reusable respirators (Figure 1), as unlike nonreusable SCBA it is impossible to ignore half-mask weight. Besides, filtering boxes have negative influence.

(a)   (b)

Figure 1. Types of respirators: (a) nonreusable (filtering layer performs function of a frame); (b) reusable (half-mask is made of elastic material).

2 THE MAIN PART

To solve given problem calculation model of the most popular reusable respirator RPA (Pulse) was designed. Taking into account construction symmetry concerning UAH surface (Figure 2a) system of external forces and reactions influencing respirator was considered.

Half-masks of this respirator are of three sizes with various geometrical parameters for more accurate fitting to man's face. Table 1 shows half-mask basic numerical characteristics of the second size as it is more popular.

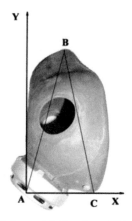
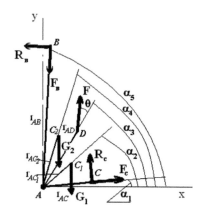

Figure 2. Calculation model of respirator RPA.

Table 1.

r_{AB}, mm	r_{AC}, mm	r_{AD}, mm	r_{AC1}, mm	r_{AC2}, mm	α_1^0	α_2^0	α_3^0	α_4^0	α_5^0
118	14	32	18	23	8	15	35	62	75

Taking into account experimental studies (Ennan & Schneider 1994), reactions along obturation bar can be reduced to two concentrated forces: in the places of nose bridge ($\overline{R_B}$) and chin ($\overline{R_C}$). Where:

$\overline{G_1}$ и $\overline{G_2}$ – weight force of mask and filtering box;

$\overline{F}$ – headblock tightening force;

$\overline{F_B}$ и $\overline{F_C}$ – friction force of sliding.

Value of friction forces is determined by corresponding reactions and friction coefficients between obturator and face skin in the places of nose bridge and chin.

System of forces influencing respirator can be expressed by static equation

$$\begin{cases} \sum M_A = 0; \\ \sum F_{ix} = 0; \\ \sum F_{iy} = 0. \end{cases}$$

$$F_B = f_B R_B \text{ and } F_C = f_C R_C \quad (1)$$

$$R_B r_{AB} + R_C r_{AC} + F r_{AD} \sin\theta - G_1 r_{AC_1} \cos\alpha_4 - G_2 r_{AC_2} \cos\alpha_2 = 0;$$
$$- F_C \cos\alpha_1 - R_C \sin\alpha_1 + F \cos(\theta + \alpha_3) + F_B \cos\alpha_5 - R_B \sin\alpha_5 = 0; \quad (2)$$
$$- F_C \sin\alpha_1 + R_C \cos\alpha_1 + F \sin(\theta + \alpha_3) + F_B \sin\alpha_5 + R_B \cos\alpha_5 - G_1 - G_2 = 0.$$

System of equations (2) describes limited balance state of respirator half-mask on man's face. Its solution will enable to determine minimum value of forces F, R_B, R_C providing relative immobility of SCBA. Analyzing solution of system (2) we come to conclusion that reaction in the place of chin (curve 2) is much more than in the place of nose bridge (curve 1) (Figure 3) that is confirmed by increase of polluted air blow-off in the place of nose bridge while using respirator (Mironov 2002).

Results of study are also confirmed by experimental data which were obtained according to methods given in (Kabiesz & Makowka 2009).

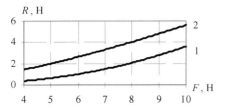

Figure 3. Reaction dependence in the places of nose bridge (curve 1) and chin (curve 2) on tightening force.

As we can see in (Figure 4) maximum pressure to face skin takes place on the chin, it is lower in the place of nose bridge. These factors confirm accuracy of given model of force distribution of respirator RPA.

Let's estimate influence of headblock tightening force on reaction along obturation bar. From the first equation of system (2) we find:

$$F = G_1 r_{AC_1} \cos\alpha_4 + G_2 r_{AC_2} \cos\alpha_2 - \frac{R_B r_{AB} R_C r_{AC}}{r_{AD} \cos\theta} . \quad (3)$$

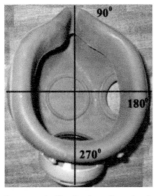 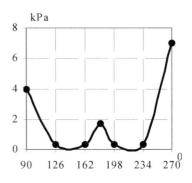

Figure 4. Axisymmetric distribution of mechanical pressure along obturation bar.

We will write the second and the third equations of system (2) taking into account (1) in the form of:

$$\begin{cases} R_C (f_C \cos\alpha_1 - \sin\alpha_1) + F \cos(\theta + \alpha_3) + R_B (f_B \cos\alpha_5 - \sin\alpha_5) = 0; \\ R_C (f_C \sin\alpha_1 + \cos\alpha_1) + F \sin(\theta + \alpha_3) + R_B (f_B \sin\alpha_5 - \cos\alpha_5) - G_1 - G_2 = 0. \end{cases} \quad (4)$$

Changing values of force F in these equations by the law (3) and introducing symbols:

$$a_{11} = -f_c \cos\alpha_1 + \sin\alpha_2 - \frac{r_{AC} \cos(\theta + \alpha_3)}{r_{AD} \sin\theta} ;$$

$$a_{12} = f \cos\alpha_5 - \sin\alpha_5 - \frac{r_{AB} \cos(\theta + \alpha_3)}{r_{AD} \sin\theta} ;$$

$$a_{21} = -f_C \sin\alpha_1 + \cos\alpha_1 - \frac{r_{AC} \sin(\theta + \alpha_3)}{r_{AD} \sin\theta} ;$$

$$a_{22} = f_B \sin\alpha_5 + \cos\alpha_5 - \frac{r_{AB} \sin(\theta + \alpha_3)}{r_{AD} \sin\theta} ;$$

$$b_1 = -\frac{\cos(\theta + \alpha_3)}{r_{AD} \sin\theta} \left[G_1 \cos\alpha_4 r_{AC_1} + G_2 \cos\alpha_2 r_{AC_2} - R_B r_{AB} - R_c r_{AC} \right];$$

$$b_2 = G_1 + G_2 - \frac{\cos(\theta + \alpha_3)}{r_{AD} \sin\theta} \left[G_1 \cos\alpha_4 r_{AC_1} + G_2 \cos\alpha_2 r_{AC_2} - R_B r_{AB} - R_c r_{AC} \right],$$

we come to equation system

$$\begin{cases} R_C a_{11} + R_B a_{12} = b_1, \\ R_C a_{21} + R_B a_{22} = b_2 \end{cases},$$

solution of which will enable to determine reactions in places of nose bridge R_B and chin R_C depending on point of force application F_5.

$$R_C = \frac{b_1}{a_{11}} - \frac{a_{12}(b_2 a_{11} - b_1 a_{21})}{a_{11}(a_{22}a_{11} - a_{12}a_{21})};$$

$$R_B = \frac{b_2 a_{11} - b_1 a_{21}}{a_{22}a_{11} - a_{12}a_{21}}. \qquad (5)$$

To increase protective properties and reduce pain along obturation bar it is required to equalize pressure in places of nose bridge σ_B and chin σ_C. Ignoring irregularity of pressure distribution along

contact area average value can be calculated by formula:

$$\sigma_B = \frac{R_B}{S_B}, \quad \sigma_C = \frac{R_C}{S_C}, \text{ kPa} \qquad (6)$$

where S_B, S_C – areas of contact zones in places of nose and chin, m².

We estimate influence of geometric sizes of respirator half-mask on the value of stresses σ_B and σ_C using dependences (5) and (6).

The range of changing geometric mask sizes is not too large and is limited by anthropometric face sizes. However, it is possible to find out such values of geometric half-mask parameters under which reactions in places of nose bridge and chin will be equal or close within this relatively small range. So, Figure 5 a, b shows graph of behaviour σ_B and σ_C depending on angles α_1, α_5 determining mask boundaries.

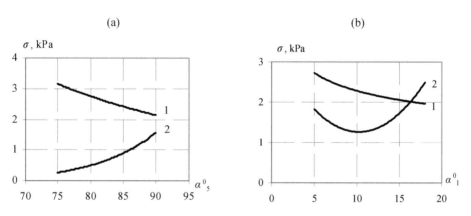

Figure 5. Dependence of stresses σ_B (curve 1) and σ_C (curve 2) on geometric mask parameters (angles α_1 and α_5).

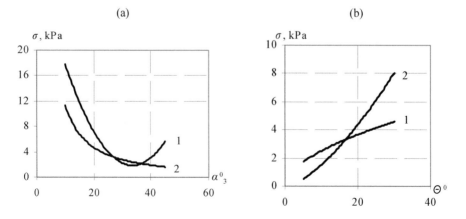

Figure 6. Dependence of stresses σ_B (curve 1) and σ_C (curve 2) on position of headblock.

320

Given dependences show that minimum difference (about 35%) between these stresses will be at the angle $\alpha_5 = 90°$. Difference between reactions in places of nose bridge and chin increases while decreasing this angle. In the place of angle α_1 change there is such its value ($\alpha_1 = 16°$) under which $\sigma_B = \sigma_C$ (Figure 5b). However, half-mask with such angle will be not comfortable.

Position of headblock greatly influences on redistribution of stresses (Figure 6 a, b).

Thus, according to given dependences built on the basis of expressions (5) and (6) under $23.5 < \alpha_3 < 33.5$ stresses σ_B and σ_C are close to each other and they are equal on the boundaries of this range. Difference between them rises sharply beyond the boundaries of this site. It explains unbalance of half-mask construction as point of tightening force application F is below center of mass. While changing headblock position in regard to fixation by varying angle θ (Figure 6b) it is possible to find out angle $\theta = 14°$ under which stresses are also equalized.

3 CONCLUSIONS

Therefore, experimental and analytical studies showed irregularity of pressure distribution along respirator obtiration bar. However, study results of constructive parameter influence on stress value determined such its values under which stresses in places of nose bridge and chin are balanced.

REFERENCES

Mironov, L.A. 2002. *Importance of leak-in of polluted air into under-mask area of filtering respirators and methods of its detection.* Work clothes, 3(15): 33-35.

Ennan, A.A., Schneider, V.G., Baidenko, N.I. & Mironov, A.A. 1994. *Interaction of protective efficiency and total respirator pressure.* Labor safety in industry, 11: 11-12.

Ennan, A.A., Baidenko, V.I., Klimova, L.V. & Belinskij, Ye.Ye. 2002 *Modeling of mathematical construction of light respirators "Snezhok".* Materials of the 1st International Scientific Conference "Environmental protection, health, safety in welding engineering". Odessa: Astroprint: 255-276.

Kabiesz, J & Makowka, J. 2009. *Empirical-analytical method for evaluation the pressure distribution in the coal seams.* Mining Science and Technology. Formerly Journal of China University of Mining & Technology. Vol. 19, #5:556-562.

Mining of Mineral Deposits – Pivnyak, Bondarenko, Kovalevs'ka & Illiashov (eds)
© 2013 Taylor & Francis Group, London, ISBN: 978-1-138-00108-4

Research of mine workings stability on volumetric models made from optically active materials

P. Ponomarenko
National Mining University, Dnipropetrovs'k, Ukraine

ABSTRACT: The results of experimental researches on volumetric models made from optically active materials. Horizontal mine working is driven along the rock. The models were investigated by "freezing" method. The isochor picture was deciphered at polariscope.

1 INTRODUCTION

To study the stress-strain state of the rock massif when a mine working stability increases it is necessary to install several measuring stations on a horizontal axis as the drift face advances. During installation of the stations and conduction of mine researches at them it is necessary to use significant material, labor and time expenses. Hence, the methods of physical modeling under the concrete conditions are the most effective and gained wide distribution during study of various problems in rock mechanics. It is explained by their large popularity and low material and time expenses for their implementation.

2 METHOD OF THE RESEARCH

One of the obvious methods is based on physical methods. In the work (Ponomarenko 2010) of the given paper's authors the essence of this method is described together with such leading figures as Frokht F., Trumbachov V.F., Filatov N.A. (Frokht 1950; Trumbachov 1962 & Filatov 1975), which caused a significant influence on the development of this method and conductions of fundamental researches during solving various tasks of rock mechanics.

In the cases when the stresses components situated along the coordinate parallel to the mine working direction are variable the volumetric task were being studied. Volumetric modeling is significantly more complicated and labor-intensive compared to the research on flat models. Testing the volumetric models has become possible with the occurrence of the epoxy resins allowing to produce large-scale blocks.

Volumetric models were made from optically active material – epoxy resin with maleic anhydrite. They looked like cylinders with diameter of 125 mm and height of 162 mm. As in the flat models the modeling scale was takes to be 1:100.

The model was placed in the pneumatic loading device for volumetric models loading in the form of cylinders with diameter and height up to 30 mm. As the pressure source for creating the load in the loading device the balloons with the compressed nitrogen and reducing valves were used.

The force scale was chosen based on the same criteria as for the flat models in the work (Ponomarenko 2010) providing the possibility to have a comfortable for measuring optical picture in the cross-section. Experimentally it was established that the load value on the modeling area that imitates the load of the above-situated strata makes up 4-6 MPa.

From several existing methods of the volumetric-stressed objects experimental research the "freezing" method was chosen. Its basic advantage compared to other methods (composite models method, dispersed light method) consists in the possibility of the information acquisition about active stresses practically for any point of the model.

Based on the above-stated the loading device with the model were installed in the oven and was "frozen" under hydrostatic load ($P = 5$ MPa) according to the methodology of the former "All-Union Scientific Research Institute of Surveying" (inter-disciplinary scientific center). Then the model with "frozen" deformations was sawn into one meridional and 10-15 transversal cuts of 2 mm. The cuts were photographed and isochor pictures in the cross-cuts were deciphered on the same devices with the same method used for the flat models. The research results on volumetric models were presented as coefficients of the stresses concentrations.

During analysis conduction of the installation technology and protection of mine workings it was marked that low rates of mine workings drivage on manganese-ore mines are considerably substantiated by frequent rock falls into the unsupported face area as a result of high stresses at the mine working contour and low bearing capacity of host rocks.

It is known that stress state of the rock massif within the face of development workings is volumetric. Under the conditions of weak rocks the formation of plastic deformation zones is possible. Analytic research of the existing stress field is hard even in the case when its components are elastic. Extremely limited data on the alike tasks solution exist in the literature. In connection with that the laboratory methods of mechanical processes modeling taking place in the massif weakened by a mine working turn out to be the simplest ones. In particular, photoelasticity methods are well developed. Their use in the considered case of complex spatial object allows to evaluate the influence of the face area and relief wells on the stress state on a high quality and quantity level, and also to provide recommendations as to their quantity and basic parameters.

In connection with that, the researches are conducted on tree volumetric models made from one group of epoxy resin. The galleries were modeled with a cylindrical hole with diameter of 30 mm and length – 57 mm. The advance well was modeled with cylindrical hole of 10 mm diameter and length of 23 mm.

A gallery with flat face was modeled in the first place. After deciphering the interference picture it was concluded that the maximal stresses concentration was observed in the corners of the gallery (abutment places of the gallery with the face).

Horizontal and radial stresses rapidly increase here. The stresses concentration in the gallery corners leads to rocks discontinuity and their fall. On the second stage the gallery with flat face and advance well was modeled. Decrease of horizontal stresses from 2 to 1.4 Pa was observed on the gallery contour but the stresses concentration in the gallery corners did not decrease.

In order to decrease the maximum of horizontal stresses in the gallery corner the spherical form of the face was proposed with remaining the advance well. As a result, the horizontal stresses coefficient K decreased at the gallery contour from 1.4 to 1.3 Pa, and its maximal value at the gallery contours lowered from 7 to 8.5 Pa.

Based on the above-stated, it can be concluded that the advance well in combination with the spherical form of the face provides more significant effect for decreasing the rocks horizontal stresses at the gallery contour.

Reduction of stresses increases the stability of unsupported area of a gallery as in this case the contour rocks are less subjected to caving. It facilitates an erection of a constant mine support, increases labor safety of the drift miners due to the reduction of rocks caving danger from the face, walls and roof of the gallery.

The first two models and one more additional made from the same resin group ЭД-16 are used by the author during research of the advance relief of rock massif.

In the patent literature the technical solutions on rock massif relief with inclined wells are described. The original attempt to research such an unloading method on flat models was not successful.

In connection with that the researches of the volumetric model were conducted. From the isochor picture around the mine working on the medial cut it is obvious that the wells do not give a significant reduction of stresses on the gallery contour.

Hence, it is needed to place the samples as close to each other as possible in order to squeeze the crosshole pillar and form the cavity similar to a relief slit under natural conditions. Besides, in weak rocks the wells can play a role of compensation cavities.

Also the stress-strain state research was conducted around the galleries with a line of horizontal, vertical relief wells, slits and cavity-slit in the gallery walls. The experiments were conducted on flat and volumetric models.

It is established that the series of horizontal and vertical wells, and slits under conditions of plastic host rocks will not provide a significant relief effect of near-the-contour part of the massif. The largest dislocations of stresses concentrations into the massif take place when the slits are located in horizontal plane.

Maximal relief effect is achieved during construction of so-called cavity-slits in the gallery walls. The inelastic deformations zone parameters changes by the following way: its linear dimensions do not exceed 1\4 of the gallery span width. Insignificant plastic zones occur in the mine roof (1\8 of the gallery span width). In general, the experiment researches testify about a considerable relief effect of near-the-contour part of the rock massif during constructing horizontal slits and additional cavities in the gallery walls.

3 CONCLUSIONS

Thus, the research method of mine workings stability made from optically active materials is quite effective when studying the stress-strain state of a single mine working.

REFERENCES

Ponomarenko, P.I. 2010. *The research of stresses around the mine working relative to a coal seam in the models made of optical material.* New Techniques and Technologies in Mining 2010.Dnipropetrovs'k: National Mining University, Ukraine: 67-169.

Frokht, MM. 1950. *Practice of practical modeling.* Novosybirs'k: Science Photoelasticity.

Trumbachov, V.F. 1962. *Study of stress state of rocks and mine workings with help of optical method*: the author's abstract. Moscow.

Filatov, N.A. 1975. *Development of modeling methods in connection with complex researches of rock pressure in underground workings.* The author's abstract. Leningrad.

Mining of Mineral Deposits – Pivnyak, Bondarenko, Kovalevs'ka & Illiashov (eds)
© 2013 Taylor & Francis Group, London, ISBN: 978-1-138-00108-4

Researches of structural-mechanical properties of coal tailings as disperse systems

O. Gayday
National Mining University, Dnipropetrovs'k, Ukraine

ABSTRACT: Preforming fuel are considered at adhesion-chemical technology preforming coal tailings and breazes in clause ways of optimization of physical-mechanical properties Researches of structural and mechanical properties of coal tailings as disperse systems by means of the methods, defining deformation properties are given: durability – limit tension on shift; elasticity module; relaxation characteristics; after-effects, etc.

1 INTRODUCTION

Fields which are presented by storages coal tailings occupy the huge space that leads to alienation of agricultural grounds and notable deterioration of an ecological situation of territories. But quantity of useful combustible components, in such storages of tailings, about 20-75% which can be treated into fuel. In this case the actual industrial and social problem of consumption in additional firm fuel and decrease in an environmental pressure of regions using underground mining of coal fields is solved.

As effective processing of energy resources and reduction to technical requirements it is offered to technology of a preforming adhesive and chemical (Gayday 2006).

2 FORMULATING THE PROBLEM

Justification of physical-mechanical parameters of preforming composite fuel requires research of structural and mechanical properties of coal tailings as disperse systems. Researches were conducted by means of methods of measurement of structural and mechanical properties of disperse systems. Coal tailings (fraction of 0-1 mm) as the disperse rocks including a disperse phase and the disperse environment, represent the system based on physical and chemical interaction of its components.

Existence of structure gives the system peculiar mechanical properties. These properties – elasticity, durability, plasticity, viscosity – depend on the chemical nature of the substances forming this system; decide by molecular forces of adhesion between structure elements, their interaction on the dispersive environment and extent of development of structure in all volume of system. Therefore the

improvement of structural and mechanical properties necessary for process of their preforming demands further research (Pahalok 1952).

For an assessment of the characteristic of mechanical properties of the structured disperse systems, the most rational methods are determination of their deformation properties: durability – limit tension on shift; elasticity module; relaxation characteristics; after-effects, etc. (Remesnikov 1955).

3 MAIN PART

In many cases for an assessment of the structured disperse system it is possible to be limited to determination of the simplest and almost important size – the mechanical durability of structure that is limit tension of shift P at small rate of deformation.

Methods of measurement of structural and mechanical properties of disperse systems (Remesnikov 1955; Lurye, Boytsova & Ravich 1957; Lurye 1956; Volarovich 1954; Pertsov 1999; Almazov & Pavlotskiy 1971) can be divided into three groups:

1) Methods of extraction of working part of the device from studied system. Here the method of extraction of the cylinder treats with sharp cutting according to Volarovich-Tolstoy (previously such cylinder – the screw – has to be completely filled or screwed in the studied system);

2) Methods of penetration of a tip (indenter) of the correct geometrical form and giving the prints similar each other (a cone, a pyramid), in almost boundless volume of studied system. These methods allow to investigate in detail the process of a plastic current of the system at a set tension and rates of deformation and to receive curve currents; this is rheological characteristics of the structured plastic

and viscous systems. Determined by the greatest immersion of a tip (a cone, a pyramid) the limit tension of shift characterizes the plastic durability of system corresponding to the top limit of fluidity;

3) Shift methods in studied system at preservation of constancy of a surface of contact of an indenter with system, this is at preservation of constancy of tension of shift given to system. The method of tangential shift of the plate placed in system on Veylera-Rehbinder belongs to methods of this group, and also a method of a twisting of the cylinder shipped in system (Shvedov's method). Valuable feature of these methods is an opportunity to define absolute values of all elastic-plastic characteristics of the structured systems.

3.1 Method of immersion of a cone

The method of immersion of a cone is allocated with the simplicity, and also strict validity of calculation, and gives the chance to make measurements at small deformations of shift, i.e. small gradients of speed with transition in a limit to an assessment of static limit stress of shift or limit pressure which for the majority of plastic and viscous systems can characterize durability of their structure. This method is described and presented in works (Remesnikov 1955; Lurye, Boytsova & Ravich 1957; Almazov & Pavlotskiy 1971).

This method consists in definition of dynamics of immersion of a cone in studied system under the influence of constant loading of F, as gives the conditional rheological characteristic – a current curve which expresses dependence of speed of immersion dh/dt on tension of the shift P which is continuously decreasing in process of immersion owing to increase in the area of contact of a cone with system. Respectively the speed of immersion of $V = dh/dt$ while it doesn't become almost equal to zero at the greatest immersion of h_m decreases also. The limit tension of shift of P_m, that is greatest of all static tension, possible in this system, is equal to the smallest value of the operating tension of P_l, corresponds to an equilibration of external force F plastic durability of structure.

The size of P_m is calculated on limit immersion of a cone to h_m caused by this loading of F, assuming that at cone immersion in system the current of a layer of system along a lateral surface of a cone takes place. This condition is carried out in rather plastic systems therefore tension of shift P causing this current, is defined by a projection of force of

F operating on a cone to l forming a cone carried to unit of area of contact of a cone with system S:

$$P = \frac{F}{S} = \frac{F \cdot cos\dfrac{\alpha}{2}}{\pi r l} \; ; \tag{1}$$

$$r = h \cdot tg\frac{\alpha}{2} \; ; \tag{2}$$

$$l = \frac{h}{cos\dfrac{\alpha}{2}} \; ; \tag{3}$$

$$P = K_\alpha \cdot \frac{F}{h^2} \; , \tag{4}$$

where $K_\alpha = \dfrac{1}{\pi}cos^2\dfrac{\alpha}{2} \cdot ctg\dfrac{\alpha}{2}$ – the constant of a cone depending only on a corner α at its top (in axial section), and, therefore

$$P_m = K_\alpha \frac{F}{h_m^2} \; . \tag{5}$$

In some disperse systems with stronger and fragile structure the current isn't possible, and at introduction of a cone plastic deformation a compression takes place. In such systems the limit pressure of P_l which pays off division of force operating on a cone of F, into a horizontal projection of S_m is defined by a method of immersion of a cone:

$$P_l = \frac{F}{S_m} = \frac{F}{S_m \cdot sin\dfrac{\alpha}{2}} = \frac{F \cdot cos\dfrac{\alpha}{2}}{\pi h_m^2 \cdot tg\dfrac{\alpha}{2} \cdot sin\dfrac{\alpha}{2}} \; ; \tag{6}$$

$$P_l = K_\alpha' \frac{F}{h_m^2} \; , \tag{7}$$

where $K_\alpha' = \dfrac{1}{\pi}ctg^2\dfrac{\alpha}{2}$ – cone constant; h_m – the greatest immersion of a cone.

Application of this or that formula is defined by invariance of results, this is independence of the calculated limit tension of shift of P_m or limit pressure of P_l from a cone corner α or F loadings.

In Figure 1 the general view of the device penetrometer is presented to LP on which measurements of dynamics of durability of structure of system "coal-binding" were made.

Penetrometer – the device for measurement of a consistence of semi-fluid materials by determination of depth of penetration of a test body of the standard sizes and weight on the tested environment. Measures penetration number – an indicator characterizing rheological properties of substances which is equal to depth of immersion of a working body penetrometer in terms of the tenth shares of millimeter. For example, if the working body penetrometer plunged on 20 mm, the number of a penetration will be equal 200.

(a)

(b)

Figure 1. Device penetrometer LP: (a) laboratory; (b) industrial.

Usually penetrometer is applied in the form of freely sliding plunger with the working body fixed on it in the form of a needle or a cone. Before measurement the edge of a working body is brought closely to a surface of the studied environment, and then the plunger is released and starts plunging on environment under own weight. Penetration depth in a definite time (penetration number) is fixed, at a certain temperature and in advance chosen mass of assembly a plunger a working body.

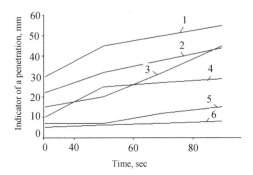

Figure 2. Dependence of an indicator of a penetration in time for systems of coal tailings with the various binding: 1 – cement; 2 – plaster; 3 – physical solute; 4 – liquid glass; 5 – lignin; 6 – not extinguished lime.

Depending on binding on the example of coal tailing of brand a dependence of an indicator of a penetration in time for disperse system (Figure 2) is presented.

3.2 Method of tangential shift of a plate

One of the most sensitive methods of determination of durability of structure – the limit tension of shift of P_t in the structured disperse systems – is the method of tangential shift of a plate.

The principle of this method offered by S.Y. Veyler and P.A. Rehbinder, consists in definition of the effort necessary for shift of a plate, shipped in studied system.

Rectangular or round very thin metal plate (50-100 micrometers) is shipped by thickness in investigate substance and suspended on a rigid thread to a quartz or glass spring. Rectangular a ditch with the examinee substance it is strengthened on a special little table. The electric motor with a reducer smoothly lowers or lifts a little table with a constant speed. Shift of a plate and stretching of a spring is counted by means of the horizontal microscope supplied with a preforming micrometer.

Shift tension in this method is calculated on stretching of a spring and the corresponding this stretching of a spring to effort of F and a lateral surface of a plate of S:

$$P = \frac{F}{2S}. \qquad (8)$$

The limit tension of shift characterizing durability of structure of system of P_m, corresponds to the greatest effort of F_m (in the absence of system sliding along a plate surface), respectively $P_m = \dfrac{F_m}{2S}$.

This method allows defining not only the limit tension of shift, but also the elasticity module, effective viscosity, to investigate relaxation process, and also to remove full deformation curves at different speeds of deformation. The method possesses big sensitivity, and is applicable in a wide interval of durability of structure, from weak structure systems to strongly-shaped systems with high-strength structure (Pertsov 1999).

3.3 Method of a twisting of the cylinder (Volarovich 1954)

The principle of a method is based on definition of elastic and plastic characteristics of the structured systems on a twisting of the cylinder suspended on an elastic thread and shipped in this system. The twisting device has an exact head in which the elastic thread is fixed. On a thread the corrugated cylinder which is completely shipped in studied system is suspended. At turn of a twisting head to a certain corner α torque it is transferred through a thread to the cylinder and causes shift deformations in the concentric layers of system surrounding the cylinder. The cylinder thus to twirl on some corner β to balance between elasticity of the twirled thread and resistance of the deformed system. The difference $\alpha - \beta$ gives a corner of twisting of a thread ω, F corresponding to a certain effort, established on preliminary graduation of a thread.

Let's consider balance between force of elasticity of the twirled thread and elastic reaction of the deformed layers of system. Let's divide system into concentric layers height to h equal to height of the cylinder, the radius of r and dr thickness. The layer adjacent to walls of a vessel, remains in rest; the layer which has been directly connected with the cylinder, twirl together with it on a corner β; all intermediate layers experience shift deformations relatively each other.

The moments of all forces applied to a surface are equal, from where there is a module of elasticity of system

$$E = \frac{K}{4\pi h}\left(\frac{1}{r^2} - \frac{1}{r_l^2}\right) \cdot \frac{\beta}{\omega}, \tag{9}$$

where $r_l = r + dr$.

At such definition of the module of elasticity of system if diameter of the internal cylinder is insignificant in comparison with its height, influence of a

bottom of the cylinder can be neglected or it can be considered by special experiences.

3.4 The experiment description in laboratory (when using SNS-2)

In a glass the examinee solute is filled in. At glass rotation solute entrains the cylinder which was in it measuring and all suspended system until the moment of a twisting of a thread doesn't become equal to a torque developed by a static stress of shift of solute on the cylinder. Static limit stress is determined by the maximum corner of a twisting of a thread.

4 CONCLUSIONS

Thus, on the basis of researches of physic mechanical properties of disperse coal tailings, it is possible to draw conclusions:

– to number of the major factors defining structural and rheological properties of disperse systems of coal tailings: force of adhesion in contacts between particles; coordination number (this is number of the contacts falling on one particle), depending on increase in concentration of a disperse phase in the dispersive environment, its dispersion and distributions of particles by the size.

– coal tailings as disperse rocks, represent the system of a firm and liquid phase based on physical and chemical interaction of its components. Therefore improvement of physical-mechanical properties, disperse coal tailings has to be based on physical and chemical influence on them. With dependence of $P = 17.853 Ln(t) + 30.816$ the disperse system coal tailings possesses the best physical-mechanical characteristics at physical-mechanical influence by cement.

REFERENCES

Gayday, O.A.2006. *Researches of strength properties of briquettes from coal tailings and breazes, received in the way of a cold performing*. Col. scien. wor. NMU, 26. V.1: 208.

Pakhalok, I.F. 1952. *The capital equipment of brown coal briquette factories*. Moscow: Ugletekhizdat: 236.

Remesnikov, I.D. 1955. *Questions of the theory of briquetting of brown coals*. Moscow: Ugletekhizdat: 320.

Lurye, L.A., Boytsova, G.F. & Ravich, B.M. 1957. *Researches on briquetting of brown coals*. Moscow: Ugletekhizdat: 264.

Lurye, L.A. 1956. *The new directions in development of briquetting of coal*. Moscow: Ugletekhizdat: 173.

Volarovich, M.P. 1954. *Research of rheological properties*

of disperse systems. Colloidal magazine, 3. V. XVI: 227-240.

Pertsov, A.V. 1999. *Methodical development to a work-shop on colloidal chemistry.* 6th edition. Moscow.

Almazov, A.B. & Pavlotskiy, I.P. 1971. *Probabilistic methods in the theory of polymers.* Moscow: Science: 151.

Mining of Mineral Deposits – Pivnyak, Bondarenko, Kovalevs'ka & Illiashov (eds)
© 2013 Taylor & Francis Group, London, ISBN: 978-1-138-00108-4

Theoretical investigation of dry frictional separation of materials on rotating cylinder

Yu. Mostyka, V. Shutov, L. Grebenyuk & I. Ahmetshina
National Mining University, Dnipropetrovs'k, Ukraine

ABSTRACT: The influence of the definite constructive and technological parameters on the particle tearing off point from the surface of the revolving cylinder (roller or barrel) has been shown. The parameters are: cylinder diameter, feeding point position, angular velocity of cylinder rotation, friction coefficient of mineral particle along the cylinder surface, motion speed of the product particles in the initial point of their contact with the cylinder. The received dependences can be used for choosing the efficient combination of parameters for dry frictional separation on rotation cylinder.

1 INTRODUCTION

The purpose of this article is the theoretical evaluation of a possibility to separate dry mixes of granular materials using the difference between friction coefficients of the mixture components along the working face of separating apparatus. Similar tasks are considered in the range of works, both theoretical and experimental, for example (Derkach 1966; Zverevich 1971; Kravetc 1986; Blagov 1984; Svoboda 1987; Tyurja 2001; Pilov 2001 & Tyuria 2003). Thereby, friction coefficient is considered either as a parameter, which influences on the process of separation along with other dividing signs of separation process, or as a general dividing sign. Dividing of friable coal and rock mixture moving down the inclined surface under the impact of gravity and friction force and counteracting the gravity has been investigated in detail in the works of P.I. Pilov and Y.I. Tyurja (Tyurja 2001; Pilov 2001; Tyuria 2003); the control of separating process is performed by adjusting inclination angles of the working surfaces. Two schemes of dry coal and rock separation by friction are investigated in the research (Pilov & Tyurja 2001): the first one is "acceleration – flight" scheme, which assumes that the dividing process can be regulated by changing the inclination angle of the accelerated zone and the angle of the descent from the working face (in the separation point); the second one is "acceleration – deceleration" scheme, which assumes varying of the inclination angles of the accelerated and decelerated zones of the working face. It appears to be reasonable (although it needs to be proved experimentally) to extend the possibilities of regulating the separation process by replacing the fixed working face

with the rotating cylinder. Thereby, the following supplementary parameters for the regulation process can be set: cylinder rotation frequency, cylinder diameter, the inclination angle of the surface tangent to the cylinder in the feeding point of the initial product.

The particles of two components of the material (if these components have different coefficients of friction along the work surface f_1 and f_2) will be tearing off the rotating cylinder with two different values φ_1, φ_2 the polar angle φ (Figure 1) and different speeds V_1 and V_2. The particles of two different components, which have friction coefficients f_1 and f_2, remove by trajectories $y_1(x)$ and $y_2(x)$, correspondingly after hey have been torn off the cylinder. Obviously, that it's necessary to provide sufficiently large divergence of curves $y_1(x)$ and $y_2(x)$ for separating components from mixture. These curves are defined by: angle φ in tearing off point, the particles speed in the tearing off point, the particles density (when the density of the particles is increasing, the influence of aerodynamic resistance on the particles moving is decreasing). That's why analytical investigation of curves divergence $y_1(x)$ and $y_2(x)$ by the range of the parameters (R_0, φ_0, $V_{\theta,0}$, n, $\rho_p\,d_p$, f) is very difficult issue. As a result, it is advisably to solve this issue by two stages: first – investigation, how these parameters influence on polar angle φ tearing off the particles from cylinder. Second stage is investigation, how these parameters influence on the trajectories $y_1(x)$ and $y_2(x)$. The general purpose of this article is investigation of the first stage of present issue.

As for second stage, this study presents approximate formula for particle trajectory (without taking aerodynamic resistance into consideration). More accurate and extensive studies of curves divergence $y_1(x)$ and $y_2(x)$ for mentioned parameters would allow to receive quantitative data for process of separation. However, this investigation remains outside the context of this article because of its big volume, and will be published later.

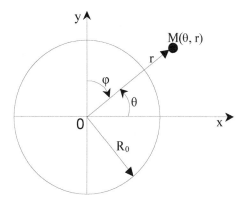

Figure 1. The cylinder cross-section and the coordinate systems, which are connected with its.

When calculating the parameters of the division process on dry rollers or cylinder magnetic separator it is usually assumed that the speed of non-magnetic particles along the working face of a roller or a cylinder is equal to the circular speed of this face. However some works, e.g. (6), show that there is frictional sliding of non-magnetic particles along the working face of a roller or a cylinder, that's why the position of the tearing off point of the particle from the roller or cylinder surface will be most probably different from that position calculated under the absence of frictional sliding.

Dependence of the particle speed from the polar angle and friction coefficient were investigated in the article (6). However the given calculations suppose that particle speed is equal to zero in the feeding point of a roller or a cylinder (the initial point). Additional influencing parameters have been considered in this work, and the position of particle tearing off point depending on the definite variables has been calculated. These variables are: angular velocity rotation of a roller or a cylinder, diameter of a roller or a cylinder, friction coefficient of a particle along the surface of a roller or a cylinder, position of feeding point (the initial motion point), speed of the particle in the initial point. The results of this research can be used firstly to define the efficient po-

sition for the divisor of magnetic and non-magnetic fractions coming off from the surface of the roller or cylinder. Secondly (and this is the subject of this work), the results of the calculations are intended for the theoretical substantiation of the possibility to divide granular materials on the separator of cylinder or roller kind using the difference between friction coefficients of the original material grains on the working face.

In the sequel the rotating cylinder will be considered to be the separation device meaning a roller or a cylinder.

2   THEORETICAL MODEL

Let us consider the cylinder cross-section, i.e. its section with plane normal to the rotation axis, and let us introduce polar coordinate system θ, r its centre is consistent with the cross-section center (Figure 1).

Let us write down the equation of particle motion along the cylinder surface in projection to the coordinate direction θ:

$$\frac{dV_\theta}{dt} = \frac{1}{m_p}\left(F_{\Sigma,\theta}\right), \tag{1}$$

where t – time; V_θ – tangential component of the particle speed (while moving along the roller surface V_θ coincides with the modulus of the particle speed vector); m_p – particle mass; $F_{\Sigma,\theta}$ – the sum of tangential components of forces influencing the particle;

$$F_{\Sigma,\theta} = F_{g,\theta} \pm F_f, \tag{2}$$

where $F_{g,\theta}$ – the tangential component of gravity; F_f – friction force of the particle along the roller surface; "+" or "–" sign depends on the correlation between speeds of the particle and the roller surface in the formula (2).

After substituting expressions $F_{g,\theta}$, F_f in (2) and then the expression $F_{\Sigma,\theta}$ in (1), and integrating by time there will be the following:

$$V_\theta(t_1) = V_{\theta,0} + \int_0^{t_1} a(t)dt, \tag{3}$$

$$a(t) = g\sin\varphi(t) + \left(signF_f\right)f\left[g\cos\varphi(t) - \frac{V_\theta^2(t)}{R_0}\right], \tag{4}$$

where $V_{\theta,0}$ – the value $V_{\theta,0}$ in the initial point;

334

g – free fall acceleration; $signF_f$ – algebraic sign "+" while $V_\theta < V_w$ and "–" while $V_\theta > V_w$; V_w – the surface circular speed of cylinder; R_w –

radius of cylinder; $V_w = \omega R_w$; $\varphi = \frac{\pi}{2} - \theta$;

$R_0 = R_w + \dfrac{d_p}{2}$ – distance from the roller axis to the center of mass of the particle; ω – the angular velocity of the rotation cylinder; f – sliding friction coefficient of the particle along the roller surface.

In the sequel we decide approximately $R_0 = R_w$.

If the initial particle speed (i.e. the speed of feeding the initial product to the cylinder) is less than the circular speed of the cylinder surface, that at the definite time moment $t_{f,0,1}$, the value $V_{\theta,0}$ will be

approach to the value V_w: $|\overline{V}_\vartheta - \overline{V}_w| < \varepsilon$, where ε is

a small value, e.g. $\varepsilon = 10^{-4}$ m / sec.

This is explained by the friction force impact to particle. Then starting from $t = t_{f,0,1}$ time moment the static friction coefficient f_0 is used instead of the sliding friction coefficient f ($f_0 > f$) for calculating the frictional force. Meantime, let us suppose $V_\theta = V_w$ instead of the formulas (3) and (4). This calculation scheme is used until the time moment $t = t_{f,0,2}$ $t = t_{f,0,2}$, starting from that moment the condition

$$F_{g,\theta} > F_{f,0} \tag{5}$$

is completed. After substitution instead $F_{g,\theta}$ and $F_{f,0}$, their expressions:

$$F_{g,0} = m_p g \sin\varphi, \tag{6}$$

$$F_{f,\theta} = m_p f_0 \left[g\cos\varphi - \frac{V_\theta^2(\varphi)}{R_0} \right] \text{ under assuming}$$

$$V_\theta(\varphi) = V_w, \tag{7}$$

we will have received:

$$\sin\varphi - f_0\left(\cos\varphi - \frac{V_w^2}{gR_0} \right) > 0;$$

$$\varphi_0 < \varphi < \pi/2; \ \varphi = \varphi(t). \tag{8}$$

The values $t_{f,0,1}$, $t_{f,0,2}$, i.e. the boundaries

$t_{f,0,1} < t < t_{f,0,2}$ of the time interval where $V_\theta = V_w$, can be defined while integrating by time (although obtaining these values explicitly is not required for calculating the angle φ_s).

The formula similar to (3) and (4), is used again for $t < t_{f,0,2}$, the lower limit of integration is equal to $t_{f,0,2}$, and $V_{\theta,0}$ is equal to V_w:

$$V_\theta(t_1) = V_w + \int_{t f0,2}^{t_1} \times$$

$$\times \left\{ g\sin\varphi(t) - f\left[g\cos\varphi(t) - \frac{V_\theta^2(t)}{R_0} \right] \right\} dt. \tag{9}$$

Let us notice that the speed of the particle in the moment t_1 is obtained as a result of integration according to the formulas (3) and (9), whereas the variable value $V_\theta(t)$ (under the integral sign in the right part of formulas (4) and (9)) runs from $V_{\theta,0}$ to $V_\theta(t_1)$ while integrating.

Integration by time can be substituted with integration by angle φ, excluding time t from (3) with the help of the following formula:

$$dt = \frac{R_0 d\varphi}{V_\theta}. \tag{10}$$

In this case speed V_θ is defined as a function from the angle φ

$$V_\theta(\varphi_1) = V_{\theta,0} + R_0 \int_0^{\varphi_1} a(\varphi) \frac{d\varphi}{V_\theta(\varphi)}. \tag{11}$$

$$a(\varphi) = g\sin\varphi + signF_f f\left[g\cos\varphi - \frac{V_\theta^2(\varphi)}{R_0} \right]. \tag{12}$$

For $V_\theta(\varphi) = 0$ formula (11) is unacceptable, then formula (3) is used.

Integrating by formulas (3), (4), or (11), (12) performed by a numerical procedure. The condition of the inseparable (without tearing) motion of particles along the cylinder surface is checked in the process of integration:

$$V_\theta^2(t_1) \le gR_0 \cos\varphi(t_1). \tag{13}$$

If this condition is not implemented, the particle tears off from the cylinder surface. If the speed is defined by (3) and (4), then the moment of time $t_s = t_1$ when the particle tears off is used as the upper integration limit in the formula to calculate the tearing off angle φ_s :

$$\varphi_s = \varphi_0 + \int_0^{t_s} \frac{V_\theta(t)}{R_0} dt ,\qquad (14)$$

where φ_s – value of polar angle corresponding to the feeding point.

If the speed is calculated according to the formulas (11) and (12), the tearing off angle φ_s is defined as the value of the upper integration limit φ_1 in (11), where particle tears off from the cylinder surface, i.e. the following condition is implemented:

$$V_\theta^2(\varphi_1) > gR_0 \cos\varphi_1 .\qquad (15)$$

After calculation of a value φ_s and an particle velocity $V_{\theta,s}$ for $\varphi = \varphi_s$, the particle trajectory after their tearing off from the cylinder can be calculated. If the particle doesn't experience electrostatic and electrodynamics interacting between the particle and the working face and the aerodynamic resistance is neglected, the formula for the particle trajectory can be written in the following way:

$$y(x) = y_s - tg\varphi_s(x - x_s) -$$

$$-\frac{g}{2V_{\theta,s}^2 \cos^2\varphi_s}(x - x_s)^2 ,\qquad (16)$$

where x, y – the current coordinates of the center of mass of the particle in rectangular coordinate system (Figure 1); x_s, y_s – tearing off point coordinates; $x_s = R_0 \sin\varphi_s$; $y_s = R_0 \cos\varphi_s$; $V_{\theta,s}$ – particle speed in the tearing off point.

3 RESULTS

The calculation of the particle speed along the cylinder contour and the angle value φ_s when the particle tears off from the cylinder surface has been performed using the formulas (3), (4) and the condition (13). The results of calculating the tearing off angle φ_s depending on the number of cylinder revolutions per minute n are presented in Figure 2a for $R_0 = 0.05$ m, $\varphi_0 = 0$: the curves 1,2 are for $V_{\theta.0} = 0$; the curves 3, 4, 5, 6 are for $V_{\theta.0} = 0.4$ m / sec; the curves 1, 3, 5 are for $f = 0.2$; the curves 2, 4, 6 are for $f = 0.6$. The dependences 1, 2, 3 and 4 have been calculated assuming that the static friction coefficient f_0 is equal to the sliding friction coefficient f, the dependences 5, 6 have been calculated assuming that $f_0 = 2f$.

(a)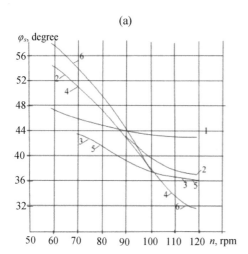

(b)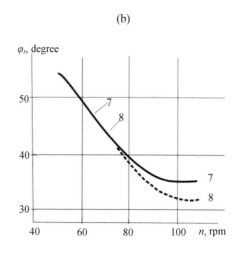

Figure 2. Dependences the particle tearing off angle φ_s on the cylinder numbers of revolution per minute n : (a) $R_0 = 0.05$ m; $\varphi_0 = 0$; $1, 2 - V_{\theta.0} = 0$; $3, 4, 5, 6 - V_{\theta.0} = 0.4$ m / sec; $1, 2, 3, 4 - f_0 = f$; $5, 6 - f_0 = 2f$; $1, 3, 5 - f = 0.2$; $2, 4, 6 - f = 0.6$; (b) $R_0 = 0.075$ m; $f = 0.6$; $f_0 = f$; $7 - V_{\theta.0} = 0$; $8 - V_{\theta.0} = 0.4$ m / sec.

Functions $\varphi_s(n)$ are decreasing for the combinations of the determining parameters (R_0, φ_0, $V_{\theta.0}$, f, f_0) under consideration, and for $n > n_*$ (in general case values n_* are different for different curves) they become constant. Apparently, the dependences of the angle φ_s from n under $f = 0.6$ are stronger than under $f = 0.2$. The dependences $\varphi_s(n)$ for $R_0 = 0.075$ m (Figure 2b) ($\varphi_0 = 0$; $f_0 = f$; $f = 0.6$; 7 – for $V_{\theta,0} = 0$; 8 – for $V_{\theta,0} = 0.4$ m / sec) have the same character.

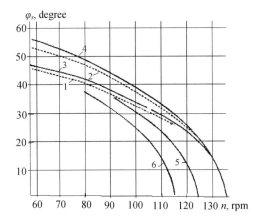

φ_s, degree

Figure 3. Dependences the particle tearing off angle φ_s on the cylinder numbers of revolution per minute n when the particle initial speed is equal the circumferential speed of cylinder work surface $V_{\theta,0} = V_w$; 1, 2, 3, 4 – have been receive for $R_0 = 0.05$ m; 1 – for $f = 0.2$; assumes $f_0 = f$; 2 – for $f = 0.6$; assumes $f_0 = f$; 3 – for $f = 0.2$; assumes $f_0 = 2f$; 4 – for $f = 0.6$; assumes $f_0 = 2f$; 5 – have been received for $R_0 = 0.06$ m; $f_0 = f = 0.4$; 6 – have been received for $R_0 = 0.07$ m; $f_0 = f = 0.4$; 5 and 6 – have been calculated under assuming $f_0 = f$.

To explain the coincidence of the curves 2 and 4 in Figure 2a diagrams of the speeds $V_\theta(\varphi)$ of the particle motion along the cylinder surface under the different number of revolutions n have been calculated. The calculation results are shown in Figure 7a for the case $V_{\theta,0} = 0$, and for $V_{\theta,0} = 0.4$ m / sec are shown in Figure 7b. The comparison of curves 1 ($n = 60$) in Figures 7a,b shows their considerable divergence only under $\varphi < 10°$ and then these curves practically coincide. By-turn, this can be explained in the following way: the particle is accelerated for $V_{\theta,0} = 0$, under the frictional force influence, but for $V_{\theta,0} = 0.4$ m / sec it is decelerated, as for $n = 60$ rpm and $R_0 = 0.05$ m, $V_w = 0.314$

m / sec. In both cases the particle acquires the speed V_w of the cylinder surface (under $\varphi \approx 10°$), that's why by the time the particle tears off its speeds are almost equal in both cases and therefore the values of the tearing off angle φ_s are almost equal too. That's why the curves 2 and 4 in Figure 2a coincide in the point $n = 60$ rpm.

Speed diagrams $V_\theta(\varphi)$ for $n = 80$ and 90 rpm (curves 2 and 3 in Figure 7a are for $V_{\theta,0} = 0$ and those in Figure 7b are for $V_{\theta,0} = 0.4$ m / sec) are also slightly different on the angles area φ corresponding to the particle speed alignment with the speed V_w, and under angles φ close to the tearing off angle φ_s. Due to it the curves 2 and 4 in Fifure 2a for $n < 90$ rpm almost coincide. However, it is not observed for $n > 90$ rpm: curves 2 and 4 in Figure 2a corresponding to $V_{\theta,0} = 0$ and 0.4 m / sec for $f = 0.6$ considerably differ.

For $f = 0.2$ the speed diagrams $V_\theta(\varphi)$ differ for all values of n (from 60 till 120 rpm) present in Figure 2a, as in this case frictional force has considerably lower influence on the particle motion, than under $f = 0.6$. Thereby, curves 1 and 3 in Figure 2a considerably differ.

Therefore, friction influence appears to be sufficient to level differences in the initial speeds $V_{\theta,0} = 0$ and 0.4 m / sec in case $f = 0.6$ and it is insufficient in case $f = 0.2$.

The dependences $\varphi_s(n)$ have been considered separately for the case when the cylinder feeding speed is equal to the circular speed of the surface with radius $R_0 : V_{\theta,0} = V_w$. The calculation results for this case for $R_0 = 0.05$ m, $\varphi_0 = 0$, are presented in Figure 3: curves 1 and 3 for $f = 0.2$; curves 2 and 4 for $f = 0.6$; curves 1 and 2 have been calculated assuming that $f_0 = f$; curves 3 and 4 were calculated assuming that $f_0 = 2f$. In this case the dependences $\varphi_s(n)$ are descending as in previous case, however they differ, as for $n \geq n_*$ the tearing off angle φ_s is equal to zero.

The value n_* is determined by the following conditions:

$$V_{\theta,0}^2 / R_0 = g ; \; V_{\theta,0} = V_w ; \; V_w = R_0 \omega_* ; \; \omega_* = \frac{\pi n_*}{30} .$$

From these conditions is followed:

$$n_* = \frac{30}{\pi}\sqrt{\frac{g}{R_0}}. \qquad (17)$$

Let us emphasise that calculation results presented in Figures 2 and 3 refer to the special case $\varphi_0 = 0$. Influence of changing the initial angle φ_0 on the tearing off angle φ_s is investigated below.

The dependences $\varphi_s(n)$ presented in Figures 2 and 3 show that inaccuracy of setting the value f_0 slightly influences on the calculation results φ_s, if the limit condition is the following: $1 \le f_0 / f \le 2$. Thus further calculations are performed assuming that $f_0 = f$.

The calculations results for the tearing off angle φ_s depending on the initial angle φ_0 for the range of different values n, $V_{\theta,0}$, f, for $R_0 = 0.05$ m are presented in Figure 4.

Let us note that the particle can slide in the direction reverse to the cylinder rotation if in area negative values φ_0 $|\varphi_0| > \alpha$ and $V_{\theta,0} = 0$ or values $V_{\theta,0}$ aren't enough large for the friction coefficient f, where $\alpha = arctg(f)$.

As the initial product can contain the particles with different friction coefficient values f, it is interesting to evaluate the influence of the value f dispersion on the tearing off angle of the particle φ_s.

The dependences $\varphi_s(f)$ calculated assuming that $f_0 = f$ are shown in Figure 5: curves 1a, b are for $n = 60$ rpm; 2a, 2b, 2c are for $n = 90$ rpm; 3a, 3b, 3c, 3d are for $n = 120$ rpm; 4a, 4b, 4c, 4d are for $n = 150$ rpm; letters a, b, c, d correspond to the values of particle initial speed $V_{\theta,0} = 0$; 0.2; 0.4 and 0.6 m / sec.

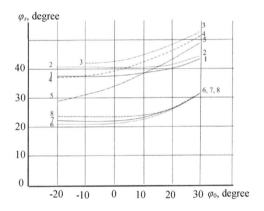

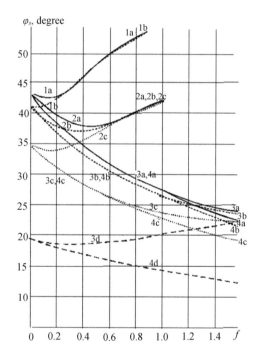

Figure 4. Dependences the particle tearing off angle φ_s on the initial angle φ. $R_0 = 0.05$ m under assuming $f_0 = f$; 1, 2 – $n = 90$ rpm; $V_{\theta,0} = V_w = 0.471$ m / sec; $f = 0.2$; 0,4; 3, 4, 5 – $n = 20$ rpm; $V_{\theta,0} = 0$; $f = 0.2$; 0.4; 0.8; 6, 7, 8 – $n = 120$ rpm; $V_{\theta,0} = V_w = 0.628$ m / sec; $f = 0.2$; 0.4; 0.8.

Figure 5. Dependences the particle tearing off angle φ_s on the sliding friction coefficient f: 1 – for $n = 60$ rpm; 2 – $n = 90$ rpm; 3 – $n = 120$ rpm; 4 – $n = 150$ rpm; a – when $V_{\theta,0} = 0$ (for all values n); b – when $V_{\theta,0} = 0.2$ m / sec; c – when $V_{\theta,0} = 0.4$ m / sec; d – when $V_{\theta,0} = 0.6$ m / sec.

Apparently, for $V_{\theta,0} = V_w$ the tearing off angle almost doesn't depend on the initial angle φ_0, (at least in the area, where $\varphi_0 < 10°$) both in case $n = 90$ rpm (curves 1 and 2) and in case $n = 120$ rpm (curves 6, 7 and 8). However the angle φ_s for zero initial speed $V_{\theta,0} = 0$ is significantly increased with increasing φ_0 (curves 3, 4, 5).

The dependences $V_{\theta,0}$ in Figure 5 on the one hand can be considered as four set of curves, for each of them $n = const$, and the initial speed $V_{\theta,0}$ varies. In this case the dependences $\varphi_s(f)$ of one set of curves have the maximum difference between them with little f values, and the dependences for different $V_{\theta,0}$ gradually approach when f is increasing. For $n = 60$ and 90 rpm the dependences $\varphi_s(f)$ coincide for different $V_{\theta,0}$ in the area $f > f_*$: for $n = 60$ rpm $f_* \approx 0.3$; for $n = 90$ rpm $f_* \approx 0.7$. For the case $n = 120$ rpm the value f_* transcends the real values of f; for $f \approx 1.6$ the dependences $\varphi_s(f)$ for $V_{\theta,0} = 0$; 0.2; 0.4 and 0.6 m / sec differ not more than 2 degrees.

From the other hand, the dependences $\varphi_s(f)$ in Figure 5 can be considered as the set of curves $V_{\theta,0} = const$, where the number of revolutions n varies. The dependences $\varphi_s(f)$ for different n and little f are close to each other in the set of curves $V_{\theta,0} = const$, and under $f \approx 0$ they coincide. While f is increasing, the dependency difference $\varphi_s(f)$ ascends for different n. Apparently, this tendency is reverse to the other observed for the set of curves $n = const$. If the revolution number n is not very big ($n = 60$ and 90 rpm), the divergences between the curves $\varphi_s(f)$ for different n may occur, starting from $f \approx 0$, as for example, curves 1a and 2a ($V_{\theta,0} = 0$, $n = 60$ and 90 rpm), and curves 1b and 2b ($V_{\theta,0} = 0.2$ m / sec, $n = 60$ and 90 rpm).

Curves divergence $\varphi_s(f)$ begins with higher values of f for different n in case of the relatively big revolution numbers ($n = 120$ and 150 rpm). For example, curves 3a and 4a ($V_{\theta,0} = 0$, $n = 120$ and 150 rpm) and curves 3b and 4b ($V_{\theta,0} = 0.2$ m / sec; $n = 120$ and 150 rpm) begin to diverge under $f \approx 0.8-1.0$. Curves 3c and 4c ($V_{\theta,0} = 0.4$ m / sec; $n = 120$ and 150 rpm) begin to diverge visibly under $f \approx 0.5$.

The calculation results presented in Figure 5 show that the dependences $\varphi_s(f)$ are not monotonous in general case, and the tendencies of their changing for variable parameters n and $V_{\theta,0}$ have different disposition in different areas of these parameters.

This fact can be explained by influence of two opposite factors on the particle speed, such as the following: frictional force influence becomes more intense while the sliding friction coefficient f is increasing, and the influence of the tangential component of gravity becomes more intense while f is decreasing. In Figures 6a,b, c, d, e the dependences $V_{\theta}(\varphi)$ of the particle speed V_{θ} from the angular coordinate φ (Figure 1) for $R_0 = 0.05$ m; $\varphi_0 = 0$ for different values of f under the condition of inseparable particle motion along the cylinder surface are shown.

Corner points of the curves $V_{\theta}(\varphi)$ in Figures 6a-e correspond to the angle φ values, when the particle speed reaches the circular speed of the cylinder surface. Figures 6d, e illustrate the weak dependency of the angle φ near the tearing off point on the friction coefficients f (for $f = 0.1$; 0.2; 0.4 and 0.6), despite the fact that diagrams $V_{\theta}(\varphi)$ are significantly different for different f in the middle part of the inseparable particle motion way.

The results of calculating $V_{\theta}(\varphi)$ for different revolution numbers n for fixed f for $R_0 = 0.05$ m, $\varphi_0 = 0$ are shown in Figure 7a ($V_{\theta,0} = 0$) and 7b ($V_{\theta,0} = 0.4$ m / sec). Similar calculation results for the case of $R_0 = 0.075$ m, are shown in Figures 8a ($V_{\theta,0} = 0$) and 8b ($V_{\theta,0} = 0.4$ m / sec).

One can see in Figures 7a, 7b, 8a, 8b that the speed diagrams $V_{\theta}(\varphi)$ of inseparable particle motion for different numbers n coincide in the area $\varphi_0 \leq \varphi \leq \varphi_w$, where φ_w is the value φ, then the particle speed becomes equal to the cylinder surface speed for this value. Under $\varphi > \varphi_w$ the particle can be fixed in regard to the cylinder surface if the tangential component of gravity for the particle is less than the static friction force, or in opposite case it can slide the cylinder surface. In both cases under $\varphi > \varphi_w$ the diagrams $V_{\theta}(\varphi)$ for different n are significantly different.

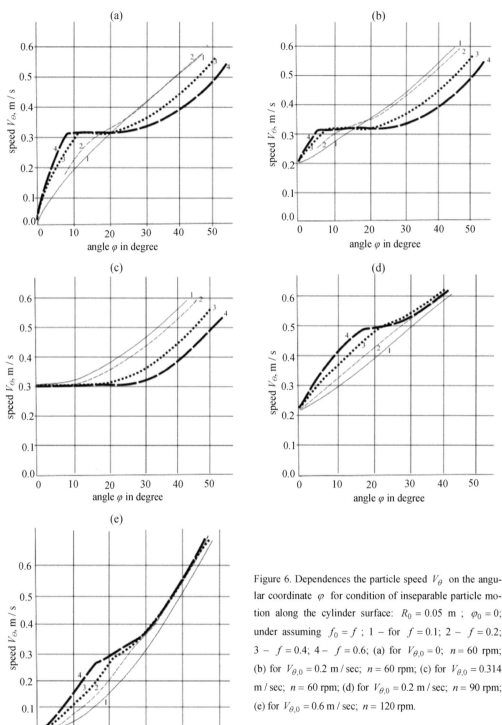

Figure 6. Dependences the particle speed V_θ on the angular coordinate φ for condition of inseparable particle motion along the cylinder surface: $R_0 = 0.05$ m ; $\varphi_0 = 0$; under assuming $f_0 = f$; 1 – for $f = 0.1$; 2 – $f = 0.2$; 3 – $f = 0.4$; 4 – $f = 0.6$; (a) for $V_{\theta,0} = 0$; $n = 60$ rpm; (b) for $V_{\theta,0} = 0.2$ m / sec; $n = 60$ rpm; (c) for $V_{\theta,0} = 0.314$ m / sec; $n = 60$ rpm; (d) for $V_{\theta,0} = 0.2$ m / sec; $n = 90$ rpm; (e) for $V_{\theta,0} = 0.6$ m / sec; $n = 120$ rpm.

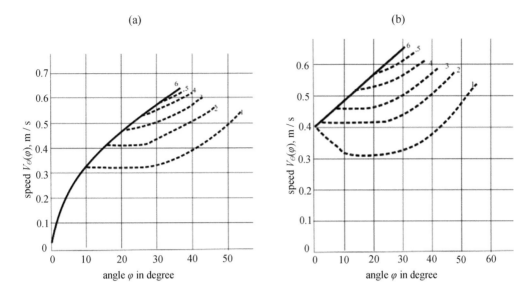

Figure 7. Dependences the particle speed V_θ on the angle φ for different the cylinder rotation speed n for $R_0 = 0.05$ m; $\varphi_0 = 0$; $f = 0.6$; assumes $f_0 = f$; (a) $V_{\theta,0} = 0$; (b) $V_{\theta,0} = 0.4$ m / sec; $1 - n = 60$ rpm; $2 - n = 80$ rpm; $3 - n = 90$ rpm; $4 - n = 100$ rpm; $5 - n = 110$ rpm; $6 - n = 120$ rpm.

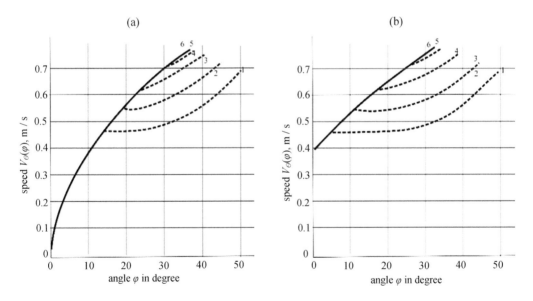

Figure 8. Dependences the particle speed V_θ from the angle φ for different the cylinder rotation speed n for $R_0 = 0.075$ m; $\varphi_0 = 0$; $f = 0.6$; assumes $f_0 = f$; (a) $V_{\theta,0} = 0$; (b) $V_{\theta,0} = 0.4$ m / sec; $1 - n = 60$ rpm; $2 - n = 70$ rpm; $3 - n = 80$ rpm; $4 - n = 90$ rpm; $5 - n = 100$ rpm; $6 - n = 110$ rpm.

341

4 CONCLUSIONS

1. In case when the particle initial speed $V_{\theta,0}$ is equal to the circular speed V_w of the cylinder working surface radius R_0 while increasing the number of cylinder revolutions per minute n in the area $n > n_*$, the tearing off angle φ_s is equal to the initial angle φ_0 ($\varphi_0 \geq 0$); value n_* doesn't depend on the coefficients of sliding friction f and static friction f_0, but depends on radius R_0. If $V_{\theta,0} < V_w$, while increasing the number n in the area $n > n_*$, the tearing off angle φ_s remains constant; in this case value n_* depends on the initial sped $V_{\theta,0}$, the initial angle φ_0, the coefficients f and f_0, and on the radius R_0.

2. The dependences $\varphi_s(\varphi_0)$ of the particle tearing off angle from the initial angle have different disposition for the cases $V_{\theta,0} = 0$ and $V_{\theta,0} = V_w$. Particularly, the following is proved for the diapasons of the initial angle values $-20° \leq \varphi_0 \leq 30°$ and the sliding friction coefficient $0.2 \leq f_0 \leq 0.8$:

a) in case $V_{\theta,0} = V_w$ the tearing off angle φ_s increases along with the increase of the initial angle φ_0 in the area $\varphi_0 > \varphi_*$, whereas the angle φ_s doesn't depend on φ_0 in the area $\varphi_0 < \varphi_*$ (the value φ_* depends on the parameters n, f);

b) in case $V_{\theta,0} = 0$ the dependency $\varphi_s(\varphi_0)$ is the increasing function in the diapason of the initial angle φ_0 under consideration.

3. The dependence $\varphi_s(f)$ of the tearing off angle φ_s from the sliding friction coefficient f for the definite combination of parameters n, $V_{\theta,0}$ can be a decreasing function, and for other combination of these parameters it can be an increasing function. In general case the dependency $\varphi_s(f)$ is not monotonous and can have the maximum point. This is explained by the influence of two opposite factors on the current velocity of the particle $V_\theta(\varphi)$: the frictional force influence while f is increasing becomes stronger, and the gravity influence decreases correspondingly.

REFERENCES

Derkach, V.G. 1966 *Special methods for mineral concentration*. Moscow.

Zverevich, V.V. & Perov, V.A. 1971. *Basis for concentration minerals*. Moscow.

Kravetc, B.N. 1986. *Special and combination concentration methods*. Moscow.

Blagov, I.S. 1984. *Guide for concentration coals*. Moscow.

Svoboda, J. 1987. *Magnetic Methods for the Treatment of Minerals*. Elsevier.

Edward, D., Hollam, P. & Kojovic, T. 1995. *The motion of the mineral sand particles on the roll in high tension separators*.

Tyurja, Yu. I. 2001. *Investigation the friction coefficient for the coal and rock along the steel*. Dnipropetrovs'k: 130-135.

Pilov. P.I. & Tyurja. Yu.I. 2001. *About increasing efficiently the coal and rock dry separation according by the friction*. Dnipropetrovs'k: 83-90.

Tyuria, Yu.I. 2003. *Regulations of distribution the coal and rock particles during tribo-gravitation*. Dnipropetrovs'k: 148-152.

Tyuria, Yu. I. 2002. *Developing device for separation according the friction*. Dnipropetrovs'k: 135-138.

Mining of Mineral Deposits – Pivnyak, Bondarenko, Kovalevs'ka & Illiashov (eds)
© 2013 Taylor & Francis Group, London, ISBN: 978-1-138-00108-4

Influence of undermined terrain mesorelief on accuracy of forecasting subsidence and deformation of earth surface

M. Grischenkov & E. Blinnikova
*Ukrainian State Research and Design Institute of Mining Geology,
Rock Mechanics and Mine Surveying, Donets'k, Ukraine*

ABSTRACT: There is considered the compliance of existing method of forecasting subsidence and deformation of earth surface to the real situation. There is developed and tested the mathematical model for forecasting subsidence and deformation of earth surface with taking into account the influence of mesorelief. Results of the model calculations are shown that errors of deformations caused by mesorelief may reach significant values, especially errors of relative deformations. These errors can significantly affect the accuracy of forecasting deformation of earth surface.

1 INTRODUCTION

The available method of forecast for subsidence and deformation of earth surface based on using in calculation of average depth of mining, i.e. some averaged plane representing earth surface. In many cases real earth surface deviates significantly from plane. In particular, majority of territories in Donets'k coal basin are characterized by mesorelief with difference of heights up to 50-100 m within the boundaries of mine allotment.

In present paper there is proposed algorithm accounting the effects of mesorelief in forecasting for subsidence and deformation of earth surface caused by underground mining.

2 ANALYSIS OF PROBLEM

Donets'k coal basin is the densely populated industrial region where underground mining is carried out under cities and towns immediately. Underground mining of coal reserves inevitably leads to subsidence of rock mass and earth surface over mine workings. Deformations of earth surface are capable of causing serious damage to buildings, constructions, engineering communications caught in the zone of influence of mining (in the subsidence throat). Therefore if excavation of coal reserves is under territories where are objects needed in protection from the influence of mining that this excavation is carried out according to the projects of undermining objects. In these projects according to special method there is forecast subsidence and de-

formations of earth surface caused by influence of mining and there is estimated the degree of their impact on protected objects. If the projected values of deformations exceed regulatory tolerances there are planned mine and constructive measures of protection or there is made the decision on preservation of objects of protective pillars.

The existing method of calculating the deformation of the earth surface is realized in industry standard "Rules of undermining..." (GSTU 101.00159 226.001 – 2003). In this method deformation in the subsidence throat is calculated for some fixed average depth of mining. This is equivalent to calculating the deformation for some horizontal plane located from the center of longwall working on distance equal to average depth of mining. As real undermined surface differs from this plane of calculating the deformation as there is obviously the emergence of inaccuracies in the determining of deformation when using existing algorithm. Thus the magnitude of these errors will be more than value of deviation of real earth surface from horizontal plane of calculation will be more. Maximum error of determination of deformation we should expect in the presence of enough major landforms on the undermined territory.

Unlike microrelief which describes small land forms and has a small amplitude of height volatility the mesorelief represents a major landforms with amplitude of height up to some tens of meters. Just mesorelief characterizes the earth surface of the vast majority of the undermined territories in Donets'k coal basin. The territory of the Central region of Donbass and East Donbass is very typical in this respect. Here there is almost everywhere the hilly terrain, increased steepness of the hill slopes, extensive structure of hollows (type gullies). The mesorelief of Donbass territory is clearly displayed on maps of scale 1:25000-1:50000 (with height of a relief cut to 5-10 m).

In a number of Donbass areas the ratio of amplitude of surface's height volatility to average depth of mining reaches to 10-15%. This fact is able to significantly reduce the accuracy of determining the deformation of the earth surface. It should be borne in mind that the most sensitive to mesorelief height volatility is determining the relative deformations (slopes, curvature, horizontal deformations) but not determining the absolute deformations. Just this circumstance is the key for developing a reliable method forecast and dictates the necessity of accounting of the mesorelief effect for calculating the deformation of the earth surface.

3 MODEL DEVELOPMENT

To take into account the mesorelief effect there was developed algorithm for calculating the deformation earth surface which includes additionally two new factors. First factor is the using digital terrain model (DTM) for describing the mesorelief of earth surface. The base for DTM formation is the determining the elevation marks of DTM points from the re-

sults of topographic survey or from results of vectorizing the topographic maps.

Second factor of the new calculating model is the concept of dynamic depth of mining. The essence of the proposed concept is visible from the following drawing (Figure 1).

For each calculation point i dynamic depth of mining H_i^d takes equal

For each i-th design point the depth is assumed to be

$$H_i^d = Z_i - Z_c = H + \Delta H_i, \qquad (1)$$

where Z_i – height mark of calculation point i; Z_c – the height mark of the center of longwall face (on seam ground); H – depth of mining in the center of longwall face; ΔH_i – the elevation point i above the plane of calculation that located from the center of longwall face on height H.

Value ΔH_i is determined from next equation

$$\Delta H_i = Z_i - (Z_c + H). \qquad (2)$$

Further the determination of all design parameters and values will be on base of dynamic depth of mining H_i^d for each calculation point i instead of average depth of mining H.

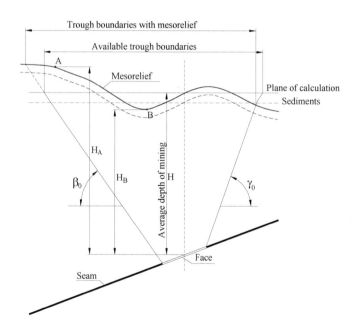

Figure 1. Scheme of mesorelief influence on the accuracy of calculating deformation of earth surface.

4 RESULTS AND DISCUSSIONS

In accordance with the given algorithm there were made changes in computer program "Undermining" which developed in UkrSRMI. This program is already operated for several years to develop the projects of undermining the different objects. Changes were made in conjunction with algorithm of deformation of mesorelief slopes. That algorithm is based on the hypothesis on interaction of topsoil and ledge rocks under earth surface undermining (Peng 1992 & Knothe 1957).

To test the proposed algorithm there were developed 24 digital terrain models. These DTM were developed for six most possible variants of undermining and four possible variants of slope of earth surface. For each variant of undermining DTM include consequently 1221, 999, 825, 675, 837 and 1023 calculation points.

The results of the test calculations demonstrate that for average depth of mining in the range of 500 up to 1000 meters height variations of mesorelief in the range of 50 up to 100 meters lead to next errors: in subsidence up to 40-60 millimeters, in slopes up to 0.3-$1.1 \cdot 10^{-3}$, in horizontal deformations up to 0.3-$0.8 \cdot 10^{-3}$.

For example, permissible values of horizontal deformations are: $2.0 \cdot 10^{-3}$ for dams and dykes; 1.0-$2.5 \cdot 10^{-3}$ for underground steel gas pipeline; $2.5 \cdot 10^{-3}$ for main railways; 1.0-$1.5 \cdot 10^{-3}$ for technological equipment of rolling shops and so on. Thus we can say that relative deformations due to the influence of mesorelief may reach very high values for a number of protected objects on undermined territory.

5 CONCLUSIONS

1. Existing method of forecasting the subsidence and deformation of earth surface based on using fixed average depth of mining, i.e. some averaged horizontal plane. The influence of mesorelief in this case is not considered.

2. In Donetsk coal basin the vast majority of undermined territories is characterized by mesorelief. In many cases difference of heights reaches 10-15% of average depth of mining in the range of mine allotment. This affects the accuracy of forecasting deformation of earth surface.

3. The forecast of deformation of earth surface with taking into account the influence of mesorelief shows that errors caused by mesorelief may reach significant values.

4. Errors caused by mesorelief are able to reach 40-80% permissible values of horizontal deformations and 28-55% permissible values of slopes.

REFERENCES

GSTU 101.00159226.001 – 2003. *Rules of undermining of buildings, constructions and natural objects under coal mining in the underground way. Instead "Rules of protection...".* Moscow: Nedra, 1981: 288; Introd. 01.01.2004. Kyiv, 2004: 128.

Peng, S.S. 1992. *Surface Subsidence Engineering.* Society for Mining, Metallurgy and Exploration. Inc. – Ann Arbor: MI: 161.

Knothe, S. 1957. *Observations of Surface Movements Under Influence of Mining and Their Theoretical Interpretation.* Proceedings European Congress Ground Movement. Leeds: 210-218.

Mining of Mineral Deposits – Pivnyak, Bondarenko, Kovalevs'ka & Illiashov (eds)
© 2013 Taylor & Francis Group, London, ISBN: 978-1-138-00108-4

Aspects of sulphurous feed extraction in Ukraine

V. Kharchenko & O. Dolgyy
National Mining University, Dnipropetrovs'k, Ukraine

ABSTRACT: Researches of scientists about an origin and hydrogen sulfide formation in the Black Sea and its stocks are resulted. Two ways of extraction hydrogen sulfide are given, its scopes in industry and the household purposes are specified.

1 INTRODUCTION

Economical independence of any state requires sufficient number of energy carriers and energy efficient technique.

Nowadays Ukraine can fully provide itself only with coal. Its extraction builds up 75-80 mln. t / year and to provide industrial and municipal spheres it has to be 110-120 mln. t / year. With current tempo of coal extraction Ukraine will have enough coal for next 100-200 years. Tempo of coal extraction increase in Ukraine is behind country needs because of different reasons.

As a result, lack of coal is balanced by its import 20 mln. t / year from Russia and Poland, 30-40 bln. m³ / year of gas and about 50 bln. t / year of oil.

This means that there is a need to search new sources of energy. Economists and energetics come to the conclusion that it is impossible to substitute nuclear energy. Although, after Chernobyl its danger is undeniable especially for countries with unstable conditions and dangerous terrorism level. Meanwhile there is an alternative to nuclear energetics. Engineer Yutkin L.A. has called the Black Sea a unique stockroom with unlimited stores, so called "El Dorado" of renewable feed sources (Golcov & Yutkin 1996). Yutkin L.A. offered a project of electrohydraulic effect of gases enrichment. The matter is that, the Black Sea waters with depth lower than 80-100 m. contain hydrogen sulfide. Especially important is the fact that unlike other anthracites, stores of hydrogen sulfide in the Black Sea are renewable. According to the researches, hydrogen sulfide is replenished by two sources: bacterium activity which in anaerobic conditions can restore sulphate sulfur to sulphide one and hydrogen sulfide, synthesized inside Caucasus Mountains and occurred from cracks of Earth crust. Concentration of hydrogen sulfide is settled by its oxidation in surface water levels. Atmospheric oxygen with dissolving in water affects hydrogen sulfide, turning it into sulfuric acid. The acid reacts with dissolved mineral salts and forms sulfates. These processes are simultaneous, that is why there is a dynamic equilibrium in the Black Sea.

2 FORMULATING THE PROBLEM

Computations showed that during one year the part of converted into sulfates hydrogen sulfide is 1 to 4. This is the result of oxidation in the Black sea.

So, it is possible to extract about 250 million tonnes of hydrogen sulfide every year from the Black Sea. Energy intensity of it is about 1012 kW / h. One consumed kilogram of hydrogen sulfide builds up about 4000 kcal. Such amount of energy is equal to the annual production of electricity in the Former Soviet Union. That means that the Black Seas hydrogen sulfide generator can fully provide need of energy for all Black Sea countries, including Ukraine.

In the air hydrogen sulfide ignites at the temperature of about +300 °C (Achmetov 2001).

Hydrogen sulfide set afire in the air is consumed according to one of the following equations:

$2H_2S + 3O_2 = 2H_2O + 2SO_2 + 1125$ kJ (with oxygen excess); $2H_2S + O_2 = 2H_2O + 2S + 531$ (with oxygen lack).

Thermal power of hydrogen sulfide is 23.7 mJ / m³ (with 0 °C and 101.3 kPa).

Critical temperature of H_2S equals to 100 °C with critical pressure 89 atm.

Dangerously explosive mixtures of hydrogen sulfide with the air containing from 4 to 45 volume % H_2S.

With expansion of hydrogen sulfide and its further combustion significant energy gain can be reached. According to energetics (combustion heat), 1 m³ of hydrogen sulfide is equal to 1.49 m³ of household gas.

Weight thermal power of hydrogen (28630 kcal) is 2.8 times bigger than the gas one. Ignition energy is 15 times less than hydrocarbonic fuel's one. Maximal propagation velocity of flame front is 8 times faster in comparison with hydrocarbon.

3 MATERIALS UNDER ANALYSIS

According to the scientists' calculations, one burned kilogram of hydrogen sulfide makes up 4000 kcal and 2 kg of sulfur. So, near Moscow station "Podzemgaz" along with their main product – gas – hydrogen sulfide was producing such chemical products as high quality brimstone in its desulfurization shop.

Experience of desulfurization shop of near Moscow station has proved technical feasibility of sulfur and its derivatives production from hydrogen sulfide. Gas production on the station was stopped in 1963. During its operating period 21.9 th. t. of sulfur were produced (Antonova and other 1990).

Foregoing direct use of hydrogen sulfide as fuel is very dangerous for environment.

Currently, environmental friendly and economical plazma-chemical process and experimental-industrial equipment aiming to decompose hydrogen sulfide and obtain sulfur and hydrogen, importance of which in modern energetics is constantly growing, were developed.

Unlike other traditional methods of hydrogen sulfide decomposition plazma-chemical one does not lead to emissions of SO_2 into the atmosphere. There is almost no loss of hydrogen (Malinin 1971).

$$H_2S + Q \downarrow = H_2 + S \ ;$$

$$2H_2 + O_2 = 2H_2O + 14Q \uparrow .$$

With decomposition of hydrogen sulfide and further combustion of hydrogen, 14 times bigger energy win can be reached.

Hydrogen can be used as environmental friendly fuel for cars.

There are different hypotheses concerning hydrogen sulfide staying in the depth. According to some scientists' opinion, hydrogen sulfide can be kept in dissolved state only by significant pressure of overlaying waters (10-20 atm.). Remove this cork – and the water will boil. Hydrogen sulfide will be released very quickly in gas form (like a can of soda).

Hydrogen sulfide can be easily oxidized in water solution and it is strong reducing agent (Figure 1).

In 1979 an engineer L.A. Yutkin offered quite real project of electrohydraulic effect. The essence

of it was built on separation and enrichment of gases.

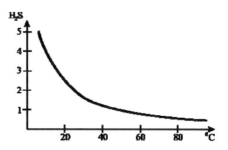

Figure 1. Solubility of hydrogen sulfide (volumes for 1 volume of water).

For this reason, Yutkin L.A. offered to rise bottom layers of water with great volume of hydrogen sulfide on the technological height, where they could be exposed to electrohydraulic charges, which provide extraction of hydrogen sulfide. Obtained gas should be liquidized and burned and produced sulfur dioxide oxidized into sulfuric acid.

The authors offer two schemes of hydrogen sulfide feed extraction technique.

The first one – with fixed installations (Figure 2) and the second one (Figure 3) with floating platform.

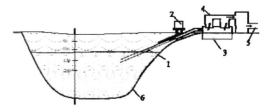

Figure 2. Fixed station with equipment sizes: 1 – input pipeline; 2 – station of reagent input; 3 – separating pool and hydrogen sulfide storage; 4 – tank for hydrogen sulfide gathering and liquidizing; 5 – delivery of hydrogen sulfide for domestic purpose (compressor house); 6 – hydrogen sulfide in sea water.

In the first case, extracted hydrogen sulfide is transported through the pipelines to the customers on shore. In the second case, extracted hydrogen sulfide is utilized in tank-containers and delivered to the customers.

The essence of the first scheme is that there will a fixed station installed on the shore which includes intake pipeline (1), station of reagents input (2), water and hydrogen sulfide separating pool (3), hydrogen sulfide collector (4), tank to gather hydrogen sulfide and its liquidizing (5), pipeline to deliver hydrogen sulfide for industrial needs. Such fixed plant should be located on shore with significant depth of the sea. Fixed plant has significant disadvantages: first of all, it is significant length of input pipeline (1) and possibility of being located near/inside towns.

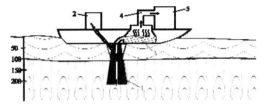

Figure 3. Floating platform with equipment sizes: 1 – input pipeline; 2 – station of reagent input; 3 – separating pool and hydrogen sulfide storage; 4 – tank for hydrogen sulfide gathering and liquidizing; 5 – tank to gather hydrogen sulfide.

That is the second option with floating platform is offered. It includes: input pipeline (1), station of reagent input (2), separating pool (3), tank to gather and liquidize hydrogen sulfide (4). Such installation has line of significant advantages:

a) extraction of hydrogen sulfide can be carried out in any area of the Black Sea;

b) tanks are filled with liquidized hydrogen sulfide and they be exchanged for the empty one, which is very efficient.

In such way hydroge sulfide is delivered in any ports of destination, besides, one of the advantages of floating platform is extraction of hydrogen sulfide in the sea, which provides safety of shore population.

4 CONCLUSIONS

Offered development will provide the customer with following:

1. Increase of the Black Sea biodiversity by means of significant increase of hydrogen sulfide concentration in sea water.

2. Receiving of cheap electricity by means of high potential heat of hydrogen sulfide processing processes utilization.

3. Possibility of receiving additional market product, chemically pure sulfur. Receiving of valuable product – sulfur makes new perspectives to develop big chemical manufacture.

4. Providence of Autonomous Republic of Crimea with drink and technically fresh water by means of sea water desalination.

5. Implementation of new working places

In should be marked that according to Ukrainian laws, hydrogen sulfide is not energy carrier. Its industrial consumption is not quoted and does not require specific budget payments.

REFERENCES

Golcov, L.I., Yutkin F.L. 1996. *Black Sea, good sea.* Moscow: Magazine. Inventor and innovator, 2.

Antonova, R.I., Bezhanishvili, A.E., Blinderman, M.S., Grabovskaya, E.P., Gusev, A.F., Kazak, V.N., Kapralov, V.K. & Fominih, V.G 1990. *Underground gasification of coal in USSR.* Moscow: CRIEI Coal.

Achmetov, N.S. 2001. *General inorganic chemistry.* Moscow: Vishaya shkola.

Malinin, K.N. 1971. *Hydrogen sulfide handler handbook.* Moscow: Chemistry.

The stress-strain state of the belt
on a drum under compression by flat plates

D. Kolosov & O. Dolgov
National Mining University, Dnipropetrovs'k, Ukraine

A. Kolosov
Moscow State University of Technology and Management, Moscow, Russia

ABSTRACT: The effect of the geometric parameters of the rubber-rope belt on its stress-strain state in the interaction with clamping elements on the driving drum winder is investigated. The obtained results determine expediency of the flat clamping elements use in hoist engines with bobbin winding and can be applied in engineering design practice.

1 INTRODUCTION

In the hoist engine reel-type member a flat belt winds in a few layers by the Archimedes spiral on a drum which is assumed to be solid. The multi-layered winding, regardless of amount of layers in a bobbin, does not change the amount of layers which interact with a drum. Only one layer interacts with a drum that does not correspond to the safety condition, requiring the presence of three coils of friction of the belt on a drum.

For the removal of the above mentioned contradiction it is necessary to make alteration in the construction of bobbin. Such a change of construction is possible if, for example, join to drum a flat belt and metallic sheet of the length equal to the first coil length. Such a sheet interacts with the first and second layer of the rope. Its longitudinal deformations are small as compared to deformations of rope. In these construction three surfaces of the first two layers interacts with the driving drum of the hoist engine both directly and through a metal sheet.

There are other engineering designs of the belt interaction with drum. In Figure 1 the amount of layers interacting with a drum can be increased by the use of additional clamping straps. Moreover, the drum of the hoist engine can have a driving pulley with ditches and belt with return ledges for prevention of displacement of belt from the axis of getting up of load.

2 PROBLEM STATEMENT

For the safe exploitation of hoist engine with the reel-type member of winding it is necessary to make alteration in the construction of this member. This alteration conduces to the change of the belt loading conditions. The estimation of impact of such change needs in available mathematical modeling and solution of stress-strain state problem of rubber layer of the belt as the least strong element. Therefore, analyze of belt stress and strain dependence on its constructive parameters, at its interaction with clamping elements on the driving drum, is an **actual scientific and technical problem.**

The influence of the belt ledges and ditches on the pulley of hoisting engine on distribution of tensions in a rubber layer is analyzed in (Belmas, Kolosov A. & Kolosov D. 2013). The mathematical stress and strain state model of the flat infinite rubber-rope belt compressed by rigid flat flags is worked out in (Kolosov D., Belous & Tantsura 2012). Such loading corresponds to the case of plane deformation.

The clamping slats shown on a Figure 1 have a limit width, so forces do not act on a belt outside slats. It allows to consider a belt in conditions near to the plane stress and to use the model developed in (Kolosov D., Belous & Tantsura 2012) with corresponding changes.

3 PROBLEM SOLUTION

Taking into account the belt regular construction it is possible to consider that rubber at any rope deforms identically. Cut out the element of unit thickness (Figure 2).

There are following limitations for such an element:

at $y = 0$, $v = \tau = 0$; $\qquad\qquad$ (1)

at $y = h$, $v = \tau = 0$;   (2)

at $x = 0$, $u = \tau = 0$,   (3)

where v, u – displacements along y- and x-axis; τ – shearing stress.

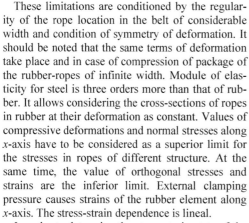

Figure 1. Joining of the flat rope to the drum.

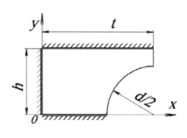

Figure 2. Fastening terms of the rubber inter-rope element.

These limitations are conditioned by the regularity of the rope location in the belt of considerable width and condition of symmetry of deformation. It should be noted that the same terms of deformation take place and in case of compression of package of the rubber-ropes of infinite width. Module of elasticity for steel is three orders more than that of rubber. It allows considering the cross-sections of ropes in rubber at their deformation as constant. Values of compressive deformations and normal stresses along x-axis have to be considered as a superior limit for the stresses in ropes of different structure. At the same time, the value of orthogonal stresses and strains are the inferior limit. External clamping pressure causes strains of the rubber element along x-axis. The stress-strain dependence is lineal.

In order to determine the stress-strain state of the rubber element, use the method of initial functions. Accordingly to the terms (1) and (2) we have

$$v_{0y} = \tau_{0y} = 0,$$

$$u_{0x} = \left[\frac{1-\mu_g}{1+\mu_g} sin(\alpha h) - \alpha h cos(\alpha h)\right] C e^{\alpha x},$$ (4)

$$\sigma_{0x} = 2\alpha\left[sin(\alpha h) - \alpha h cos(\alpha h)\right] C e^{\alpha x},$$

where C – constant; $\alpha = \alpha_n = \dfrac{n\pi}{h}$; n – any positive integer.

Finally, taking into account terms (3), functions describing the stress-strain state of the rubber element have a form:

a) displacements along x-axis

$$u = A_0 + \frac{1-\mu_g}{2}B\,x - \sum_{n=1}^{\infty}(-1)^n\, h\, cos(\alpha_n y)\left[\alpha_n sh(\alpha_n x)A_n + \left(\frac{2\mu}{1+\mu}sh(\alpha_n x) + \alpha_n x\, ch(\alpha_n x)\right)D_n\right],$$

b) displacements along y-axis

$$v = \sum_{n=1}^{\infty}(-1)^n\, h\, sin(\alpha_n y)\left[\alpha_n ch(\alpha_n x)A_n + \left(\frac{3+\mu}{1+\mu}ch(\alpha_n x) + \alpha_n x\, sh(\alpha_n x)\right)D_n\right],$$

c) normal stresses

$$\sigma_y = E\mu B + 2E\sum_{n=1}^{\infty}(-1)^n \alpha_n h\, cos(\alpha_n y)\left[\alpha_n ch(\alpha_n x)A_n + \left(3ch(\alpha_n x) + \alpha_n x\, sh(\alpha_n x)\right)D_n\right],$$

$$\sigma_x = E\,B - 2E\sum_{n=1}^{\infty}(-1)^n\,\alpha_n h\,cos(\alpha_n y)\Big[\alpha_n ch(\alpha_n x)A_n + \big(ch(\alpha_n x)+\alpha_n x\,sh(\alpha_n x)\big)D_n\Big], \qquad (5)$$

d) shearing stress

$$\tau = 2G\sum_{n=1}^{\infty}(-1)^n\,\alpha_n h\,sin(\alpha_n y)\Big[\alpha_n sh(\alpha_n x)A_n + \big(2sh(\alpha_n x)+\alpha_n x\,ch(\alpha_n x)\big)D_n\Big],$$

where A_i, B, D_i – unknown constants of integration. Formulate boundary conditions. In accordance with Figure 2, on the surfaces

$$x(y) = \begin{cases} t \ \ when \ \ y \ge \left(\dfrac{d}{2}\right), \\[4mm] t - \sqrt{\left(\dfrac{d}{2}\right)^2 - y^2} \ \ when \ \ y < \left(\dfrac{d}{2}\right) \end{cases} \qquad (6)$$

the following conditions are valid:

$$u = \delta, \qquad (7)$$

$$\tau = 0 \ if \ y \ge \left(\frac{d}{2}\right) \ \ and \ \ v = 0 \ if \ y < \left(\frac{d}{2}\right). \qquad (8)$$

Fourier series for function (6), taking into account symmetry condition, is

$$x(y) = a_0 + \sum_{n=1}^{\infty} a_n\,cos(\alpha_n y). \qquad (9)$$

Obey (7), then

$$A_0 + \frac{1-\mu_n}{2}B\left[a_0 + \sum_{m=1}^{\infty} a_m\,cos(\alpha_m y)\right] - \sum_{n=1}^{\infty}(-1)^n\,h\,cos(\alpha_n y)\times$$

$$\times\left[\begin{array}{l} \alpha_n sh\!\left(\alpha_n\!\left[a_0 + \sum_{m=1}^{\infty} a_m\,cos(\alpha_m y)\right]\right)A_n + \\[5mm] + \left(\begin{array}{l}\dfrac{2\mu}{1+\mu}sh\!\left(\alpha_n\!\left[a_0 + \sum_{m=1}^{\infty} a_m\,cos(\alpha_m y)\right]\right)+ \\[5mm] +\alpha_n\!\left[a_0 + \sum_{n=1}^{\infty} a_n\,cos(\alpha_n y)\right]ch\!\left(\alpha_n\!\left[a_0 + \sum_{m=1}^{\infty} a_m\,cos(\alpha_m y)\right]\right)\end{array}\right)D_n \end{array}\right] = \delta. \qquad (10)$$

In (10) there are components depending on y, and constants. Thus we have

$$A_0 + \frac{1-\mu_g}{2}Ba_0 = \delta.$$

Constant A_0 takes into account displacement of element as a solid. Take it equal to zero.

Obey (8) taking the first N members in sums for corresponding number of equations for points on the surface of inter-rope element:

$$x(y), \ y_n = \frac{n}{N} \ (n = 1, 2, \dots, N).$$

As a result one can obtain a set of algebraic equations of $2N$ order:

$$\begin{pmatrix} Q_1\beta_{1,1} & \cdots & Q_{2\,N}\beta_{1,2\,N} \\ \cdots & \cdots & \cdots \\ Q_1\beta_{2\,N,1} & \cdots & Q_{2\,N}\beta_{2,2\,N} \end{pmatrix} = \begin{pmatrix} b_1 \\ \cdots \\ b_{2\,N} \end{pmatrix}. \qquad (11)$$

There are hyperbolic functions in equations (10). Their values increase in proportion to that of n and coordinate y. Thereafter, the values of elements in matrix β differs essentially. As a result, some algebraic accuracy decreases. For the purpose of elimination of this fault the set (11) is transformed as following:

$$\begin{pmatrix} q_1\dfrac{\beta_{1,1}}{\gamma_1} & \cdots & q_{2\,N}\dfrac{\beta_{1,2\,N}}{\gamma_{2\,N}} \\ \cdots & \cdots & \cdots \\ q_1 & \cdots & q_{2\,N} \end{pmatrix} = \begin{pmatrix} b_1 \\ \cdots \\ b_{2\,N} \end{pmatrix}, \qquad (12)$$

where $q_i = Q_i\gamma_i$; γ – subsidiary vector.

Components of the vector γ are taken equal to maximum absolute values of the corresponding β matrix column. The values of the vector components A_i and D_i are determined from set (12).

353

These values were used for further calculus.

For the analysis of influence of geometrical parameters of rubber layer in a rubber-rope belt with the rope of 7 mm diameter, the values of tensions and deformations for various ratios of $2t$ (thickness of belt) and $2h$ (a step of rope laying) are determined for the case of compression on $\delta = t/100$. The results of calculus are given in Table 1-3 for Poisson's ratio equal to 0.48, which corresponds to rubber.

Table 1. Impact of parameter $2h/d$ (at $2t/d = 1.1$) on stresses and strains in rubber element of flat rubber rope belt compressed by two plates.

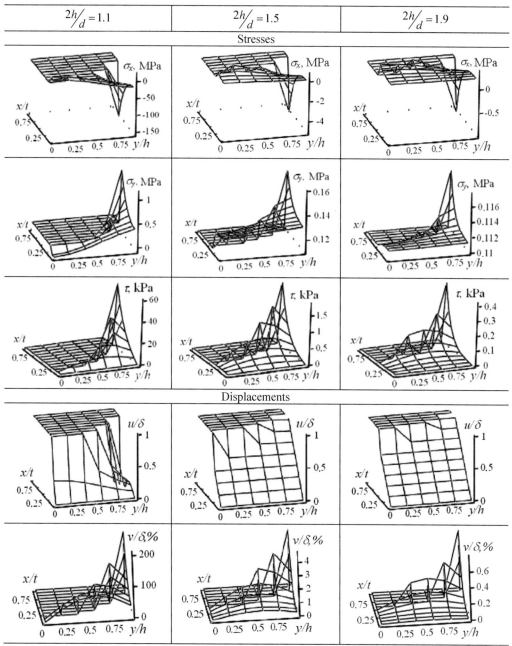

354

Table 2. Impact of parameter $2t/d$ (at $2h/d = 1.1$) on stresses and strains in rubber element of flat rubber rope belt compressed by two plates.

$2t/d = 1.1$	$2t/d = 1.5$	$2t/d = 1.9$
Stresses		
σ_x, MPa; 0, -50, -100, -150; x/t 0.75 0.25; 0 0.25 0.5 0.75 y/h	σ_x, MPa; 0, -20, -40, -60; x/t 0.75 0.25; 0 0.25 0.5 0.75 y/h	σ_x, MPa; 0, -5, -10, -15; x/t 0.75 0.25; 0 0.25 0.5 0.75 y/h
σ_y, MPa; 1, 0.5, 0; x/t 0.75 0.25; 0 0.25 0.5 0.75 y/h	σ_y, MPa; 0.5, 0.4, 0.3, 0.2, 0.1; x/t 0.75 0.25; 0 0.25 0.5 0.75 y/h	σ_y, MPa; 0.18, 0.16, 0.14, 0.12; x/t 0.75 0.25; 0 0.25 0.5 0.75 y/h
τ, kPa; 40, 20, 0; x/t 0.75 0.25; 0 0.25 0.5 0.75 y/h	τ, kPa; 20, 10, 0; x/t 0.75 0.25; 0 0.25 0.5 0.75 y/h	τ, kPa; 4, 2, 0; x/t 0.75 0.25; 0 0.25 0.5 0.75 y/h
Displacements		
u/δ; 1, 0.5, 0; x/t 0.75 0.25; 0 0.25 0.5 0.75 y/h	u/δ; 1, 0.5, 0; x/t 0.75 0.25; 0 0.25 0.5 0.75 y/h	u/δ; 1, 0.5, 0; x/t 0.75 0.25; 0 0.25 0.5 0.75 y/h
v/δ,%; 200, 100, 0; x/t 0.75 0.25; 0 0.25 0.5 0.75 y/h	v/δ,%; 60, 40, 20, 0; x/t 0.75 0.25; 0 0.25 0.5 0.75 y/h	v/δ,%; 10, 5, 0; x/t 0.75 0.25; 0 0.25 0.5 0.75 y/h

Obtained graphic dependencies indicate that the ratio of rope step laying, their diameters and thickness of the belt affects its stress and strain condition caused by compression of flat plates. At a constant value of the strain, increasing of the portion of rubber in the cross section of the belt reduces the stress in it. The measure of impact of changes in the rope step and belt thickness is different. Increasing of the

355

ratio of step laying to the diameter, results in a more significant reduction in the maximum stresses than the ratio of the belt thickness to the diameter of ropes. The values of the maximum normal and shear stresses are also different. The latter are almost three orders less.

Table 3. Impact of parameter $2t/d = 2h/d$ on stresses and strains in rubber element of flat rubber rope belt compressed by two plates.

$\dfrac{2t}{d} = \dfrac{2h}{d} = 1.1$	$\dfrac{2t}{d} = \dfrac{2h}{d} = 1.5$	$\dfrac{2t}{d} = \dfrac{2h}{d} = 1.9$
σ_x, MPa	σ_x, MPa	σ_x, MPa
σ_y, MPa	σ_y, MPa	σ_y, MPa
τ, kPa	τ, kPa	τ, kPa
Displacements		
u/δ	u/δ	u/δ
v/δ, %	v/δ, %	v/δ, %

356

Stress redistribution is specified by the changing nature of deformation of rubber, which depends on the shape of the layer disposed between the cables. Gradient of strains, coinciding with the direction of deformation of the belt, decreases at increase both of thickness and rope step laying. The values of the maximum orthogonal displacements are much more sensitive to changes in step than the thickness of the belt. This leads to a decrease in shear stress.

The force transmission from the drum to the cables of the belt, including through clamping bars, are provided by shearing stresses. For rubber-rope belt the most dangerous (after breaking the cable) is break of the adhesive bond of rubber and cables. Accordingly, it can be argued that the reduction of step laying of cables, from the point of view of the strength of the connection of the cable and the rubber, is impractical. Joint increase of step laying of the ropes and its thickness is insignificant effect on the above-mentioned features of the stress-strain state of the belt compressed by bars.

However, note that the conditional increase in thickness of the belt may be provided by joint winding of the rope and rubber strip of the same width. This winding will reduce the uneven distribution of stresses in the belt without changing its parameters, including the mass per unit length. Note also that the rubber in the belt construction affects the dependence of the deformation of the belt from the pressing forces of the metal clamping elements. Therefore, for ensuring a given clamping force it is necessary to take into account the results of the above analysis.

Deformation of the element depends not only on its shape and loading conditions, but also on the material properties. Poisson's ratio is one of these characteristics. These studies show that the Poisson's ratio have an influence on the stress and strain state of the belt, but practically only at small thicknesses of the belt (up to $2h/d = 1.1$) and step laying of ropes (up to $2t/d = 1.1$).

The adequacy of the obtained solutions was examined by determining the stress state of the element of the same size, but with a nominal diameter of the rope, equal to zero. The results are consistent with the terms of the compression of the rectangular element.

4 CONCLUSIONS

The ratio of the step arrangement of ropes, their diameters and thicknesses have an influence on their stress and strain state caused by compression of a flat plate. At a constant value of the strain, increasing the portion of rubber in the section of the belt reduces the stress in it. Reduction of the step laying of ropes from the point of view of the strength the connection of rope and rubber is not rational.

The use of materials with reduced Poisson's ratio is impractical because it leads to an increase in the maximum shear stress. The stress and strain state of infinitely wide flat rubber-rope belt compressed between two flat plates is the same as for the package of compressed by plates ropes which confirms the adequacy of the obtained results.

The use of wide clamping planks pressing the rope reduces the maximum shear stress, which determines the appropriateness of their use in hoisting machines with bobbin winding of the belt.

The obtained results can be used in engineering practice and designing of the hoist engines with bobbin winding.

REFERENCES

Belmas, I.V., Kolosov, A.L. & Kolosov, D.L. 2013. *Determination of tensions in the rubber cover of rubber-rope belt with a ledge.* All-Russian scientific conf. "Irreversible processes in nature and technique": Labours. Moscow: Bauman Moscow State Technical University Edition. Part II: 236-240.

Kolosov ,D.L., Belous, E.I. & Tantsura, A.I. 2012. *Model and determination of stress and strain conditions of rubber-robe belt compressed by rigid flat flags.* Collection of scientific works of the Kerch state marine technological university and Dniprodzerzhyns'k state technical university. Issue 13: 64-68.

Multifactorial mathematical model of mechanical drilling speed

M. Moisyshyn, B. Borysevych & R. Shcherbiy
Ivano-Frankivs'k National Oil and Gas University, Ivano-Frankivs'k, Ukraine

ABSTRACT: Aiming at determining the multifactorial mathematical model of mechanical drilling speed, which takes into account the overall influence of operating parameters and the drilling tool characteristics on the energy output, we have used the method of rational planning of experiments. According this method the combination of variable factors, including the axle static load F_{stat}, the drilling bit rotation rate n, the severity C and the damping factor β of the drilling tool occurs only once. According to this the overall multidimensional function can be presented in the form of the gain of separate dependencies on the variable factors – $V_{mech} = B_{av} \cdot f(F_{stat}) \cdot f(n_b) \cdot f(\beta) \cdot f(C)$ The constant factors when conducting the controlled experiment were the consumption of flushing fluid (water), the type and the diameter of the three-roller drill bit cutter and the rock hardness by the stamp.

According to the controlled experiment results, the equation of the multifactorial mathematical model of mechanical drilling speed looks as follows:

$$V_{mech} = 1.6703 \cdot 10^{-3} \cdot F_{stat}^{1.565623} \cdot n_b^{0.64089277} \cdot \beta^{-0.03314602} \cdot C^{-4.550764 \cdot 10^{-2}}.$$

1 INTRODUCTION

One of the main tasks of bore-hole drilling is the increase in the technical-economic indices of drilling: advance per bit h and the drilling rate V_{mech}. According to the results of experimental research conducted in test-bench and industrial conditions (Allikvander 1969; Belikov 1972 & Potapov 1961) correlations (mathematical models) of the drilling rate to the axle static load on the drilling bit F_{stat} and the drilling bit rotation rate n were suggested. Starting from the 1960 in order to reduce the harmful influence of shaking on the operation of the drill-string in the bottom part of its construction arrangement the antivibration devices (AVD), called buffers, have been used. When used at drilling sites in the USA the influence of AVD on the main indicators of drilling (drilling rate and advance per bit) was revealed (Usage of near-the-face…1974). According to the results of this research it has been determined that the use of the buffer in the bottom part of the construction arrangement of the drill-string reduced the cost of one meter of drilling by $14-23. The research into the influence of AVD on drilling rate was conducted both in the USA and the USSR. According to these investigations no definitive conclusion about their positive influence on V_{mech} has

been drawn. According to some sources (Nazarov 1985) when using AVD both the increase and the decrease in the drilling rate was observed. Some investigations (Yamaltdinov 1986) provide the results of testing buffers in "Kransodarneftegas" in Western Siberia and in "Ukrzahidnaftegas", according to which the use of the same type of buffer depending on the operational conditions differently influences drilling indicators. This signifies the necessity of a substantiate choice of the characteristics of the buffer (severity C and the damping ratio β) for the specific conditions of drilling, which may be done only on the basis of experimentally tested empirical interrelations between the drilling rate and these characteristics. Some investigations (Scherbiy 2012) according to the results of experimental tests determined the empirical correlations between the damping ratio of the drilling tool and the drilling rate. The probability of the existence of the correlation is bigger than 0.95.

2 FORMULATING THE PROBLEM

Taking into account the above mentioned, in order to determine the multifactorial mathematical model of drilling rate V_{mech}, which includes the overall influence of operating parameters and the drilling tool characteristics, (Borysevych 2009) conducted experimental tests on the drilling stand (see Figure 1) with the use of the method of rational planning of experiments (Protodyakonov 1970 & Yaremiychuk 1966). During the conduct of these experiments, devices, whose construction allows changing severity and the drilling tool damping ratio of the elastic element irrespective of each other, were used for the change of severity and the drilling tool damping ratio (Borysevych 2009). The change of severity within the interval of 225-2500 kN / m was conducted by means of using coil springs of compression of different sizes as a elastic element of the device. The damping ratio during the use of these springs is practically unchanged.

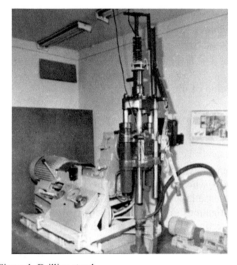

Figure 1. Drilling stand.

3 MATERIALS UNDER ANALYSIS

In order to change the damping ration of the drilling tool construction hydraulic absorbers of vibrations КВЗ-ЛПЖТ were used. The use of these absorbers with a differing number of valve lashes changed the damping ration within the interval of 40-90 kN·s / m.

The investigation was conducted in accordance with the planned four-factorial experiment by the levels of the change of the factors (F_{stat} , n , β , C) in the blocks of Vorotyshche sandstone sunk, with the severity by the stamp of p_{st} = 2050 MPa. The matrix of the planned experiment with the experimental values of the drilling rate is presented in Table 1.

Table 2 presents the results of the experiment, averaged by operating conditions, and Table 3 – averaged by the drilling tool characteristics.

The selection of partial empirical dependencies between the variable factors and the drilling rate according to the experiment results of Table 1 is given in Table 4.

According to Tables 1 and 2 the most significant factor in the change of V_{mech} is the axle static load on the drilling bit. When the load was changed from 10 to 25 kN the rate was increased by 4.06 times. The probability P of the existence of empirical interdependency between the energy output and the static axle load according to the curvelinear regression is bigger than 0.95.

Table 1. Results of the investigation according to the planned four-factorial experiment.

# test.	Rotation rate n_b , min^{-1}	Axle static load on the drilling bit F_{stat} , kN	Drilling tool se- verity C , kN / m	Damping ratio β , kN s / m	Drilling rate V_{mech} , m / hour
1	82	10	400	0.1	1.05
2	82	15	2500	40	1.46
3	82	20	800	70	2.77
4	82	25	1700	90	4.41
5	133	10	800	90	1.47
6	133	15	1700	70	2.05
7	133	20	400	40	3.2
8	133	25	2500	0.1	5.78
9	188	10	1700	40	1.92
10	188	15	800	0.1	3.02
11	188	20	2500	90	4.84
12	188	25	400	70	7.44
13	285	10	2500	70	1.83
14	285	15	400	90	3.64
15	285	20	1700	0.1	7.54
16	285	25	800	40	7.91

Table 2. Average results of the planned experiment by operating conditions.

F_{stat} , kN / n_b , min^{-1}	82	133	188	285	Total	Average
10	1.05	1.47	1.92	1.83	6.27	1.5675
15	1.46	2.05	3.02	3.64	10.17	2.5425
20	2.77	3.20	4.84	7.54	18.35	4.5875
25	4.41	5.78	7.44	7.91	25.54	6.385
Total	9.69	12.5	17.22	20.92	60.33	
Average	2.4225	3.125	4.305	5.23		

Table 3. Average results of the planned experiment by the drilling tool characteristics.

C , kN / m / β , kN·s / m	400	800	1700	2500	Total	Average
0,1	1.05	3.02	7.54	5.78	17.39	4.3475
40	3.20	7.91	1.92	1.46	14.49	3.6225
70	7.44	2.77	2.05	1.83	14.09	3.5225
90	3.64	1.47	4.41	4.84	14.36	3.59
Total	15.33	15.17	15.92	13.91	60.33	
Average	3.8325	3.7925	3.98	3.4775		

Table 4. Results of the selection of partial empirical dependencies according to experimental data.

Variable	Partial empirical dependencies	Empirical value of the correlation ratio	Average quadratic deviation
Axle static load	$V_{mech} = 0.04 \cdot F_{stat}^{1.565623}$	0.9906 $P > 0.95$	0.2040
Drilling bit rotation rate	$V_{mech} = 0.142 \cdot n^{0.6409277}$	0.9707 $P > 0.95$	0.1652
Damping ratio of the drilling tool	$V_{mech} = 4.054 \cdot \beta^{-0.03021951}$	−0.9917 $P > 0.95$	0.0398
Severity of the drilling tool	$V_{mech} = 4.647 \cdot C^{-0.03013011}$	−0.5542 $P < 0.8$	0.1912

In order to increase the influence of other variables on the drilling rate we neutralize the influence of the load. This neutralization is conducted in two ways: according to Protodyakonov method (Protodyakonov 1970) and Yaremiychuk-Reihert method (Yaremiychuk 1966). According to Protodyakonov method all experimental results of $V_{mech.i}$ in Table 1 are corrected (recalculated) according to the formula:

$$V_{mech.cor.1.i} = V_{mech.i} + \left[f(F_{stat.av}) - f(F_{stat.i}) \right], \quad (1)$$

where $f(F_{stat.av})$ – values of energy output, determined by the formula $V_{mech} = 0.04 \cdot F_{stat}^{1.565623}$, and the average value of $F_{stat} = 17.5$ kN; $f(F_{stat.i})$ – values of drilling rate, determined by the formula $V_{mech} = 0.04 \cdot F_{stat}^{1.565623}$, and the values of $F_{stat} = 10, 15, 20$ and 25 kN.

According to Yaremiychuk-Reihert method all experimental results of $V_{mech.i}$ in Table 1 are corrected according to the formula:

$$V_{mech.cor.2.i} = V_{mech.i} \frac{f(F_{stat.av})}{f(F_{stat.i})}. \quad (2)$$

The results of the experiment corrected according to Protodyakonov method and averaged by operation conditions and the drilling tool parameters are presented in Tables 5 and 6, and according to Yaremiychuk-Reihert method – in Tables 7 and 8.

The results of the selection of partial gradual empirical dependencies between the variable factors and the drilling rate adjusted in accordance with the corrected results, received after neutralizing the influence of the axle static load according to Protodyakonov and Yaremiychuk methods are provided in Tables 9 and 10.

The results of the selection of partial dependencies $V_{mech} = f(n)$ by experimental and corrected data are presented in Table 11.

According to Table 11 the better partial dependency $V_{mech} = f(n)$ is the dependency selected by experimental data – $V_{mech} = 0.142 \cdot n^{0.6409277}$. It has the lowest quadratic deviation.

Table 5. Corrected results of the experiment (Cor. 1).

F_{stat} , kN / n_b , min^{-1}	82	133	188	285	Total	Average
10	3.11	3.53	3.98	3.89	14.51	3.6275
15	2.23	2.82	3.79	4.41	13.25	3.3125
20	1.95	2.38	4.02	6.72	15.07	3.7675
25	0.77	2.14	3.8	4.27	10.98	2.745
Total	8.06	10.87	15.59	19.29	53.81	
Average	2.015	2.7175	3.8975	4.8225		

Table 6. Corrected results of the experiment (Cor. 1).

C , kN / m _____ β , kN·s / m	400	800	1700	2500	Total	Average
0.1	3.11	3.79	6.72	2.14	15.76	3.94
40	2.38	4.27	3.98	2.23	12.86	3.215
70	3.8	1.95	2.82	3.89	12.46	3.115
90	4.41	3.53	0.77	4.02	12.73	3.1825
Total	13.7	13.54	14.29	12.28	53.81	
Average	3.425	3.385	3.5725	3.07		

Table 7. Corrected results of the experiment (Cor. 2).

F_{stat} , kN / n_b , min^{-1}	82	133	188	285	Total	Average
10	2.52	3.53	4.61	4.40	15.06	3.765
15	1.87	2.62	3.87	4.66	13.02	3.255
20	2.25	2.60	3.93	6.12	14.9	3.725
25	2.52	3.31	4.26	4.52	14.61	3.6525
Total	9.16	12.06	16.67	19.7	57.59	
Average	2.29	3.015	4.1675	4.925		

Table 8. Corrected results of the experiment (Cor. 2).

C , kN / m _____ β , kN·s / m	400	800	1700	2500	Total	Average
0.1	2.52	3.87	6.12	3.31	15.82	3.955
40	2.6	4.52	4.61	1.87	13.6	3.4
70	4.26	2.25	2.62	4.4	13.53	3.3825
90	4.66	3.53	2.52	3.93	14.64	3.66
Total	14.04	14.17	15.87	13.51	57.59	
Average	3.51	3.5425	3.9675	3.3775		

Table 9. Results of the selection of partial empirical dependencies.

Variable	Partial empirical dependencies according to the corrected results (Protodyakonov method)	Empirical value of the correlation ratio	Average quadratic deviation
Axle static load	$V_{mech} = 6.012 \cdot F_{stat}^{-0.2093885}$	-0.6213 $P > 0.8$	0.3778
Drilling bit rotation rate	$V_{mech} = 0.082 \cdot n^{0.7234297}$	0.9947 $P > 0.95$	0.1732
Damping ratio of the drilling tool	$V_{mech} = 3.649 \cdot \beta^{-0.03314602}$	-0.9904 $P > 0.95$	0.0366
Severity of the drilling tool	$V_{mech} = 4.318 \cdot C^{-0.03593761}$	$-0,5758$ $P < 0.8$	0.1904

Table 10. Results of the selection of partial empirical dependencies.

Variable	Partial empirical dependencies according to the corrected results (Yaremiychuk-Reihert method)	Empirical value of the correlation ratio	Average quadratic deviation
Axle static load	$V_{mech} = 6.012 \cdot F_{stat}^{-0.2093885}$	0.07302 $P > 0.8$	0.2343
Drilling bit rotation rate	$V_{mech} = 0.137 \cdot n^{0.6389839}$	0.9787 $P > 0.95$	0.1916
Damping ratio of the drilling tool	$V_{mech} = 3.767 \cdot \beta^{-0.01887814}$	−0.8529 $P > 0.8$	0.1438
Severity of the drilling tool	$V_{mech} = 3.342 \cdot C^{-0.01033095}$	−0.0140 $P < 0.8$	0.2532

Table 11. Results of the selection of partial dependencies $V_{mech} = f(n)$.

Input data	Partial empirical dependencies	Empirical value of the correlation ratio	Average quadratic deviation
Experimental	$V_{mech} = 0.142 \cdot n^{0.6409277}$	0.9707 $P > 0.95$	0.1652
Corrected (Cor. 1) $F \approx const$	$V_{mech} = 0.082 \cdot n^{0.7234297}$	0.9947 $P > 0.95$	0.1732
Corrected (Cor. 2) $F \approx const$	$V_{mech} = 0.137 \cdot n^{0.6389839}$	0.9787 $P > 0.95$	0.1916

According to the accepted partial dependency $V_{mech} = f(n)$ we neutralize the influence of the second most important factor for the drilling rate change– rotation rate according to experimental data (see Tables 1 and 2). When the rotation rate was changed from 82 to 285 min⁻¹ the drilling rate increased from 2.42 m / hour to 5.23 m / hour, that is by 2.16 times. According to Table 7 the probability of the existence of empirical dependency between the rotation rate and drilling rate is bigger than 0.95.

According to Protodyakonov method all experimental data $V_{mech.i}$ of Table 1 are recalculated according to the formula:

$$V_{mech.cor.3.i} = V_{mech.i} + \left[f(n_{av}) - f(n_i) \right], \qquad (3)$$

where $f(n_{av})$ – values of energy output, determined by the formula $V_{mech} = 0.142 \cdot n^{0.6409277}$, and the average values of $n = 183.5$ min⁻¹; $f(n_i)$ – values of energy output, determined by the formula $V_{mech} = 0.142 \cdot n^{0.6409277}$, and the average values of $n = 82, 133, 188$ and 285 min⁻¹.

According to Yaremiychuk-Reihert method all experimental data $V_{mech.i}$ of Table 1 are recalculated according to the formula:

$$V_{mech.cor.4.i} = V_{mech.i} \frac{f(n_{av})}{f(n_i)}. \qquad (4)$$

The results of the experiment corrected according to Protodyakonov method and averaged by operation conditions and the drilling tool parameters are presented in Table 12, and according to Yaremiychuk-Reihert method – in Table 13.

The results of the selection of partial gradual empirical dependencies between the drilling tool parameters and the drilling rate, received during the neutralization of the influence of the rotation rate according to Protodyakonov method are presented in Table 14, and according to Yaremiychuk-Reihert method – in Table 15.

Table 16. Provides the results of the selection of partial dependencies $V_{mech} = f(\beta)$ according to experimental and corrected data.

According to Table 16 the better partial dependency is the dependency selected on the basis of the results corrected according to Protodyakonov method (see Table 6) – $V_{mech} = 3.649 \cdot \beta^{-0.03314602}$. It has the lowest quadratic deviation.

According to the accepted partial dependency $V_{mech} = f(\beta)$ we neutralize the influence of the third most important factor for the drilling rate change – damping ration according to the corrected results of the experiment (Cor. 1). During the change of the damping ratio from 0.1 to 90 kN·s / m the drilling rate decreased from 3.94 to 3.18 m / hour, that is by 1.24 times.

Table 12. Corrected results of the experiment (Cor. 3).

$\dfrac{C\,,\,kN/m}{\beta\,,\,kN\cdot s/m}$	400	800	1700	2500	Total	Average
0.1	2.67	2.96	6.23	6.53	18.39	4.5975
40	3.95	6.6	1.86	3.08	15.49	3.8725
70	7.38	4.39	2.8	0.52	15.09	3.7725
90	2.33	2.22	6.03	4.78	15.36	3.84
Total	16.33	16.17	16.92	14.91	64.33	
Average	4.0825	4.0425	4.23	3.7275		

Table 13. Corrected results of the experiment (Cor. 4).

$\dfrac{C\,,\,kN/m}{\beta\,,\,kN\cdot s/m}$	400	800	1700	2500	Total	Average
0.1	1.76	2.98	5.68	7.11	17.53	4.3825
40	3.94	5.96	1.89	2.45	14.24	3.56
70	7.33	4.65	2.52	1.38	15.88	3.97
90	2.74	1.81	7.4	4.77	16.72	4.18
Total	15.77	15.4	17.49	15.71	64.37	
Average	3.9425	3.85	4.3725	3.9275		

Table 14. The results of the selection of partial empirical dependencies.

Variable	Partial empirical dependencies according to the corrected results (Protodyakonov method)	Empirical value of the correlation ratio	Average quadratic deviation
Damping ratio of the drilling tool	$V_{mech} = 4.305 \cdot \beta^{-0.02841417}$	-0.9947 $P > 0.95$	0.040
Severity of the drilling tool	$V_{mech} = 4.888 \cdot C^{-0.02815838}$	-0.5542 $P < 0.8$	0.1912

Table 15. The results of the selection of partial empirical dependencies.

Variable	Partial empirical dependencies according to the corrected results (Yaremiychuk-Reihert method)	Empirical value of the correlation ratio	Average quadratic deviation
Damping ratio of the drilling tool	$V_{mech} = 4.175 \cdot \beta^{-0.01585461}$	-0.9947 $P > 0.95$	0.280
Severity of the drilling tool	$V_{mech} = 3.348 \cdot C^{-0.02606999}$	0.2843 $P < 0.8$	0.2184

Table 16. The results of the selection of partial dependency $V_{mech} = f(\beta)$.

Input data	Partial empirical dependencies	Empirical value of the correlation ratio	Average quadratic deviation
Experimental	$V_{mech} = 4.054 \cdot \beta^{-0.03021951}$	-0.9917 $P > 0.95$	0.03984
Corrected (Cor. 1) $F \approx const$	$V_{mech} = 3.649 \cdot \beta^{-0.03314602}$	-0.9904 $P > 0.95$	0.0366
Corrected (Cor. 2) $F \approx const$	$V_{mech} = 3.767 \cdot \beta^{-0.01887814}$	-0.8529 $P > 0.8$	0.1438
Corrected (Cor. 3) $n \approx const$	$V_{mech} = 4.305 \cdot \beta^{-0.02841417}$	-0.9947 $P > 0.95$	0.03995
Corrected (Cor. 4) $n \approx const$	$V_{mech} = 4.175 \cdot \beta^{-0.01585461}$	-0.9947 $P > 0.95$	0.280

According to Table 16 the probability of the existence of empirical dependency between the damping ratio and drilling rate is bigger than 0.95.

The results corrected according to Protodyakonov

method and averaged according to the drilling tool parameters are presented in Table 17, and according to Yaremiychuk-Reihert method – in Table 18.

Table 19 provides the results of the selection of partial dependencies $V_{mech} = f(C)$ according to experimental and corrected data.

According to Table 19 the better partial dependency is the dependency selected on the basis of the results corrected at first according to Protodyakonov, method and then according to Yaremiychuk-Reihert method, $V_{mech} = 4.4 \cdot C^{-0.04550764}$. The average quadratic deviation is the lowest for this dependency, and the probability of the existence of empirical dependency between the severity and drilling rate is bigger than 0.9.

Table 17. Corrected results of the experiment (Cor. 5).

C, kN / m — β, kN·s / m	400	800	1700	2500	Total	Average
0.1	2.39	3.07	6	1.42	12.88	3.22
40	2.37	4.26	3.97	2.22	12.82	3.205
70	3.85	2	2.87	3.94	12.66	3.165
90	4.49	3.61	0.85	4.1	13.05	3.2625
Total	13.1	12.94	13.69	11.68	51.41	
Average	3.275	3.235	3.4225	2.92		

Table 18. Corrected results of the experiment (Cor. 6).

C, kN / m — β, kN·s / m	400	800	1700	2500	Total	Average
0.1	2.57	3.09	5.39	1.75	12.8	3.2
40	2.37	4.27	3.96	2.18	12.78	3.195
70	3.86	1.98	2.86	3.95	12.65	3.1625
90	4.52	3.62	0.79	4.12	13.05	3.2625
Total	13.32	12.96	13.0	12.0	51.28	
Average	3.33	3.24	3.25	3.0		

Table 19. the results of the selection of partial dependency $V_{mech} = f(C)$.

Input data	Partial empirical dependencies	Empirical value of the correlation ratio	Average quadratic deviation
Experimental	$V_{mech} = 4.647 \cdot C^{-0.03013011}$	-0.5542 $\beta < 0.8$	0.1912
Corrected (Cor. 1) $F \approx const$	$V_{mech} = 4.318 \cdot C^{-0.03593761}$	-0.5758 $\beta < 0.8$	0.1904
Corrected (Cor. 2) $F \approx const$	$V_{mech} = 3.342 \cdot C^{-0.01033095}$	-0.0140 $\beta < 0.8$	0.2532
Corrected (Cor. 3) $n \approx const$	$V_{mech} = 4.888 \cdot C^{-0.02815838}$	-0.5542 $\beta < 0.8$	0.1912
Corrected (Cor. 4) $n \approx const$	$V_{mech} = 3.348 \cdot C^{-0.02606999}$	0.2843 $\beta < 0.8$	0.2184
Corrected (Cor. 5) $\beta \approx const$	$V_{mech} = 4.177 \cdot C^{-0.037719}$	-0.5758 $\beta < 0.8$	0.1904
Corrected (Cor. 6) $\beta \approx const$	$V_{mech} = 4.4 \cdot C^{-0.04550764}$	0.8912 $\beta > 0.9$	0.0816

If each combination of the values of variable factors occurs only once, then the overall multidimensional function can be presented in the form of the gain of separate dependencies on the variable factors – $V_{mech} = B_{av} \cdot f(F_{stat}) \cdot f(n) \cdot f(\beta) \cdot f(C)$. Partial dependencies between each factor and the drilling rate are presented in Table 20.

Table 21 presents the values of coefficient B, determined according to the formula:

$$B_i = \frac{V_{mech.i}}{f(F_i) \cdot f(n_i) \cdot f(\beta_i) \cdot f(C_i)}, \qquad (5)$$

where $V_{mech.i}$ – the value of the drilling rate according to the data of Table 1, which corresponds consequently to the experiment;

$f(F_{stat}) \cdot f(n_i) \cdot f(\beta_i) \cdot f(C_i)$ – gain of partial empirical dependencies of variable factors, whose values correspond to the conditions and consequently to the experiment of Table 20.

Table 20. Partial dependencies $V_{mech} = f(F_{stat})$, $V_{mech} = f(n)$, $V_{mech} = f(\beta)$, $V_{mech} = f(C)$.

F_{stat} , kN	10	15	20	25
Exp. value V_{mech} , m / hour	1.57	2.54	4.59	6.39
$V_{mech} = 0.04 \cdot F_{stat}^{1.565623}$	1.47	2.78	4.35	6.18
$F_{stat}^{1.565623}$	**36.781**	**69.393**	**108.874**	**154.400**
n_b , min^{-1}	82	133	188	285
Exp. value V_{mech} m / hour	2.42	3.13	4.31	5.23
$V_{mech} = 0.142 \cdot n_b^{0.64089277}$	2.39	3.26	4.07	5.32
$n_b^{0.64089277}$	**16.848**	**22.97**	**28.742**	**37.436**
β , kN·s / m	0.1	40	70	90
Cor. value V_{mech} , m / hour	3.94	3.22	3.12	3.18
$V_{mech} = 3.649 \cdot \beta^{-0.03314602}$	3.94	3.23	3.17	3.14
$\beta^{-0.03314602}$	**1.0793**	**0.8849**	**0.8686**	**0.8614**
C , kN / m	400	800	1700	2500
Cor. value V_{mech} , m / hour	3.33	3.24	3.25	3.0
$V_{mech} = 4.4 \cdot C^{-0.04550764}$	3.35	3.25	3.14	3.08
$C^{-4.550764 \cdot 10^{-2}}$	**0.7614**	**0.7377**	**0.7128**	**0.7004**

Table 21. The value of coefficient в for all tests of the planned experiment.

# test	1	2	3	4	5	6
B	$2.0619 \cdot 10^{-3}$	$2.0149 \cdot 10^{-3}$	$2.3567 \cdot 10^{-3}$	$2.2036 \cdot 10^{-3}$	$2.1853 \cdot 10^{-3}$	$2.0772 \cdot 10^{-3}$
# test	7	8	9	10	11	12
B	$2.0136 \cdot 10^{-3}$	$2.1559 \cdot 10^{-3}$	$2.3607 \cdot 10^{-3}$	$1.9682 \cdot 10^{-3}$	$2.5636 \cdot 10^{-3}$	$2.535 \cdot 10^{-3}$
# test	13	14	15	16		
B	$2.1846 \cdot 10^{-3}$	$2.1364 \cdot 10^{-3}$	$2.4046 \cdot 10^{-3}$	$2.0964 \cdot 10^{-4}$		

According to Table 21 the average value of coefficient B is $1.6703 \cdot 10^{-3}$. Table 22 presents the values of the drilling rate determined according to the equation of multifactorial mathematical model (6) and the ranges of relative error (RE) between these values and the experimental results of the test.

$$V_{mech} = 1.6703 \cdot 10^{-3} \cdot F_{stat}^{1.565623} \cdot n_b^{0.64089277} \times$$
$$\times \beta^{-0.03314602} \cdot C^{-4.550764 \cdot 10^{-2}}. \qquad (6)$$

The average value of the relative error between the experimental results and the drilling rate values calculated according to the equation of multifactorial mathematical model is 6.6%.

Partial empirical dependencies $V_{mech} = f(F_{stat})$, $V_{mech} = f(n)$ $V_{mech} = f(C)$ and $V_{mech} = f(\beta)$ according to the results of Table 19 are depicted in Figure 2.

Table 22. Values of the drilling rate.

# test	Rotation rate n_b, min⁻¹	Axle static load on the drilling bit F_{stat}, kN	Drilling tool severity C, kN / m	Damping ratio β, kN s / m	Drilling rate V_{mech}, m / hour	V_{mech}, m / hour acc. to formula	RE, %
1	82	10	400	0.1	1.05	1.12	-6.7
2	82	15	2500	40	1.46	1.60	-9.6
3	82	20	800	70	2.77	2.59	6.5
4	82	25	1700	90	4.41	4.42	-0.2
5	133	10	800	90	1.47	1,48	-0.7
6	133	15	1700	70	2.05	2.18	-6.3
7	133	20	400	40	3.2	3.51	-9.7
8	133	25	2500	0.1	5.78	5.95	-2.9
9	188	10	1700	40	1.92	1.8	6.3
10	188	15	800	0.1	3.02	3.39	-12.2
11	188	20	2500	90	4.84	4.17	13.8
12	188	25	400	70	7.44	6.48	12.9
13	285	10	2500	70	1.83	1.85	-1.1
14	285	15	400	90	3.64	3.76	-3.3
15	285	20	1700	0,1	7.54	6.92	8.2
16	285	25	800	40	7.91	8.33	-5.3

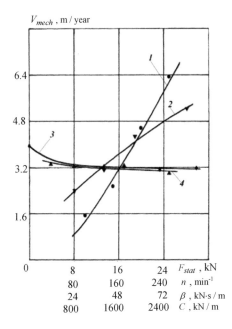

Figure 2. Partial dependencies $V_{mech} = f(F_{stat})$, $V_{mech} = f(n_b)$ $V_{mech} = f(C)$ and $V_{mech} = f(\beta)$ obtained as a result of the planned experiment:

$1(\bullet) - V_{mech} = 0.04 \cdot F_{stat}^{1.565623}$;

$2(\blacktriangledown) - V_{mech} = 0.142 \cdot n^{0.64089277}$;

$3(\blacklozenge) - V_{mech} = 3.649 \cdot \beta^{-0.03314602}$;

$4(\blacktriangle) - V_{mech} = 4.4 \cdot C^{-0.04550764}$.

4 CONCLUSIONS

1. With the help of the rational planning method, the multifactorial mathematical model between the drilling rate and axle static load on the bottomhole, the drilling bit rotation rate, the severity and the damping ratio of the drilling tool has been determined. This model looks as follows:

$$V_{mech} = 1.6703 \cdot 10^{-3} \cdot F_{stat}^{1.565623} \cdot n^{0.64089277} \times$$

$$\times \beta^{-0.03314602} \cdot C^{-4.550764 \cdot 10^{-2}} .$$

The maximum value of the relative error while using the mathematical model amounts to 13.8%, the average value – 6.6%.

2. Partial dependencies $V_{mech} = f(F_{stat})$, $V_{mech} = f(n_b)$ and $V_{mech} = f(\beta)$, received as a result of the planned experiment, with the probability higher than 0.95 are approximated with empirical gradual de-

pendencies, and the partial dependency $V_{mech} = f(C)$ is approximated with empirical gradual dependency with the probability higher than 0.9.

REFERENCES

Allikvander, E. 1969. *Modern deep drilling.* Moscow: Nedra: 232.

Belikov, V. & Postash, S. 1972. *Rational development and wear resistance of roller cone bits.* Moscow: Nedra: 149.

Borysevych, B., Moisyshyn, V. & Scherbiy, R. 2009. *Drilling stand for rock breaking researching and dynamic of drilling device.* Prospect and development of oil and gas deposits, 3(32): 23-29

Borysevych, B., Moisyshyn, V. & Scherbiy, R. 2009. *Devices of drilling dynamic changes on drilling stand.* Prospect and development of oil and gas deposits, 4(33): 18-23.

Nazarov, V. 1985. *Dampeners for borehole sinking.* Drilling.

Usage of near-the-face vibration dumpers during drilling oil and gas boreholes in USA. 1974. Oil and gas industry, 3: 50-52.

Potapov, Y. & Simonov, V. 1961. *Rock breaking by three cone drill bits with small diameter.* Moscow: 86.

Protodyakonov, M. 1970. *Method of rational experiment planning.* Moscow: 76.

Scherbiy, R., Moisyshyn, V. & Liskanych, O. 2012. *Establishment of correlation empiric dependence between damping coefficient of drilling device and mechanical drilling velocity of oil and gas boreholes.* Prospect and development of oil and gas deposits, 1(42): 79-88.

Yamaltdinov, A. & Mavlyotov, M. 1986. *Increasing of prop stability of roller cone bits during borehole sinking.* Moscow: 35.

Yaremiychuk, R. & Rayhert, L. 1966. *Drilling of high diameter shafts.* Moscow: Nedra: 174.

Mining of Mineral Deposits – Pivnyak, Bondarenko, Kovalevs'ka & Illiashov (eds)
© 2013 Taylor & Francis Group, London, ISBN: 978-1-138-00108-4

Elastic waves influence upon enhancement of shale rocks fracturing

Ya. Bazhaluk, O. Karpash, I. Kysil, Ya. Klymyshyn, O. Gutak & Yu. Voloshyn
Ivano-Frankivs'k National Technical University of Oil and Gas, Ivano-Frankivs'k, Ukraine

ABSTRACT: The aim of study is enhancement of gas yield while shale deposits development. The methodology being used is based upon impulse-wave action, special mode of which let us increase shale rock inner surface due to creation of new fractures. Peculiarity of elastic oscillations use is that enough energy is created for fatigue destruction of rock, which will lead to shale gas desorption activating and decrease of energy consumption in the course of formation hydraulic fracturing.

1 INTRODUCTION

Creation of additional network of fractures and correspondingly inner surface in shale rock can be essential for the enhancement of shale stratum hydraulic fracturing effectiveness. It can be done by application of variable pressures, that can be created in the stratum by special technical means in the course of elastic oscillations propagation. In case of sufficient intensity of elastic oscillations as well as certain operating time, additional fractures appear in the formation (Sobolev et al. 2003). They appear only if asymmetrical variable pressure parameters reach beyond rock fatigue strength limit.

Preliminary treatment of formation with elastic oscillations aimed at enhancement of its fracturing will result in increase of shale rock inner surface in comparison with the surface, created while hydraulic fracturing without treatment. As a result shale gas desorption will become more intense and power consumption will be decreased while hydraulic fracturing. This method can be also used to intensify gas yield processes during well operation.

The objective of this scientific paper is to evaluate elastic oscillations parameters (intensity and frequency), necessary to create fatigue fractures in shale stratum at a given distance from borehole generator of fatigue fractures and to stipulate technical requirements for borehole generator.

2 CALCULATION OF ELASTIC OSCILLATIONS PARAMETERS AND GENERATOR ACOUSTIC CAPACITY, NECESSARY TO CREATE FATIGUE FRACTURES IN A FORMATION

For further calculations the following is assumed:

– cylindrical wave is propagating in the stratum;
– direction of wave propagation in the horizontal part of a well is vertical to shale rock layers;
– strata thickness is a constant value;
– for the evaluation of rock fatigue strength σ_a cyclic loads of stratum with variable pressures are taken into account;
– strength (strength limit $\sigma_в$) and acoustic (density ρ, speed of longitudinal oscillations propagation in stratum c_n, coefficient of damping k of elastic oscillations in the specified range of frequencies f) characteristics of shale rock are constant.

Elastic oscillations in rock occur as a result of influence of hydraulic pressure impulses in borehole (Bazhaluk et al. 2012). In this case generator of oscillations is an acoustic system which consists of one or several pressure impulse generators, well liquid medium and a part of perforated casing string (Figure 1).

Hydraulic pressure impulses actuate train of elastic oscillations in reservoir environment. In rocks, high-frequency constituents of oscillations trains decline at a little distance from the generator. If distances exceed 20 m, oscillations in the range of 30-80 Hz are dominating (Bazhaluk et al. 2012).

Assume that fatigue strength limit of rock σ_a is on the level of $\alpha(n)$ from its strength limit $\sigma_в$ (Sobolev et al. 2003), thus

$$\sigma_a = \alpha(n)\sigma_в, \qquad (1)$$

where $\alpha(n)$ – parameter, linking fatigue strength with static strength; n – number of load cycles. For the conditions of all-around compression, static strength parameter is evaluated with the help of well-known correlations by Huber-Mises, Mohr, Hoeke, Schleicher-Nadai etc. (Myslyuk et al. 2004). For instance, in accordance with Hoeke criterion

$$\sigma_{\text{B}} = a\sigma^b + c, \qquad (2)$$

where a, b, c – empirical constants; σ – average or hydrostatic pressure.

Then from the condition of reservoir fatigue fracturing

$$p_x \geq \alpha(n)\sigma_{\text{B}} \qquad (3)$$

for the given number n of load cycles, value of variable pressure amplitude p_x is evaluated. Oscillation strength, necessary for creation of variable pressure amplitude p_x in reservoir, can be determined by formula (Riznichenko 1985)

$$I_x = \frac{p_x^2}{2\rho c_n}, \qquad (4)$$

where I_x – oscillation strength in reservoir at the distance x from borehole generator.

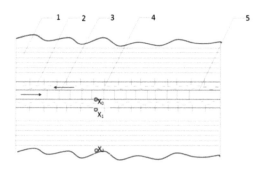

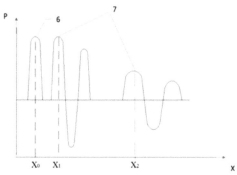

Figure 1. Schematic illustration of borehole oscillator of cylindrical wave and elastic oscillations trains in a formation: 1 – stratum; 2 – perforated casing string; 3 – pressure impulse generator; 4 – working fluid; 5 – tubing string; 6 – hydraulic pressure impulse in a well; 7 – elastic oscillations trains in a formation.

For the evaluation of oscillation strength I_0 on the surface of oscillator, necessary to obtain strength I_x in reservoir, expression for determination of intensity change of cylindrical wave depending on distance to the generator can be used (Riznichenko 1985):

$$I_0 = I_x \frac{x}{x_0} exp(2kx), \qquad (5)$$

where k – coefficient of elastic oscillations damping in the given range of frequencies.

Let's evaluate intensity on the surface of oscillator for fatigue micro fracturing (stratum depth is 2500 m and average overburden pressure is 58.9 MPa), using formulae (1) – (5). For plastic rocks $\alpha(n) = 0.5...0.7$ (Myslyuk et al. 2004, Rzevskaya & Yamschykov 1984-1991), static strength parameter is $\sigma_{\text{B}} = 3$ MPa and correspondingly $\sigma_a = 1.6$ MPa. Let's assume that (Rzevskaya & Yamschykov 1984-1991) is $\rho = 2700$ kg / m^3, $c_n = 4500$ m / s and for oscillation frequency $f \leq 100$ Hz coefficient of elastic oscillations damping is $k = 10^{-4}$ m^{-1}. If strata thickness is 60 m, distance from borehole generator will be $x = 30$ m. After calculations we have the following value of $I_0 = 279.4 \cdot 10^4$ W / m^2. For intensity equal to $279.4 \cdot 10^4$ W / m^2 and with average radiation area of pressure impulses generator of 300 cm^2, acoustic capacity of generator mustn't be less than 83.84 kW. Hydraulic capacity of such a generator is determined on the basis of its acoustic capacity of 83.84 kW.

It is known that capacity N, which is created by hydraulic downhole equipment can be determined by differential pressure values Δp on appliances and bulk consumption Q of working fluid through the appliances (Eksner et al. 2003)

$$N = \eta Q \Delta p,$$

where η – coefficient of efficiency of appliances.

Maximum differential pressure across a hydrogenerator of GKP-56M type, created by "Intex" company is 7 MPa if hydraulic fluid consumption is 0.03 m^3 / s. Thus, generated hydraulic capacity is 210 kW. If coefficient of hydraulic energy conversion into acoustic energy is assumed to be equal to 0.5, then acoustic strength of generator will be equal to 105 kW. On the radiation surface equal to 300 cm^2, oscillation intensity in borehole is $315 \cdot 10^4$ W / m^2. Taking into consideration acoustic energy losses in the course of transition from liquid borehole environment into reservoir, acoustic energy transmission coefficient is assumed to be 0.9 (Riznichenko 1985).

Then oscillations intensity on reservoir entering (on the surface of cylindrical oscillator) will be equal to $283.5 \cdot 10^4$ W / m².

The above mentioned calculations confirm possibility of creation of variable pressures with an amplitude exceeding 1 MPa in reservoir at 30 m distance from borehole hydrogenerator.

Let's evaluate maximum time, necessary for reservoir treatment. Considering frequency of pressure impulses repetition $f = 50$ Hz for the number of cycles $N = 10^6$ (Bajburova 2008), time of reservoir treatment is 5.5 hours.

In the above mentioned influence of surfactants upon rocks fracturing in the course of cyclic loads has not been taken into account, particularly Rehbinder effect (Latyshev et al. 2008). Decrease of surface energy of rock under surfactants influence provides the possibility to decrease fatigue strength limit $\sigma_{\text{в}}$ up to 40% (Latyshev et al. 2008).

Technologically process of impulses and waves influence upon reservoirs can be carried out in the following way:

1. The following equipment is lowered down into perforation zone on tubing string: lower wave reflector – hydraulic generator of pressure impulses with amplitude-frequency characteristics, determined according to geological and technical well data – upper wave reflector – tubing up to wellhead.

2. Lines are connected for closed working fluid circulation according to the following scheme: pumping unit – tubing string with hydraulic generator and wave reflectors – annular space – sectional container – pumping unit.

3. In the injection line electronic fast-response pressure gage with necessary technical characteristics is installed.

4. Necessary amplitude and frequency of hydraulic generator pressure impulses repetition are adjusted by regulating pumping unit capacity (on condition that casing valve is closed). Amplitude value of pressure impulses on electronic pressure gage are installed on the level of 0.8 from amplitude value of pressure impulses on the surface of cylindrical oscillator in borehole, taking into account pressure losses while wave front propagation in tubing.

5. Reservoir treatment is carried out during estimated time with open casing valve. After treatment, reservoir fracturing enhancement is tested by injection of working fluid into reservoir with closed casing valve. Meanwhile casing pressure testing values are not exceeded. Fluid losses speak for increase of reservoir fracturing.

3 CONCLUSIONS

1. The results of conducted experimental and theoretical investigations point to the possibility of creation of pressures in shale stratum, exceeding fatigue strength of shale rock at the distances of up to 30 m from borehole generator of pressure impulses.

2. To increase shale rock fracturing further experimental investigations of elastic oscillations influence should be carried out using surfactants to define their influence upon gas yield efficiency.

3. Obtained results can be used to investigate influence of cyclic stresses upon increase of coal strata fracturing with the aim of their decontamination and also in oil and gas wells development.

REFERENCES

Sobolev, V.V., Skobenko, A.V. & Ivanchyshyn S.Ya. 2003. *Petrophysics: Textbook for higher educational establishments.* Dnipropetrovs'k: Poligraphist.

Bazhaluk, Ya.M., Karpash, O.M., Klymyshyn, Ya.D., Gutak, A.I. & Khudin, N.V. 2012. *Oil production increase due to formation stimulation with the help of mechanical oscillations train.* Electronic scientific journal "Oil and Gas Business", Issue 3: 199-210.

Myslyuk, M.A., Rybchych, I.Yo. & Jaremijchuk, R.S. 2004. *Drilling of wells: Reference book in five parts.* Part 5: Complications, Emergency Situations, Ecology. Kiev: Interpres Ltd.: 376.

Riznichenko, Yu.V. 1985. *Seismic exploration of layered media.* Moscow: Nedra: 162.

Rzevskaya, S.V., Yamshchykov, V.S. 1984-1991. *Encyclopedia of mining engineering.* Moscow: Soviet Encyclopedia.

Eksner, Kh., Freitag, R., Lang, R. et al. 2003. *Hydraulic power drive systems. Fundamentals and components.* Bosch Rexroth Bosch Group: 92.

Bajburova, M.M. 2008. *Solution of strength and destruction tasks of anisotropic materials and rocks: Abstract of a Thesis for the Degree of a Doctor of Technical Sciences.* Almetyevs'k State Oil Institute. Almetyevs'k.

Latyshev, O.G., Synbulatov, V.V. & Osypov, Y.S. 2008. *Saturation kinetics of rock mass with surfactants solutions while borehole and blastholes drilling.* Higher educational establishments News. Journal of Mining Engineering. Ural State Mining University: 123-129.